BRADY

EMT–Paramedic National Standards Review Self Test Third Edition

Charly D. Miller, NREMT-P

Based on the
Emergency Medical Technician–Paramedic
National Standard Curriculum
as set forth by the
U.S. Department of Transportation,
National Highway Traffic Safety Administration

Keyed to
Brady's *Paramedic Emergency Care*
3rd Edition, © 1997

Brady
Prentice Hall
Upper Saddle River, New Jersey 07458

Library of Congress Cataloging-in-Publication Data

Miller, C. D. (Charly D.), (date)-
EMT-paramedic national standards review self test / Charly D. Miller. — 3rd ed.
p. cm.
"Based on the Emergency medical technician-paramedic national standard curriculum as set forth by the U.S. Department of Transportation, National Highway Traffic Safety Administration."
"Keyed to Brady's Paramedic emergency care, 3rd edition."
At head of title: Brady.
ISBN 0-8359-5102-2
1. Medical emergencies—Examinations, questions, etc.
2. Emergency medical personnel—Examinations, questions, etc.
I. Bledsoe, Bryan E., (date)- Brady Paramedic emergency care.
II. United States. National Highway Traffic Safety Administration.
III. Title.
[DNLM: 1. Emergencies—examination questions. 2. Emergency Medical Services—United States—examination questions. 3. Emergency Medical Technicians—education—examination questions.]
RC86.7.B596 1997 Suppl.
616.02'5'076—DC20

96-38269
CIP

Director of Production and Manufacturing: **Bruce Johnson**
Managing Production Editor: **Patrick Walsh**
Interior Design and Layout: **Lightworks Design/Stephen Hartner**
Cover Design: **TK**
Production Manager: **Ilene Sanford**
Publisher: **Susan Katz**
Editorial Assistant: **Carol Sobel**
Printer/Binder: **Banta Company, Harrisonburg, VA**

Printed in the United States of America
10 9 8 7 6

ISBN 0-8359-5102-2

Prentice-Hall International (UK) Limited, *London*
Prentice-Hall of Australia Pty. Limited, *Sydney*
Prentice-Hall Canada Inc., *Toronto*
Prentice-Hall Hispanoamericana, S.A., *Mexico*
Prentice-Hall of India Private Limited, *New Delhi*
Prentice-Hall of Japan, Inc., *Tokyo*
Simon & Schuster Asia Pte. Ltd., *Singapore*
Editora Prentice-Hall do Brasil, Ltda., *Rio de Janeiro*

Contents

Preface

Dear Reader,

This text is designed to assist you, the EMT-Paramedic, in your preparation for any written examination. It is based on the EMT-Paramedic National Standard Curriculum as set forth by the National Highway Traffic Safety Administration's Department of Transportation (DOT). Its self-test format is designed to challenge you, to pinpoint the subjects in which you require additional study, and to test your skill at reading and responding to test questions.

The answer keys are accompanied by reference page numbers that correspond to Brady's *Paramedic Emergency Care, 3rd Edition* and the question's subject title. If you don't have the aforementioned text (and cannot purchase it), use the index of the text you do have to seek out information on the subjects you had difficulty with.

This text is specifically designed to assist you in refreshing your memory and studying prior to any examination. Although some questions appear to have obvious answers ("refresher" questions), the majority are designed to be more demanding than the average test question. If you can pass these self tests with a score of 90 percent or better, you should be able to achieve an even better score on any written exam you may face.

Be sure to read the sections that provide tips on mental and physical preparation for taking written and practical exams. All too frequently, students fail tests not because they don't know the material, but because they don't read carefully or are ill-prepared to successfully perform in the testing environment.

I hope this text will assist you to excel in your examination. I do not wish you good luck, because luck is not a factor. Your dedicated preparation and review will help you to achieve the high scores you seek.

Sincerely yours,

Charly D. Miller

Acknowledgments

The author wishes to express her continued gratitude and appreciation to Natalie Anderson and Susan B. Katz, her extraordinary editors.

The author also extends her appreciation to the staff of Rocky Mountain Adventist Healthcare, Prehospital Services for their assistance in obtaining the ECG strips used in this text.

Lastly, to both sets of her parents (Raymond and Carol Miller, James and Carolyn White), the author wishes to present a resounding, "Thank you!"

About the Author

Charly D. Miller is a nationally-known patient-care educator, who continues to work as a field-operating paramedic for Denver General Hospital's Paramedic Division. She has been involved in EMS since 1983, when she received her EMT-Basic certificate in the state of Nebraska. In addition to "working the streets," she has been an in-hospital Psychiatric Technician-EMT, and served as a helicopter medic with the Army National Guard in Nebraska and Wyoming. She received her paramedic training at Creighton University in Omaha, Nebraska and has been a Nationally Registered EMT-Paramedic since 1986. Her first Brady book, *Home Meds: A Paramedic's Pocket Guide to Prescription Medications*, was published in the fall of 1991. Since then, she has authored several EMS texts and journal articles.

Suggestions for Written Examination Preparation and Execution

Written examinations are generally as delightful as a visit to the dentist. However, anesthesia is an inappropriate solution.

Test preparation should start well before the exam day. Use this text to determine your strengths and weaknesses, and to practice your test-taking skills.

First, read each question carefully. Many written examinations are failed by knowledgeable and experienced Paramedics *simply because they don't read the question well.* Don't read more into the question than what is presented. Is the question asking you to identify a correct answer, an incorrect answer, or the "best" answer? Read each answer carefully. After making your selection, go back and reread the question, inserting that answer. Does it still seem correct?

In an exam situation, a guess is better than leaving the answer completely blank (a blank answer is an error). When taking the self tests, however, *do not guess* at the answers. If you cannot answer a question with confidence, you ought to review that particular subject. If you *do* guess, you may answer correctly by pure accident. There is no guarantee that you will guess in the same manner when faced with a similar question on your actual exam. Skip that question, circle its number on your answer sheet, and move on.

When you have completed the self test section, compare your answers to the key provided. Write down the subject and/or reference page number for each incorrect or skipped answer. Then refer to your text and study that material.

After you have studied, retake the self test. If you again have difficulty with some subjects, return to your text and repeat your review. The more often you are able to repeat this process, the better you will fare on your examination.

As you are retaking the self tests, begin to practice timing yourself. Most examinations allow one minute per question: 60 questions, one-hour time limit; 150 questions, 2 ½-hour time limit, and so on. Timed practice sessions will actually help you to increase your test-taking speed. This will afford you extra time for the questions that require extra thought.

Learn not to allow yourself to become "bogged down" on a difficult or confusing question. You can skip it and come back. When you do skip a question, however, make sure you also skip the corresponding answer sheet entry! Although everyone will warn you about making "unnecessary stray marks or erasures," it is better to *lightly* circle the number of a skipped question than to cause all your subsequent questions to be incorrectly answered. Make sure that you erase all extra marks before submitting your sheet.

If you have co-workers or friends who will be taking the same exam, get together and do group study sessions during the week or two before the test (especially in preparation for practical examinations). Sometimes "teaching" is the best way to study.

Okay. You've done all that "good student" stuff. You've done your homework and practiced your skills, but you're still fearful and uncertain. What *more* can you do? You can relax.

Be sure to get a good night's sleep before the test. If you stay up all night studying you'll certainly do more harm than good. If you've delayed your review and can't avoid last-minute studying, you *still* should go to bed early! Then get up early and study the morning of the test, ideally over a good breakfast. If you are one of those "night shift" people who cannot mentate in the early hours of morning, you will need to arrange for a good sleep the previous afternoon and evening.

Let us pause here, and discuss the definition of a "good breakfast." Guess what? Contrary to popular EMT behavior (at all levels), a good breakfast does not consist of coffee and doughnuts. Whether or not you adopt good nutrition in your normal day-to-day living, it will be to your great advantage to do so on the morning of an event as stressful as an examination.

First, the sugar contained in the doughnuts will quickly peak, resulting in a rebound lethargy, while the caffeine in the coffee will provide artificial stimulation. Do you truly want to approach an exam situation *sideways*?

A breakfast high in complex carbohydrates along with some protein is the key to the sustained energy levels you'll need during a high-stress situation. Lean meats, eggs, and milk products (such as yogurt, cheeses, and low-fat milk) are good sources of protein. Carbohydrates are found in things like fresh fruits and juices, whole-grain breads, pancakes, and rice products.

A truly "good" breakfast will fuel your body, consume energy slowly and efficiently, and help you to perform at your best. An important point to remember, however: the largest shunting of blood supply to the stomach (instead of to the brain) is immediately after ingestion of food. Allow yourself a couple of hours between breakfast and test-taking. In this way, your body will be sending energy and nutrients to your brain at the time you sit for the test.

Structured review, adequate rest, and efficient nutrition will give you an undeniable advantage and improve your performance in any testing situation.

Suggestions for Practical Examination Preparation and Execution

Throughout your entire EMS career you will periodically be subjected to the dreaded practical skills examination stations. They accompany all levels of EMS courses, certification exams, recertification exams, and the National Registry examination. More and more often, practical stations also accompany continuing education workshops (such as Critical Trauma Care or Prehospital Trauma Life Support). Even the most veteran of EMS providers tremble in their boots when faced with the ordeal of performing in the situational exam station environment.

Why?

Skills and performance, knowledge, and physical abilities are placed under the strictest scrutiny. It is unnerving to be observed and evaluated, especially when your livelihood is threatened should you fail. Even the best simulations contain components that must be recognized and remembered although you never actually see, feel, hear, or smell them. You frequently are either alone or teamed with strangers. The equipment you are given is rarely in the configuration you are accustomed to. Consequently, you may fumble around, hunting for instruments that your hands would normally find on their own. Often when you do find the equipment, it is of a different brand, make, or model, and is unfamiliar. In addition to all the above, the evaluators themselves may appear to be "out to get you." In their effort to remain unbiased and objective, even the friendliest evaluators suddenly appear cold and perhaps hostile.

What can you do about these dilemmas?

Of course, regardless of the volume of calls you actually respond to, it is imperative that you remain skilled at operating all EMS equipment, and that you frequently refresh yourself regarding patient assessment and treatment protocols (especially ACLS protocols). Periodic practice and review are essential. Volunteering your time for the testing of others is helpful in developing awareness of common performance mistakes. But the biggest improvement between past and present performances can be made by changing your approach to the situational or practical exam station itself.

When appearing for a written exam, you bring your own pencil and eraser, do you not? A similar rule applies to any practical exam station. At the very least, wear your own, old-faithful holster. This puts vital items like scissors, penlight, pens, and the like at your fingertips, just where you are accustomed to finding them. Your own stethoscope ought to be draped, bundled, or strapped about whatever it is usually draped, bundled, or strapped.
Actually changing into your uniform before going through the stations is ideal. This may appear "gung-ho" to others, but these others are not grading your performance. Indeed, if you have your own emergency care kit, bring it. Your goal should be to clothe and equip yourself in a manner that makes you feel comfortable and makes your surroundings as familiar as possible.

The previous suggestions are easily applied to any practical exam station. The "assessment" or "situation" stations present a greater challenge, however. Here, the most difficult and most vital improvement you can make is in your performance style itself.

RULES FOR SITUATIONAL PRACTICAL SKILLS STATIONS

1. Fix your eyes and your attention on the evaluator. As you enter this station, do not preoccupy yourself with eyeing the patient or the scene to get "a head start" on the situation. Fix your eyes and your attention on the evaluator. Concentrate on the instructions and information she/he is providing.

2. If you are offered time to examine the equipment, do it. Look at it carefully, especially if it differs from the equipment you are accustomed to using. If possible, arrange it in a manner that is familiar to you. Do not leave it in a messy pile beside the patient.

3. When the evaluator describes the scenario, be alert for indications of the mechanism of injury or descriptions of the scene that cannot be simulated. If you are told that the steering wheel is bent, or the windshield fractured, or the furniture broken, it means something. Take time to consider the information these clues provide.
If this type of information is not offered, you must ask for it. What is the condition of the patient's apartment? Is it tidy? Messy? What does it smell like? What does the car look like? Where was the point of impact, and where was the patient sitting? Was the patient wearing a seatbelt? Is there compartmental intrusion involving the patient's space? Is the steering wheel bent or broken? Is there glass damage?

4. As you are directed to begin, no matter what the situation, the first words you utter are "Is the scene safe?" In real life you can see smoke on approach, smell a gas leak, hear a domestic altercation, or observe that the police have not yet arrived. For situational exams, however, the first words out of your mouth must be "Is the scene safe?"

5. From this moment on, there is no longer any reason for you to look at the evaluator. This is very important. Looking back and forth between the evaluator and the patient is distracting. It will interrupt your concentration and continuity. Focus your eyes and attention on the scene and the patient. Your eyes and hands should never leave the patient (except to obtain equipment). The evaluator doesn't need to see your face. She/he is listening to what you say and watching your skills performance as much as possible.

6. Never stop talking. Whether you are questioning the evaluator, questioning the patient, or describing your actions, you should never stop talking. Frequently, evaluators miss actions or skills performed by the testing party because they are making notes on a skills performance sheet. Verbalize every single thing you are doing, everything you are thinking.

7. Don't forget to talk to your patient. This may sound difficult, but it can be easily incorporated with the previous suggestion. Since you should be addressing your attention to the patient but verbalizing all your thoughts and actions for the evaluator, tell the patient what you are thinking and doing. As often as you are able, address all your questions to the patient.

8. If the patient doesn't have the answers, ask the evaluator, but do not remove your eye contact from the patient. The evaluator is in the same room. She/he can hear you without your needing to look at her/him.

What follows is a written example of what a situational practical skills station should sound like:

EVALUATOR: You may begin.

PERFORMER: Is the scene safe?

EVALUATOR: The scene is safe.

PERFORMER: Then I am observing the scene as I approach.
Is the mechanism of injury apparent?

I can see that my patient has some blood on his left thigh. Is the blood spurting?

First, I place my hand on his head and assess his level of consciousness.

Sir? Sir? Can you hear me?

Hi. My name is Mork. I'm a paramedic and I'm here to help you. I need you to keep very still and not move your head.

Any pain in your neck as I run my fingers down it?

My invisible partner, Mindy, is going to hold your head to help you keep it absolutely still. She will not stop holding your head until we have you secured to a long backboard.

What's your name?

Okay, Endor, first I'd like to check your airway. I'm going to look inside and make sure it's clear. Anything loose in there, Endor?

Good. Are you having any difficulty breathing?

When I listen to his chest with this stethoscope, can I hear any unusual noises when he breathes?

Does your chest hurt at all when I compress it?

It feels even in excursion and I don't feel any crepitus or see any wounds or deformities.

My invisible partner, Mindy, will observe your airway and respiratory effort as she continues to immobilize your C-spine, and alert me to any changes while I examine you further.

I'm going to check his pulses now. Do I feel his radial pulse?

What quality and rate do I feel?

Is he sweaty?

What is his skin color?

Temperature?

Endor, I'm going to give you some oxygen to help you feel better. This is a nasal cannula and may tickle your nose a bit, but I'll run it at four liters per minute and it will help you.

Yes, an evaluator could be scoring this performance over the telephone.

A continuous narration of your thoughts and activities assists your score in a number of ways. It keeps you focused on your patient and your task, ensuring that you proceed without forgetting things. This extra degree of attention focusing also helps to calm you down, improving your physical and mental concentration. Perhaps most importantly, verbalizing everything you do and think will make it nearly impossible for the evaluator to miss what you have done. Without this technique, the evaluator may miss your sweeping check of the patient's clothing for gross bleeding because she/he was making notes on the skills performance sheet and didn't actually observe the check.

GROUP APPROACHES TO THE SITUATIONAL EXAM STATION

All the previous suggestions apply here as well. However, now you are working with other participants, who frequently were total strangers only moments ago. The old adage, "Too many cooks spoil the broth," applies here in quadruplicate. Few things are as debilitating and disastrous as having two, three, or four paramedics crawling all over each other, trying to treat a patient at the same time.

The secret is organization and assignment of tasks. Each group member must have an assigned task, with an assigned group leader in charge. As each new station is approached, the tasks must be clearly reassigned to allow each participant an opportunity to rotate through each different task performance.

Groups of two are easy. For trauma situational exam stations, one partner is Group Leader and the other is the C-spine/Airway Monitor. The Group Leader introduces self and partner, directing partner to maintain immobilization of the C-spine continuously and to monitor for changes in airway status after the initial examination. The Group Leader examines the patient and performs all necessary treatment.

The C-spine/Airway Monitor maintains spinal immobilization and observes airway/respiratory status no matter how tempted she/he is to help with other treatment. While doing so, however, she/he may verbally assist whenever the Group Leader seems to have forgotten something.

For groups of two in the medical situational exam stations the C-spine/Airway Monitor becomes an "Equipment Operator." The Equipment Operator manages oxygen equipment, takes vital signs, applies the ECG monitor, positions the backboard and/or pram, and so on. The Group Leader directs the activity of the Equipment Operator and is still responsible for the patient examination.

Groups of three responders are broken down into assignments of the Group Leader, the C-spine/Airway Monitor, and the Equipment Operator.

Groups of four responders may be broken down into assignments of the Group Leader, the C-spine Immobilizer, the Airway Monitor, and the Equipment Operator. If active airway maintenance is not required, the Airway Monitor becomes a secondary Equipment Operator. And the C-spine Immobilizer resumes observation of the airway.

If you have the misfortune of operating in a five-or-more-member team, send the fifth-or-more members to direct traffic.

The most important key to group performance, no matter how you divide the tasks, is that the assignment's responsibilities are clearly understood by each member.

In summary:

1. Wear your own equipment holster and stethoscope. Bring your own kit if you have one, and wear your uniform. Make the testing environment more comfortable and familiar.

2. As you first enter the station, fix your eyes and attention on the evaluator. Don't "jump the gun." Listen to the clues that the evaluator provides. Take the time to check the equipment provided. Ask for scene or situation information if it is not offered.

3. "Is the scene safe?"

4. After beginning your performance, never let your eyes or hands leave the patient.

5. Never stop talking. Talk to the patient, ask questions, and describe everything you are thinking and doing. Don't look away from the patient to ask questions of the evaluator; she/he does not need to see your face to answer you.

6. Clearly assign specific tasks to group members and take turns with task performance.

7. Do not physically deviate from your assigned task. However, you may verbally remind partners of business if they appear to have forgotten something.

8. Above all, try to enjoy yourself. Take pride in the performance level you have worked so hard to achieve and have confidence in your abilities.

1

Test Section One

Test Section One covers the following subjects:

* Roles and Responsibilities
* EMS Systems
* Medical/Legal Considerations
* EMS Communications
* Rescue
* Major Incident Response
* Stress Management
* Medical Terminology
* General Patient Assessment

EMT - Paramedic
National Standards Review Self Test
Third Edition

1. Which of the following statements regarding ethics is false?
   (a) Ethics are principles governing the conduct of the paramedic.
   (b) Ethics deal with the relationship of a paramedic to her/his patient and the patient's family.
   (c) Ethics deal with the relationship of a paramedic to her/his peers and society at large.
   (d) Ethics set standards regarding right and wrong human conduct, but do not address morality.
   (e) If the paramedic places personal considerations above all else when providing medical care, she/he will rarely have to worry about committing an unethical act.

2. Which of the following statements regarding professionalism is false?
   (a) A professional receives monetary compensation for her/his work on a regular basis.
   (b) A professional has certain special skills and knowledge in a specific area.
   (c) A professional conforms to the standards of conduct and performance in a specific field of knowledge.
   (d) Professionalism promotes quality in patient care.
   (e) Professionalism instills pride in the profession and earns the respect of the medical team.

3. The role of a paramedic includes all of the following, except
   (a) initiating and continuing emergency care under medical control and providing appropriate invasive and noninvasive treatments.
   (b) exercising personal judgment in case of immediate life-threatening conditions or interruption in medical direction.

(c) attempting rescue despite lack of personal training during incidents where adequately trained rescue personnel are not available, and the patient is in immediate life-threat.
(d) directing the maintenance and preparation of emergency care equipment and supplies.
(e) recording the details related to the incident and the patient's emergency care.

4. The process by which an agency or association grants recognition to an individual who has met certain predetermined qualifications as specified by that agency is called
(a) reciprocity.
(b) certification.
(c) licensure.
(d) Both answers (a) and (b).
(e) Answers (a), (b), and (c).

5. The process by which a governmental agency grants permission to an individual to engage in a given occupation upon review of the applicant's degree of competency is called
(a) reciprocity.
(b) certification.
(c) licensure.
(d) Both answers (a) and (b).
(e) Answers (a), (b), and (c).

6. The mutual exchange of privileges or permissions by separate governmental agencies is called
(a) reciprocity.
(b) certification.
(c) licensure.
(d) Both answers (a) and (b).
(e) Answers (a), (b), and (c).

7. Which of the following statements regarding the requirement of continuing education is false?

(a) Skills and knowledge acquired in a paramedic course may not be used with great frequency, and will quickly decay without periodic refreshment.
(b) Maintenance of continuing education assures the public and medical community that quality patient care continues to be delivered.
(c) Maintenance of continuing education is a basis for reciprocity between many states.
(d) Continuing education is a requirement that has little worth to the active street paramedic.
(e) Continuing education keeps providers informed of changes in medical protocol and the development of new skills and equipment.

8. Major benefits of subscribing to professional journals include all of the following, except that

(a) they are a tax deduction.
(b) they are a source of continuing education.
(c) they provide an opportunity for the paramedic to publish articles.
(d) they are an information source whereby paramedics can learn about other local, state, regional, or national advancements and/or issues.
(e) they encourage professional growth and awareness.

9. The EMS system includes

(a) citizen-initiated care and paramedic-initiated care.
(b) care received in the emergency department.
(c) preincident planning and incident follow-up.
(d) Answers (a) and (b) only.
(e) Answers (a), (b), and (c).

10. Standing orders regarding patient management guidelines are examples of

(a) direct medical control.
(b) indirect medical control.
(c) intermittent medical control.
(d) All of the above.
(e) None of the above.

11. Verbal orders regarding patient management guidelines are examples of

(a) direct medical control.
(b) indirect medical control.
(c) intermittent medical control.
(d) All of the above.
(e) None of the above.

12. Major incident protocols are examples of

(a) direct medical control.
(b) indirect medical control.
(c) intermittent medical control.
(d) All of the above.
(e) None of the above.

13. Training, education, and chart review requirements are examples of

(a) direct medical control.
(b) indirect medical control.
(c) intermittent medical control.
(d) All of the above.
(e) None of the above.

14. Emergency medical control at the scene should go to

(a) the responder who arrives at the scene first.
(b) any licensed physician on scene, despite a lack of emergency medical knowledge.
(c) the responder with the most knowledge and experience in prehospital delivery of care.
(d) the fire chief.
(e) the police department.

15. The law classification that deals with domestic or contractual wrongs committed by one individual against another is called

(a) criminal law.
(b) litigation.
(c) civil (tort) law.
(d) All of the above.
(e) None of the above.

16. The law classification that defines crimes and associated punishments is called

(a) criminal law.
(b) litigation.
(c) civil (tort) law.
(d) All of the above.
(e) None of the above.

17. A malpractice suit is an example of

(a) criminal law.
(b) litigation.
(c) civil (tort) law.
(d) All of the above.
(e) None of the above.

18. A paramedic practicing medicine without a license is liable for _______ proceedings.

(a) criminal law
(b) litigation
(c) civil (tort) law
(d) All of the above.
(e) None of the above.

19. A medical practice act

(a) may differ from state to state.
(b) defines the scope of practice and required licensure or certification of the paramedic.
(c) deals with medical control, protocols, and communications.
(d) Answers (a) and (b) only.
(e) Answers (a), (b), and (c).

20. Which of the following statements regarding the Good Samaritan act is true?

(a) The Good Samaritan act differs from state to state.
(b) In the United States, the Good Samaritan act provides protection to a person who renders care without performing procedures she/he is not trained for, but only when she/he does not receive payment for that care.

(c) The Good Samaritan act provides protection to persons who provide care, even when they perform procedures without previous training, but only when the situation can be proven to involve a threat to human life.
(d) All of the above are true.
(e) None of the above is true.

21. Which of the following statements regarding motor vehicle laws is true?

(a) Motor vehicle laws vary considerably from state to state.
(b) It is mandatory, in all states, that the paramedic be familiar with appropriate statutes regarding operation of emergency vehicles.
(c) Lights and sirens do not give the paramedic ambulance the right-of-way.
(d) All of the above are true.
(e) None of the above is true.

22. In the majority of states, the paramedic has an obligation to report all of the following, except

(a) abuse or neglect of children and the elderly.
(b) rape.
(c) gunshot wounds.
(d) alcohol or drug abuse.
(e) animal bites.

23. You are called to a private residence, where you are met by a woman who tells you her husband has "passed on." She further states that the patient had terminal cancer and had signed papers requesting no resuscitation attempts when he died. These papers would be an example of what is known as

(a) a death wish.
(b) a living will.
(c) suicidal ideation.
(d) All of the above.
(e) None of the above.

24. Unfortunately, the woman cannot find the papers. But she assures you they were written by her husband's lawyer and even notarized. You assess the patient and note him to be jaundiced, emaciated, apneic, and pulseless. You should

(a) initiate all aspects of ACLS except intubation, and transport the patient in the absence of a legal document.
(b) allow the wife a minimum of 10 more minutes to locate the papers before initiating resuscitation.
(c) use the "quick-look" paddles and initiate resuscitation only if the patient is in coarse V fib.
(d) recognize that the patient was suffering from terminal illness, respect his wishes by withholding resuscitation, but have his wife sign the trip sheet as having refused care.
(e) initiate BLS resuscitation, call medical control, and discuss your findings with your physician.

25. The paramedic is held responsible for performing a standard of care equal to that of

(a) any experienced EMT–Basic.
(b) any experienced EMT–Intermediate.
(c) any paramedic of similar experience.
(d) All of the above.
(e) None of the above.

26. Conduct failing to meet the standard of paramedic care is called

(a) willful disobedience.
(b) negligence.
(c) abandonment.
(d) Any of the above.
(e) None of the above.

27. The elements that must be proved in order to find a paramedic guilty of malpractice in a court of law include all of the following, except

(a) that the patient asked the paramedic for help.
(b) that the paramedic had a duty to perform.
(c) that a breach of duty to perform occurred.
(d) that damages resulted from the breach of duty.
(e) that the paramedic's action (or lack of action) was the proximate cause of the damage that occurred.

28. A patient who is unconscious may be treated under the principle of

(a) informed consent.
(b) expressed consent.
(c) implied consent.
(d) Either answer (a) or (b).
(e) Either answer (b) or (c).

29. A patient who is alert and oriented to person, place, and time may be treated only when there is

(a) informed consent.
(b) expressed consent.
(c) implied consent.
(d) Either answer (a) or (b).
(e) Either answer (b) or (c).

30. When a patient gives verbal or written consent to treatment without prior information regarding the treatment, it is called

(a) informed consent.
(b) expressed consent.
(c) implied consent.
(d) Either answer (a) or (b).
(e) Either answer (b) or (c).

31. You have transferred a nursing home resident to the emergency department for evaluation of a urinary tract infection. As you arrived, your dispatcher notified you of a 911 call waiting. All of the emergency room nurses are busy, but the admissions clerk listens to your report and assures you that she will inform the receiving nurse. You leave the patient on a hall bed with the clerk and respond to the emergency call. Which of the following statements regarding this situation is true?

(a) Your emergency room transfer was legally completed when the admissions clerk received the patient.
(b) Emergency (911) calls have legal priority over nursing home or interhospital transfers.
(c) You have abandoned your patient and may now be liable for a negligence conviction.
(d) If the clerk has EMT–Basic qualifications, she may legally relieve you of care.
(e) If the clerk has EMT–Intermediate qualifications, she may legally relieve you of care.

32. Which of the following statements regarding refusal of treatment is true?

(a) A patient who is alert and oriented to person, place, and time may refuse treatment only after the potential consequences of refusal have been explained.
(b) Any patient who requires treatment should receive it regardless of attempts to refuse.
(c) In the absence of a parent or legal guardian, a minor may refuse treatment.
(d) Any patient who signs a "release from liability" statement may refuse treatment.
(e) A person with an altered level of consciousness may refuse treatment only if over 18 years of age.

33. Which of the following is an example of a potential criminal offense?

(a) assault
(b) battery
(c) false imprisonment
(d) All of the above.
(e) None of the above.

34. Which of the following is an example of a potential civil offense?

(a) assault
(b) battery
(c) false imprisonment
(d) All of the above.
(e) None of the above.

35. Touching a patient without consent is called

(a) assault.
(b) battery.
(c) false imprisonment.
(d) All of the above.
(e) None of the above.

36. Creating apprehension of immediate bodily harm is called

(a) assault.
(b) battery.
(c) false imprisonment.
(d) All of the above.
(e) None of the above.

37. Intentional and unjustifiable detention is called
(a) assault.
(b) battery.
(c) false imprisonment.
(d) All of the above.
(e) None of the above.

38. Injuring a person's character, name, or reputation by false and malicious spoken words is called
(a) misconduct.
(b) libel.
(c) negligence.
(d) slander.
(e) malpractice.

39. The paramedic who breaches patient confidentiality by verbal discussion of the call with others may be charged with
(a) misconduct.
(b) libel.
(c) negligence.
(d) slander.
(e) malpractice.

40. Injuring a person's character, name, or reputation by false and malicious writing is called
(a) misconduct.
(b) libel.
(c) negligence.
(d) slander.
(e) malpractice.

41. A paramedic's best protection when she or he is involved in malpractice litigation is
(a) a good EMS lawyer.
(b) accurate, detailed completion of the trip report.
(c) a photographic memory.
(d) repeatedly insisting, "I do not remember."
(e) police testimony.

42. Which of the following statements regarding malpractice insurance is true?

(a) Malpractice insurance may not cover the individual while not on duty.
(b) Malpractice insurance is frequently very limited in coverage.
(c) If your employer carries malpractice insurance for you, there is no need for additional, personal policies.
(d) Both answers (a) and (b) are true.
(e) Answers (a), (b), and (c) are true.

43. A hand-held communications device with single or multiple channels is called

(a) a base station.
(b) a mobile two-way radio.
(c) a portable radio.
(d) an encoder.
(e) a decoder.

44. A communications device mounted within a vehicle, operating on single or multiple channels is called

(a) a base station.
(b) a mobile two-way radio.
(c) a portable radio.
(d) an encoder.
(e) a decoder.

45. A transmitter and receiver housed in the same cabinet that usually can only transmit on one channel at a time is called

(a) a base station.
(b) a mobile two-way radio.
(c) a portable radio.
(d) an encoder.
(e) a decoder.

46. The ______ receives a transmission from a low-power portable or mobile radio on one frequency and retransmits it at a higher power on another frequency.

(a) an antenna
(b) transmitter
(c) repeater
(d) encoder
(e) decoder

47. When activated, the _______ sends pulses or tones over the air.

(a) antenna
(b) transmitter
(c) repeater
(d) encoder
(e) decoder

48. When _______ receives the correct "code" of pulses or tones, the audio circuits of the receiver are turned on.

(a) an antenna
(b) a transmitter
(c) a repeater
(d) an encoder
(e) a decoder

49. Radio frequencies are designated by cycles per second. One cycle per second is called a

(a) kilohertz (KHz).
(b) hertz (Hz).
(c) gigahertz (GHz).
(d) microhertz (MCz).
(e) megahertz (MHz).

50. One thousand cycles per second is called a

(a) kilohertz (KHz).
(b) hertz (Hz).
(c) gigahertz (GHz).
(d) microhertz (MCz).
(e) megahertz (MHz).

51. One million cycles per second is called a

(a) kilohertz (KHz).
(b) hertz (Hz).
(c) gigahertz (GHz).
(d) microhertz (MCz).
(e) megahertz (MHz).

52. One billion cycles per second is called a
(a) kilohertz (KHz).
(b) hertz (Hz).
(c) gigahertz (GHz).
(d) microhertz (MCz).
(e) megahertz (MHz).

53. Which of the following statements regarding VHF is true?
(a) VHF stands for "very high frequency."
(b) VHF has a somewhat longer range than UHF.
(c) VHF has somewhat better penetration in dense urban areas than UHF.
(d) Both answers (a) and (b) are true.
(e) Both answers (a) and (c) are true.

54. Which of the following statements regarding UHF is true?
(a) UHF stands for "ultrahigh frequency."
(b) UHF has a somewhat longer range than VHF.
(c) UHF has somewhat better penetration in dense urban areas than VHF.
(d) Both answers (a) and (b) are true.
(e) Both answers (a) and (c) are true.

55. Radio equipment used for EMS communications typically employs ______, which is less susceptible to interference than ______.
(a) FM (faster modulation)/AM (altered modulation)
(b) AM (amplitude modulation)/FM (frequency modulation)
(c) AM (altered modulation)/FM (faster modulation)
(d) AM (amplitude modulation)/FM (faster modulation)
(e) FM (frequency modulation)/AM (amplitude modulation)

56. Which of the following statements regarding biotelemetry of ECGs is false?
(a) ECG voltage changes are converted to audio tones and sent over the air to the hospital receiver, which converts the audio signal back to voltage changes to reproduce the ECG.
(b) When biotelemetry is available, the paramedic is no longer legally responsible for mistakes made in care due to the misinterpretation of ECG findings.

(c) Telemetry interference can be caused by simultaneous transmission of voice.
(d) Telemetry interference can be caused by loose ECG electrodes or muscle tremor.
(e) Telemetry interference can be caused by 60-cycle hum or fluctuations in transmitter power.

57. A _______ transmission system uses one frequency and does not allow for simultaneous two-way communications.

(a) simplex
(b) duplex
(c) multiplex
(d) All of the above.
(e) None of the above.

58. Simultaneous two-way communication, much like a telephone conversation, requires two frequencies and is called a _______ transmission system.

(a) simplex
(b) duplex
(c) multiplex
(d) All of the above.
(e) None of the above.

59. The combination of the signals (ECG/voice) for transmission simultaneously on one channel is called a _______ transmission system.

(a) simplex
(b) duplex
(c) multiplex
(d) All of the above.
(e) None of the above.

60. Responsibilities of the dispatcher include all of the following, except

(a) knowing the location of all vehicles (ALS and BLS).
(b) directing the appropriate emergency vehicle(s) to the correct address.
(c) recognizing a "prank" call and disconnecting the caller.
(d) instructing the caller in measures that should be taken until assistance arrives.
(e) maintaining written and recorded records.

61. Which of the following statements regarding radio codes is false?

(a) "Plain English" often works as well or better than codes.
(b) Codes are useless unless everyone in the system understands them.
(c) Medical information radio codes will speed delivery of care by minimizing the time required to obtain medication orders.
(d) Codes can shorten radio air time while still providing clear information.
(e) Codes enable transmission of information in a format not understood by the patient, family, or bystanders.

62. Which of the following statements regarding radio communication techniques is true?

(a) Listen to the channel before transmitting, then "key the mike" just as you begin to speak.
(b) Avoid speaking too close to the radio, keeping the microphone 5–10 inches away, or speaking above the microphone level.
(c) Do not use the patient's name over the radio.
(d) Do not waste air time with the once-popular "echo" procedure; if you do not require clarification of the instructions, simply acknowledge by saying "copy."
(e) Use the radio rather than the telephone to protect patient privacy.

63. Radio communication of patient information frequently should include all of the following, except

(a) unit call name and number or name of the paramedic.
(b) the patient's name, sex, ethnic group.
(c) the patient's chief complaint and associated symptoms.
(d) brief pertinent medical history, medications, and allergies.
(e) physical exam findings.

64. Radio communication of physical exam findings include all of the following, except

(a) whether the patient is lethargic, obtunded, or awake.
(b) general appearance and degree of distress.
(c) vital signs, neuro exam, ECG (if applicable).
(d) significant positive findings.
(e) significant negative findings.

65. All of the following statements regarding paramedic communications with the physician from the field are true, except

(a) Always "echo" back orders given by the physician.
(b) Always question orders that are not clear.
(c) Always question orders that do not seem appropriate for the patient's condition.
(d) Always contact the physician unless transport of the injured patient is not deemed necessary.
(e) Always consult with the physician whenever you are uncertain of what course to take.

66. Which of the following statements regarding ECG telemetry is true?

(a) Use of telemetry alone, without field interpretation of ECG, is appropriate.
(b) The paramedic should always request the base physician's ECG interpretation prior to initiating care.
(c) Field interpretation alone, without telemetry, is acceptable.
(d) With advanced telemetry equipment, ECG distortion during transmission does not occur.
(e) Continuous telemetry transmission is advisable, to allow the base physician to monitor the patient's status en route.

67. Which of the following statements regarding written communications (EMS forms) is false?

(a) EMS forms are a written record of the patient's initial condition, and remain at the hospital after the paramedics have left.
(b) EMS forms are a legal record of medical treatment rendered in the patient's prehospital phase of care.
(c) The EMS form can provide documentation of a patient's refusal of care and/or transportation.
(d) EMS forms are used for medical audits, quality control, data collection, and billing.
(e) EMS forms must be complete, legible, and signed by the receiving admissions clerk.

68. Rescue is defined as

(a) risk of personal well-being to free another from confinement.
(b) risk of personal safety to free another from danger or evil.
(c) to free from confinement, danger, or evil.
(d) Both answers (a) and (b).
(e) None of the above.

69. Which of the following statements regarding safety during rescue efforts is false?

(a) The safety of the rescuer must be considered prior to any action on the part of the rescuer.
(b) When needed, the rescuer should have prior access to protective headgear, coats, and pants or coveralls.
(c) When needed, eye protection, gloves, and protective boots or shoe covers should be worn.
(d) Hearing protection is rarely needed and does not need to be considered by the paramedic.
(e) A breathing apparatus should be worn only by those rescuers trained to operate such equipment.

70. Which of the following statements regarding rescue efforts is false?

(a) Protection of the patient from the environment and scene hazards must be considered prior to any action on the part of the rescuer.
(b) When needed, protective blankets, shields, and headgear should be provided to the patient.
(c) Hearing protection for the patient does not need to be considered by the paramedic.
(d) All of the above are false.
(e) None of the above is false.

71. Place the following phases of rescue operations in appropriate order:

(1) transportation
(2) emergency care
(3) removal
(4) gaining access
(5) assessment
(6) disentanglement

(a) 6, 4, 2, 3, 5, 1
(b) 5, 4, 2, 6, 3, 1
(c) 2, 5, 6, 3, 4, 1
(d) 4, 5, 2, 6, 3, 1
(e) 3, 6, 4, 5, 2, 1

72. Scene hazards include
(a) chemical spills, gases, possibilities of fire, electrical hazards, and environmental hazards.
(b) sharp objects, compartment collapse, vehicle instability, and traffic.
(c) nonemergency personnel, bystanders, and crowds.
(d) Answers (a) and (b) only.
(e) Answers (a), (b), and (c).

73. A multiple-casualty incident (MCI) is described as
(a) any situation that will stress local EMS resources.
(b) any situation involving more patients that can be handled by responding units.
(c) any situation involving multiple sites or community-wide disasters.
(d) any situation that necessitates establishment of an Emergency Operations Center (EOC).
(e) Any of the above.

74. Upon first recognition of a major multiple-patient or mass-casualty incident, the most important act is to
(a) quickly obtain a "head count" of potential casualties.
(b) begin treatment of the patient closest to the ambulance.
(c) designate a medical control officer.
(d) immediately activate the EMS incident command system.
(e) cordon off areas for use as a morgue, a triage area, and a staging sector.

75. The most important aspect of any major EMS incident response plan is to
(a) quickly obtain a "head count" of potential casualties.
(b) designate a single incident commander.
(c) designate a treatment officer.
(d) designate a radio control officer.
(e) quickly triage the "walking wounded."

76. Functions of the incident commander include all of the following, except
(a) requesting other EMS assistance as appropriate.
(b) coordinating the overall actions at the scene.
(c) triaging patients as they are extricated.
(d) controlling radio communication between the scene and medical control.
(e) identifying her/himself as incident commander.

77. Which of the following statements regarding transfer of EMS command is true?

(a) The first EMS unit to arrive on scene activates the major EMS incident response plan and awaits the arrival of an incident commander before taking any further action.
(b) It is vitally important that the initial incident commander remain in command until the incident is resolved (or she/he is too exhausted to continue).
(c) When a higher-ranking official arrives, regardless of her/his level of training or experience, command must be transferred to the higher rank.
(d) The transfer of command may not occur until the officer assuming command has been fully briefed concerning the activities that have already occurred.
(e) None of the above is true.

78. Locating a suitable treatment sector area and evaluating resources required for patient treatment is the responsibility of the

(a) extrication sector personnel.
(b) treatment sector personnel.
(c) staging sector personnel.
(d) triage/treatment sector officer.
(e) incident commander.

79. Removal of the victims from the immediate area of the incident to the treatment sector is accomplished by

(a) extrication sector personnel.
(b) treatment sector personnel.
(c) staging sector personnel.
(d) triage officer.
(e) incident commander.

80. On-scene patient care is the responsibility of the

(a) extrication sector personnel.
(b) treatment sector personnel.
(c) staging sector personnel.
(d) triage officer.
(e) incident commander.

81. As additional units arrive, they await transportation assignments in the

(a) extrication sector.
(b) treatment sector.

(c) staging sector.
(d) triage sector.
(e) incident command sector.

82. *Stress* is defined as
(a) an acute hostile reaction to an unusual or disturbing event.
(b) a loss of emotional control when faced with unexpected events.
(c) a nonspecific mental or physical response of the body to any demand made upon it.
(d) Any of the above.
(e) None of the above.

83. Any agent or situation that causes stress is called
(a) an aggressor.
(b) a problem.
(c) a hazard.
(d) an aggravator.
(e) a stressor.

84. Which of the following statements regarding causes of stress is false?
(a) Stress is caused by loss of something that is of value to an individual, frustrations of drives, or ineffective coping abilities.
(b) Pleasant situations may cause stress.
(c) Poor health or nutrition may cause stress.
(d) All of the above are false.
(e) None of the above is false.

85. The stress stage wherein the individual begins to adapt and physiologic parameters of the body return to normal is called
(a) alarm reaction stage of stress.
(b) stage of resistance.
(c) stage of acceptance.
(d) stage of hostility.
(e) stage of exhaustion.

86. The stress stage that follows long, continued exposure to the cause of stress is called the

(a) alarm reaction stage of stress.
(b) stage of resistance.
(c) stage of acceptance.
(d) stage of hostility.
(e) stage of exhaustion.

87. Physiologic and emotional response is greatest during the

(a) alarm reaction stage of stress.
(b) stage of resistance.
(c) stage of acceptance.
(d) stage of hostility.
(e) stage of exhaustion.

88. Defense mechanisms

(a) are adaptive functions of the personality that assist us in adjusting to stressful situations.
(b) are healthy unless overused to the degree that they distort reality.
(c) are employed to seek relief of stress and may be conscious efforts or unconscious and automatic.
(d) All of the above.
(e) None of the above.

89. The defense mechanism that involves a conscious or unconscious attempt to overcome real or imagined shortcomings by developing individual skills or traits to make up for those shortcomings is called

(a) repression.
(b) regression.
(c) projection.
(d) rationalization.
(e) compensation.

90. The defense mechanism that involves the return to an earlier level of emotional adjustment is called

(a) repression.
(b) regression.
(c) projection.
(d) rationalization.
(e) compensation.

91. The defense mechanism that involves involuntary banishment of unacceptable ideas or impulses into the unconscious is called

(a) repression.
(b) regression.
(c) projection.
(d) rationalization.
(e) compensation.

92. The defense mechanism that involves attributing another person or object with those thoughts, feelings, motives, or desires which are really one's own unacceptable traits is called

(a) repression.
(b) regression.
(c) projection.
(d) rationalization.
(e) compensation.

93. The defense mechanism that involves inappropriately childish behavior in an adult is called

(a) repression.
(b) regression.
(c) projection.
(d) rationalization.
(e) compensation.

94. The defense mechanism that involves anger with oneself but may be manifested by aggression toward others is called

(a) repression.
(b) regression.
(c) projection.
(d) rationalization.
(e) compensation.

95. The defense mechanism that involves a way of "explaining" our behavior which may be self-deceiving is called

(a) repression.
(b) regression.
(c) projection.
(d) rationalization.
(e) compensation.

96. The defense mechanism that involves an unconscious, retrospective process of ascribing worthwhile motives to feelings, thoughts, or behaviors that really have other unrecognized motives is called

(a) repression.
(b) regression.
(c) projection.
(d) rationalization.
(e) compensation.

97. The defense mechanism that involves the replacement of an unattainable or unacceptable activity by one that is attainable or acceptable is called

(a) reaction formation.
(b) sublimation.
(c) denial.
(d) substitution.
(e) isolation.

98. The defense mechanism that involves the redirection of an emotion from the original object to a more acceptable replacement object is called

(a) reaction formation.
(b) sublimation.
(c) denial.
(d) substitution.
(e) isolation.

99. The defense mechanism that involves the diversion of unacceptable instinctual drives into socially acceptable channels is called

(a) reaction formation.
(b) sublimation.
(c) denial.
(d) substitution.
(e) isolation.

100. The defense mechanism that involves conscious retention of the stressful memory, but not the feeling that accompanied it, is called

(a) reaction formation.
(b) sublimation.
(c) denial.
(d) substitution.
(e) isolation.

101. The defense mechanism that involves the unconscious disavowal of thoughts, feelings, wishes, or needs that are consciously unacceptable is called

(a) reaction formation.
(b) sublimation.
(c) denial.
(d) substitution.
(e) isolation.

102. The defense mechanism that involves laughing and joking in order to avoid crying and feelings of remorse or profound sadness is called

(a) reaction formation.
(b) sublimation.
(c) denial.
(d) substitution.
(e) isolation.

103. Anxiety is defined as

(a) an emotional state caused by stress that is a key ingredient in the coping process.
(b) a loss of emotional control in response to the stress of an unexpected event.
(c) an acute hostile reaction to the stress of an unusual or disturbing event.
(d) All of the above.
(e) None of the above.

104. Which of the following statements regarding anxiety is false?

(a) Anxiety that is caused by internal conflicts between learned personal expectations and current motivations can be very helpful.
(b) Anxiety alerts a person to impending danger and maintains all potential resources (body and mind) in readiness for emergencies.
(c) Anxiety is a consequence of each individual's personal perceptions of the environment.
(d) Anxiety is a consequence of each individual's internalized psychological processes.
(e) Anxiety that is caused by external events and people around us is very harmful.

105. Which of the following statements regarding normal anxiety is true?

(a) Whether or not anxiety is "normal" is determined by what we get anxious about and is relative in intensity.
(b) Normal anxiety is adaptive, helping us cope by narrowing and focusing our field of attention.
(c) Normal anxiety enables us to increase tolerance for stress by developing coping mechanisms and defenses.
(d) All of the above are true.
(e) None of the above is true.

106. Detrimental reactions to anxiety or stress include all of the following, except

(a) reactionary coping behaviors that are disproportionate to the actual danger.
(b) increase in sympathetic discharge with release of epinephrine and norepinephrine.
(c) interference with thought processes.
(d) interference with performance of daily activity.
(e) physical dysfunction or illness.

107. Physiologic effects of anxiety or stress

(a) are mediated by the autonomic nervous system and involve intense sympathetic discharge and increased corticosteriod production.
(b) include increased blood pressure, heart rate, and blood glucose.
(c) include shunting of blood to the muscles, pupil dilation, and reduced GI peristalsis.
(d) Answers (b) and (c) only.
(e) Answers (a), (b), and (c).

108. Common physiologic effects of anxiety or stress that can be consciously felt include all of the following, except

(a) anorexia, nausea, vomiting, abdominal cramps, flatulence, or "butterflies."
(b) heart palpitations, dyspnea, tachypnea, dry mouth, chest tightness, or chest pain.
(c) increased blood pressure, increased blood glucose, and reduced GI peristalsis.
(d) body temperature fluctuations, diaphoresis, urgency and frequency of urination, dysmennorhea, or decreased sexual drive.
(e) backache, headache, aching muscles or joints.

109. Which of the following statements regarding appropriate management of stress is false?

(a) Learn to accept what cannot be changed if it is beyond your control to change it.
(b) Remember that mild stress is protective and improves performance.
(c) Lack of rest and improper diet will increase the potential harmfulness of stress.
(d) Physical exertion will increase the harmfulness of stress.
(e) Discussion of stressful events with peers or professional counselors (in private or in groups) will provide reduction of stress.

110. When dealing with death and dying, which of the following statements regarding the stage of denial and isolation is true?

(a) Denial and isolation are used by almost all patients and are a temporary stage.
(b) Denial and isolation are temporary but may be recurrent.
(c) Denial and isolation are healthy, acting as a buffer between the shock and dealing with it.
(d) Both answers (a) and (c) are true.
(e) Answers (a), (b), and (c) are true.

111. When dealing with death and dying, which of the following statements regarding the anger stage is true?

(a) The patient and/or family may displace their feelings of anger by being hostile with the paramedic.
(b) A show of tolerance in response to this hostility will encourage escalation of the anger.
(c) Avoid discussion of the patient's and/or family's anger and discreetly summon the police for protection.
(d) All of the above are true.
(e) None of the above is true.

112. The bargaining stage of death and dying

(a) is an attempt to postpone the inevitable.
(b) is a normal defense mechanism.
(c) is an attempt to reach an agreement or "bargain" that will prevent death.
(d) All of the above.
(e) None of the above.

113. The depression stage of death and dying
(a) is an abnormal response to the inevitable.
(b) is accompanied by repressed anger and the question, "Why me?"
(c) involves a great sense of loss that will not be alleviated by any amount of reassurance.
(d) Both answers (a) and (b).
(e) Both answers (a) and (c).

114. When dealing with death and dying, which of the following statements regarding the stage of acceptance is false?
(a) At this stage, the family is more in need of help, understanding, and support than the patient.
(b) The patient frequently develops a sensation of happiness and becomes more involved with loved ones in preparation for death.
(c) The patient is without fear or despair, becoming void of feelings.
(d) All of the above are false.
(e) None of the above is false.

115. When dealing with the death of a patient, the paramedic
(a) may go through the same grief stages as the family.
(b) may need to express feelings of guilt or helplessness.
(c) should recognize that expressions of emotions felt for a stranger's death are inappropriate.
(d) Answers (a) and (b) only.
(e) Answers (a), (b) and (c).

116. Which of the following statements regarding provision of care for a terminally ill or injured patient is true?
(a) Avoid discussing the terminal nature of the patient's condition.
(b) Smile frequently, in a reassuring manner, and insist that the patient will soon be much better.
(c) Acknowledge that the patient is very ill or injured, but do not tell the patient that you believe he is dying if the question is asked.
(d) All of the above are true.
(e) None of the above is true.

117. When dealing with the family of a patient who is dead, the paramedic should remember

(a) that it is not within her/his job parameters to deal with the psychological needs of the patient's family; refer them to a professional.
(b) never to use the word "dead," instead using the phrases "passed away," or "has expired."
(c) that it is natural to feel uncomfortable in this situation.
(d) Answers (a) and (b) only.
(e) Answers (b) and (c) only.

118. The prefix *trans-* means

(a) across.
(b) between.
(c) above.
(d) within.
(e) by the side of.

119. The prefix *supra-* means

(a) across.
(b) between.
(c) above.
(d) within.
(e) by the side of.

120. The prefix *para-* means

(a) across.
(b) between.
(c) above.
(d) within.
(e) by the side of.

121. The prefix *intra-* means

(a) across.
(b) between.
(c) above.
(d) within.
(e) by the side of.

122. The prefix *inter-* means
(a) across.
(b) between.
(c) above.
(d) within.
(e) by the side of.

123. The term *neuralgia* means
(a) tumor of a nerve.
(b) nerve paralysis.
(c) pain along a nerve.
(d) diminished sensation.
(e) disease of the nerves.

124. The term *leukocyte* means
(a) red blood cell.
(b) white blood cell.
(c) nerve cell.
(d) blood cell.
(e) disease of the blood cells.

125. The term *anemia* means
(a) tumor of the blood.
(b) white blood cell.
(c) red blood cell.
(d) lack of blood.
(e) disease of the blood.

126. The term *neuroma* means
(a) tumor of a nerve.
(b) nerve cell.
(c) pain along a nerve.
(d) diminished sensation.
(e) disease of the nerves.

127. The term *neuropathy* means
(a) tumor of a nerve.
(b) nerve cell.
(c) pain along a nerve.
(d) diminished sensation.
(e) disease of the nerves.

128. The prefix _____ means "pertaining to a joint."

(a) angio-
(b) ante-
(c) anti-
(d) arter-
(e) arthro-

129. The prefix _____ means "blood vessel."

(a) angio-
(b) ante-
(c) anti-
(d) arter-
(e) arthro-

130. The prefix _____ means "against, opposed to."

(a) angio-
(b) ante-
(c) anti-
(d) arter-
(e) arthro-

131. The suffix -*centesis* means

(a) creation of an opening.
(b) a cutting out.
(c) gradual decline; weakening.
(d) puncturing.
(e) repair of; tying of.

132. The suffix -*ectomy* means

(a) creation of an opening.
(b) a cutting out.
(c) gradual decline; weakening.
(d) puncturing.
(e) repair of; tying of.

133. The suffix -*ostomy* means

(a) creation of an opening.
(b) a cutting out.
(c) gradual decline; weakening.
(d) puncturing.
(e) repair of; tying of.

134. The suffix *-plasty* means
(a) creation of an opening.
(b) a cutting out.
(c) gradual decline; weakening.
(d) puncturing.
(e) repair of; tying of.

135. The suffix *-lysis* means
(a) creation of an opening.
(b) a cutting out.
(c) a gradual dissolution, decline, or weakening.
(d) a puncturing.
(e) repair of; tying of.

136. The prefix _____ means "pertaining to bile."
(a) chole-
(b) chondr-
(c) hepat-
(d) nephr-
(e) my-

137. The root word _____ means "liver."
(a) chole-
(b) chondr-
(c) hepat-
(d) nephr-
(e) my-

138. The prefix _____ means "muscle."
(a) chole-
(b) chondr-
(c) hepat-
(d) nephr-
(e) myo- or mye-

139. The prefix _____ means "pertaining to the kidney."
(a) chole-
(b) chondr-
(c) hepat-
(d) nephr-
(e) my-

140. The root word _____ means "cartilage."

(a) chole-
(b) chondr-
(c) hepat-
(d) nephr-
(e) my-

141. The prefixes *a-* or *an-* mean

(a) below, after.
(b) pertaining to the head.
(c) without, lack of.
(d) pertaining to bladder or fluid-filled sac.
(e) little.

142. The prefix *cephal-* means

(a) below, after.
(b) pertaining to the head.
(c) without, lack of.
(d) pertaining to bladder or fluid-filled sac.
(e) little.

143. The prefix *cyst-* means

(a) below, after.
(b) pertaining to the head.
(c) without, lack of.
(d) pertaining to bladder or fluid-filled sac.
(e) little.

144. The prefix *infra-* means

(a) below, after.
(b) pertaining to the head.
(c) without, lack of.
(d) pertaining to bladder or fluid-filled sac.
(e) little.

145. The prefix *olig-* means

(a) below, after.
(b) pertaining to the head.
(c) without, lack of.
(d) pertaining to bladder or fluid-filled sac.
(e) little.

146. The suffix -*algia* means

(a) condition of.
(b) weakness.
(c) pain.
(d) blood.
(e) inflammation of.

147. The root word -*asthenia* means

(a) condition of.
(b) weakness.
(c) pain.
(d) blood.
(e) inflammation of.

148. The suffix -*emia* means

(a) condition of.
(b) weakness.
(c) pain.
(d) blood.
(e) inflammation of.

149. The suffix -*itis* means

(a) condition of.
(b) weakness.
(c) pain.
(d) blood.
(e) inflammation of.

150. The suffix -*osis* means

(a) condition of.
(b) weakness.
(c) pain.
(d) blood.
(e) inflammation of.

151. The term *retroflexion* means

(a) bending sideways.
(b) bending backwards.
(c) bending foreword.
(d) breaking.
(e) rotating.

152. The term *unilateral* means
(a) one-sided.
(b) double-sided.
(c) on both sides.
(d) on the affected side.
(e) opposite to the affected side.

153. The term *rhinitis* means
(a) broken nose.
(b) congested nose.
(c) runny nose.
(d) inflammation of the nose.
(e) None of the above.

154. The term *circumoral* means
(a) protruding from the mouth.
(b) occluding the mouth.
(c) pertaining to the mouth.
(d) within the mouth.
(e) around the mouth.

155. The term *intercostal* means
(a) above the vertebrae.
(b) between the vertebrae.
(c) between the ribs.
(d) on the ribs.
(e) above the ribs.

156. The prefix *ecto-* means
(a) pertaining to the intestines.
(b) upon, on.
(c) faster.
(d) within.
(e) out from.

157. The prefix *endo-* means
(a) pertaining to the intestines.
(b) upon, on.
(c) faster.
(d) within.
(e) out from.

158. The prefixes *enter-* or *entero-* mean

(a) pertaining to the intestines.
(b) upon, on.
(c) faster.
(d) within.
(e) out from.

159. The prefix *epi-* means

(a) pertaining to the intestines.
(b) upon, on.
(c) faster.
(d) within.
(e) out from.

160. The suffix _____ means "weakness."

(a) -esthesia
(b) -paresis
(c) -plegia
(d) -phagia
(e) -phasia

161. The suffix _____ means "eating."

(a) -esthesia
(b) -paresis
(c) -plegia
(d) -phagia
(e) -phasia

162. The suffix _____ means "sensation."

(a) -esthesia
(b) -paresis
(c) -plegia
(d) -phagia
(e) -phasia

163. The suffix _____ means "speech."

(a) -esthesia
(b) -paresis
(c) -plegia
(d) -phagia
(e) -phasia

164. The suffix _____ means "paralysis."

(a) -esthesia
(b) -paresis
(c) -plegia
(d) -phagia
(e) -phasia

165. The prefix *ab-* means

(a) against, opposite.
(b) half.
(c) surrounding.
(d) away from.
(e) two.

166. The prefix *bi-* means

(a) against, opposite.
(b) half.
(c) surrounding.
(d) away from.
(e) two.

167. The prefix *contra-* means

(a) against, opposite.
(b) half.
(c) surrounding.
(d) away from.
(e) two.

168. The prefix *hemi-* means

(a) against, opposite.
(b) half.
(c) surrounding.
(d) away from.
(e) two.

169. The prefix *peri-* means

(a) against, opposite.
(b) half.
(c) surrounding.
(d) away from.
(e) two.

170. The symbol $\bar{p}$ means
(a) before.
(b) with.
(c) after.
(d) without.
(e) every.

171. The symbol $\bar{a}$ means
(a) before.
(b) with.
(c) after.
(d) without.
(e) every.

172. The symbol $\bar{c}$ means
(a) before.
(b) with.
(c) after.
(d) without.
(e) every.

173. The symbol $\bar{s}$ means
(a) before.
(b) with.
(c) after.
(d) without.
(e) every.

174. The symbol $\bar{q}$ means
(a) before.
(b) with.
(c) after.
(d) without.
(e) every.

175. The abbreviation b.i.d. means
(a) once a day.
(b) three times a day.
(c) four times a day.
(d) every other day.
(e) twice a day.

176. The abbreviation q.i.d. means
(a) once a day.
(b) three times a day.
(c) four times a day.
(d) every other day.
(e) twice a day.

177. The abbreviation t.i.d. means
(a) once a day.
(b) three times a day.
(c) four times a day.
(d) every other day.
(e) twice a day.

178. Place the following in correct order of activity:
(1) respiratory assessment
(2) major hemorrhage control
(3) EMT and patient safety considerations
(4) scene size up
(5) airway/C-spine management
(6) secondary survey
(7) vital signs
(a) 4, 3, 2, 7, 5, 1, 6
(b) 1, 4, 5, 7, 3, 2, 6
(c) 4, 5, 2, 1, 7, 3, 6
(d) 1, 4, 2, 5, 6, 3, 7
(e) 4, 3, 5, 1, 2, 6, 7

179. The primary survey mnemonic consists of the letters
(a) ABC.
(b) ABCD.
(c) ABCDE.
(d) ABCDEF.
(e) None of the above.

180. Paralysis of both lower extremities is called
(a) hemiplegia.
(b) quadriplegia.
(c) paraplegia.
(d) All of the above.
(e) None of the above.

181. Paralysis of all four extremities is called
(a) hemiplegia.
(b) quadriplegia.
(c) paraplegia.
(d) All of the above.
(e) None of the above.

182. Paralysis of the same side's arm and leg is called
(a) hemiplegia.
(b) quadriplegia.
(c) paraplegia.
(d) All of the above.
(e) None of the above.

183. The primary survey care of a patient includes
(a) assessing respiratory rate and volume.
(b) exposure of the chest to observe for retractions during respiration.
(c) treatment of pneumothorax.
(d) All of the above.
(e) None of the above.

184. The presence of a radial pulse suggests a systolic blood pressure of at least
(a) 50 mmHg.
(b) 60 mmHg.
(c) 70 mmHg.
(d) 80 mmHg.
(e) 100 mmHg.

185. If the radial pulse is absent, the presence of a femoral pulse suggests a systolic blood pressure of at least
(a) 50 mmHg.
(b) 60 mmHg.
(c) 70 mmHg.
(d) 80 mmHg.
(e) 100 mmHg.

186. If the femoral pulse is absent, the presence of a carotid pulse suggests a systolic blood pressure of at least
(a) 50 mmHg.
(b) 60 mmHg.
(c) 70 mmHg.

(d) 80 mmHg.
(e) 100 mmHg.

187. Which of the following statements regarding the primary assessment of a patient's level of consciousness is true?
(a) Apply the Glasgow Coma Scale of AVPU classification systems.
(b) Determine whether the patient is semi-conscious or unconscious.
(c) Determine whether the patient is obtunded or confused.
(d) All of the above are true.
(e) None of the above are true.

188. In the level of consciousness mnemonic "AVPU," the "A" stands for
(a) average response.
(b) adequate response.
(c) absent response.
(d) apneic.
(e) alert.

189. In the level of consciousness mnemonic "AVPU," the "V" stands for
(a) violent response to stimulation.
(b) responds to verbal stimuli.
(c) responds to various stimuli.
(d) vague response to stimulation.
(e) varied response to stimulation.

190. In the level of consciousness mnemonic "AVPU," the "P" stands for
(a) responds to painful stimulation.
(b) partial response to stimulation.
(c) responds to pinching stimulation.
(d) pulses absent.
(e) probable response to stimulation.

191. In the level of consciousness mnemonic "AVPU," the "U" stands for
(a) unable to provide stimulation.
(b) unwilling to respond to stimulation.
(c) uneven response to stimulation.
(d) unresponsive to stimulation.
(e) unreliable response to stimulation.

192. Visual examination is called
- (a) percussion.
- (b) palpation.
- (c) auscultation.
- (d) inspection.
- (e) evaluation.

193. Listening to sounds with a stethoscope is called
- (a) percussion.
- (b) palpation.
- (c) auscultation.
- (d) inspection.
- (e) evaluation.

194. Striking an area to elicit sounds or vibrations is called
- (a) percussion.
- (b) palpation.
- (c) auscultation.
- (d) inspection.
- (e) evaluation.

195. The examination technique of touching and feeling the patient is called
- (a) palpitation.
- (b) crepitation.
- (c) auscultation.
- (d) percussion.
- (e) palpation.

196. Evaluation of a patient's complaints of pain includes all of the following, except
- (a) aggravating and alleviating factors.
- (b) time of onset and activity at onset.
- (c) duration of complaint.
- (d) quality and intensity of pain.
- (e) family history of illness.

197. Which of the following statements regarding "pertinent negatives" is false?
- (a) When a patient denies the presence of a complaint, this denial is called a *pertinent negative.*
- (b) When the paramedic doesn't note signs or symptoms normally accompanying a complaint, this lack of finding is called a *pertinent negative.*

(c) When the paramedic neglects to perform a specific examination or ask a pertinent question, this is called a *pertinent negative.*
(d) All of the above are false.
(e) None of the above is false.

198. In the history mnemonic "AMPLE," the "A" stands for
(a) associated symptoms.
(b) aggravation and alleviation.
(c) allergies.
(d) All of the above.
(e) None of the above.

199. In the history mnemonic "AMPLE," the "M" stands for
(a) mechanism of injury.
(b) medications.
(c) movement/motor.
(d) All of the above.
(e) None of the above.

200. In the history mnemonic "AMPLE," the "L" stands for
(a) last meal.
(b) last oral intake.
(c) last drink.
(d) All of the above.
(e) None of the above.

The answer key to Section One begins on page 365.

2

Test Section Two

Test Section Two covers the following subjects:

* Anatomy and Physiology
* Airway and Ventilation
* Pathophysiology of Shock

EMT - Paramedic
National Standards Review Self Test
Third Edition

1. The study of the body structure is called
(a) pathophysiology.
(b) physiology.
(c) physics.
(d) anatomy.
(e) biophysics.

2. The study of body functions is called
(a) pathophysiology.
(b) physiology.
(c) physics.
(d) anatomy.
(e) biophysics.

3. The ______ is/are the source of energy production in the cell and is/are involved in synthesis of protein and metabolism of lipids.
(a) nucleus
(b) mitochondria
(c) cell membranes
(d) cytoplasm
(e) tissue

4. Genetic material (DNA) is contained within the ______ of the cell.
(a) nucleus
(b) mitochondria
(c) cell membranes
(d) cytoplasm
(e) tissue

5. A group or collection of similar cells is called
(a) a nucleus.
(b) a mitochondria.
(c) cell membranes.
(d) cytoplasm.
(e) tissue.

6. The electrolyte and fluid balance within the cells of the body is maintained by
   (a) the nucleus.
   (b) the mitochondria.
   (c) cell membranes.
   (d) cytoplasm.
   (e) tissue.

7. Body surfaces and structures are lined with
   (a) muscle tissue.
   (b) epithelial tissue.
   (c) nerve tissue.
   (d) connective tissue.
   (e) smooth tissue.

8. The majority of body tissue is
   (a) muscle tissue.
   (b) epithelial tissue.
   (c) nerve tissue.
   (d) connective tissue.
   (e) smooth tissue.

9. Adipose (fat) tissue is an example of
   (a) muscle tissue.
   (b) epithelial tissue.
   (c) nerve tissue.
   (d) connective tissue.
   (e) smooth tissue.

10. Pathways that transmit electrical impulses throughout the body are composed of
    (a) muscle tissue.
    (b) epithelial tissue.
    (c) nerve tissue.
    (d) connective tissue.
    (e) smooth tissue.

11. Tissue that contracts in response to stimulus is called
    (a) muscle tissue.
    (b) epithelial tissue.
    (c) nerve tissue.
    (d) connective tissue.
    (e) smooth tissue.

12. Cartilage is an example of

(a) muscle tissue.
(b) epithelial tissue.
(c) nerve tissue.
(d) connective tissue.
(e) smooth tissue.

13. Mucous membranes are examples of

(a) muscle tissue.
(b) epithelial tissue.
(c) nerve tissue.
(d) connective tissue.
(e) smooth tissue.

14. The brain and spinal cord are composed of

(a) muscle tissue.
(b) epithelial tissue.
(c) nerve tissue.
(d) connective tissue.
(e) smooth tissue.

15. The different types of muscle tissues are

(a) skeletal muscle.
(b) cardiac muscle.
(c) smooth muscle.
(d) Answers (a) and (b) only.
(e) Answers (a), (b), and (c).

16. The only muscle types with the ability to contract independently of an external stimulus are

(a) skeletal muscle.
(b) cardiac muscle.
(c) smooth muscle.
(d) Answers (a) and (b) only.
(e) Answers (a), (b), and (c).

17. The intestines and blood vessels are composed of

(a) skeletal muscle.
(b) cardiac muscle.
(c) smooth muscle.
(d) Answers (a) and (b) only.
(e) Answers (a), (b), and (c).

18. An organ is defined as a group of
(a) skeletal and smooth tissues functioning together.
(b) epithelial and connective tissues functioning together.
(c) smooth and nervous tissues functioning together.
(d) connective and muscle tissues functioning together.
(e) any tissue types functioning together.

19. Heat production is one of the functions of the
(a) cardiovascular system.
(b) respiratory system.
(c) genitourinary system.
(d) muscular system.
(e) skeletal system.

20. Protection of vital organs from trauma is one of the functions of the
(a) cardiovascular system.
(b) respiratory system.
(c) nervous system.
(d) muscular system.
(e) skeletal system.

21. Transportation of nutrients and gases is one of the functions of the
(a) cardiovascular system.
(b) respiratory system.
(c) nervous system.
(d) muscular system.
(e) skeletal system.

22. Control of bodily functions is accomplished by the
(a) cardiovascular system.
(b) respiratory system.
(c) nervous system.
(d) muscular system.
(e) skeletal system.

23. The alveoli are a portion of the
(a) cardiovascular system.
(b) respiratory system.
(c) nervous system.
(d) muscular system.
(e) skeletal system.

24. The spleen is a portion of the
(a) gastrointestinal system.
(b) urinary system.
(c) reproductive system.
(d) lymphatic system.
(e) endocrine system.

25. Hormones are the chemical messengers of the
(a) gastrointestinal system.
(b) urinary system.
(c) cardiovascular system.
(d) lymphatic system.
(e) endocrine system.

26. The pituitary, thyroid, and adrenal glands are portions of the
(a) gastrointestinal system.
(b) urinary system.
(c) reproductive system.
(d) lymphatic system.
(e) endocrine system.

27. The liver, gall bladder, and pancreas are portions of the
(a) gastrointestinal system.
(b) urinary system.
(c) reproductive system.
(d) lymphatic system.
(e) endocrine system.

28. The prostate is a portion of the
(a) gastrointestinal system.
(b) urinary system.
(c) reproductive system.
(d) lymphatic system.
(e) endocrine system.

29. Regulation of fluid and electrolytes is one of the functions of the
(a) gastrointestinal system.
(b) urinary system.
(c) reproductive system.
(d) lymphatic system.
(e) endocrine system.

30. *Homeostasis* is defined as

(a) the stability of the internal environment.
(b) the state of equilibrium that the body attempts to maintain.
(c) a constant internal balance required for the successful function of all body systems.
(d) All of the above.
(e) None of the above.

31. The term *medial* indicates a direction or area

(a) on or toward the front of the body.
(b) on or toward the back of the body.
(c) away from the middle of the body.
(d) toward the middle of the body.
(e) at the middle of the body.

32. The term *lateral* indicates a direction or area

(a) on or toward the front of the body.
(b) on or toward the back of the body.
(c) away from the middle of the body.
(d) toward the middle of the body.
(e) at the middle of the body.

33. The term *ventral* indicates a direction or area

(a) on or toward the front of the body.
(b) on or toward the back of the body.
(c) away from the middle of the body.
(d) toward the middle of the body.
(e) at the middle of the body.

34. The term *dorsal* indicates a direction or area

(a) on or toward the front of the body.
(b) on or toward the back of the body.
(c) away from the middle of the body.
(d) toward the middle of the body.
(e) at the middle of the body.

35. The term *posterior* indicates a direction or area

(a) on or toward the front of the body.
(b) on or toward the back of the body.
(c) away from the middle of the body.
(d) toward the middle of the body.
(e) at the middle of the body.

36. The term *anterior* indicates a direction or area

(a) on or toward the front of the body.
(b) on or toward the back of the body.
(c) away from the middle of the body.
(d) toward the middle of the body.
(e) at the middle of the body.

37. The term *midline* indicates a direction or area

(a) on or toward the front of the body.
(b) on or toward the back of the body.
(c) away from the middle of the body.
(d) toward the middle of the body.
(e) at the middle of the body.

38. The term *superior* indicates a direction or area

(a) toward the tail, or in a posterior direction.
(b) toward the front, or in an anterior direction.
(c) below, or toward the feet.
(d) above, or toward the head.
(e) at the feet.

39. The term *inferior* indicates a direction or area

(a) toward the tail, or in a posterior direction.
(b) toward the head, or in an anterior direction.
(c) below, or toward the feet.
(d) above, or toward the head.
(e) at the feet.

40. A structure or area closer to the trunk than a given point of reference is considered to be

(a) superficial.
(b) palmar.
(c) distal.
(d) proximal.
(e) deep.

41. A structure or area farther from the trunk than a given point of reference is considered to be

(a) superficial.
(b) palmar.
(c) distal.
(d) proximal.
(e) deep.

42. A structure or area close to the surface of the body is considered to be
(a) superficial.
(b) palmar.
(c) distal.
(d) proximal.
(e) deep.

43. Movement away from the body is called
(a) extension.
(b) flexion.
(c) aversion.
(d) adduction.
(e) abduction.

44. The act of bending at a joint is called
(a) extension.
(b) flexion.
(c) aversion.
(d) adduction.
(e) abduction.

45. The movement of a joint toward a straight position is called
(a) extension.
(b) flexion.
(c) aversion.
(d) adduction.
(e) abduction.

46. Movement toward the body is called
(a) extension.
(b) flexion.
(c) aversion.
(d) adduction.
(e) abduction.

47. Horizontally resting on the ventral surface of the body is called the
(a) Trendelenburg position.
(b) prone position.
(c) laterally recumbent position.
(d) semi-Fowler's position.
(e) supine position.

48. Lying on the posterior surface of the body with the lower body elevated higher than the head is called the

(a) Trendelenburg position.
(b) prone position.
(c) laterally recumbent position.
(d) semi-Fowler's position.
(e) supine position.

49. Lying on the left side is called the left

(a) Trendelenburg position.
(b) prone position.
(c) laterally recumbent position.
(d) semi-Fowler's position.
(e) supine position.

50. Lying on the posterior surface of the body with the head elevated less than 45 degrees is called the

(a) Trendelenburg position.
(b) prone position.
(c) laterally recumbent position.
(d) semi-Fowler's position.
(e) supine position.

51. Horizontally resting on the dorsal surface of the body is called the

(a) Trendelenburg position.
(b) prone position.
(c) laterally recumbent position.
(d) semi-Fowler's position.
(e) supine position.

52. Lying on the right side of the body is called the right

(a) Trendelenburg position.
(b) prone position.
(c) laterally recumbent position.
(d) semi-Fowler's position.
(e) supine position.

53. Another name for the hypopharynx is the

(a) nasopharynx.
(b) oropharynx.
(c) laryngopharynx.
(d) All of the above.
(e) None of the above.

54. Filtering, warming, and humidification of inspired air occurs in the
(a) nasopharynx.
(b) oropharynx.
(c) laryngopharynx.
(d) All of the above.
(e) None of the above.

55. The portion of the pharynx extending from the posterior soft palate to the upper end of the esophagus is called the
(a) nasopharynx.
(b) oropharynx.
(c) laryngopharynx.
(d) All of the above.
(e) None of the above.

56. The portion of the pharynx extending from the hyoid bone at the base of the tongue to the trachea (anteriorly) and the esophagus (posteriorly) is called the
(a) nasopharynx.
(b) oropharynx.
(c) laryngopharynx.
(d) All of the above.
(e) None of the above.

57. The wall that divides a chamber into two cavities is called
(a) a nare.
(b) a selea.
(c) a separation.
(d) a cilia.
(e) a septum.

58. Hairlike fibers projecting from epithelial cells, which propel mucus, pus, and dust particles, are called
(a) nares.
(b) selea.
(c) separation.
(d) cilia.
(e) septum.

59. The tongue is attached to the
(a) mandible.
(b) hyoid bone.
(c) posterior oropharynx.
(d) Both answers (a) and (b).
(e) Both answers (b) and (c).

60. The epiglottis
(a) is a thin leaf-shaped structure located immediately posterior to the base of the tongue.
(b) covers the entrance of the larynx when the individual swallows, to prevent aspiration of food or liquids.
(c) is connected to the hyoid bone.
(d) Answers (a) and (b) only.
(e) Answers (a), (b), and (c).

61. Because of the amount of vagal innervation, stimulation of the larynx by a laryngoscope or endotracheal tube may cause
(a) tachycardia, and increased respiratory rate.
(b) tachycardia and hypotension.
(c) bradycardia and hypertension.
(d) bradycardia, and decreased respiratory rate.
(e) Any of the above combinations.

62. At the level of the fifth or sixth thoracic vertebra the trachea bifurcates into the right and left main-stem bronchus. This point is called the
(a) corona.
(b) septum.
(c) carina.
(d) cornea.
(e) silia.

63. The _______ is lined with respiratory epithelium, containing hairlike structures that trap foreign particles and propel them through the airway.
(a) trachea
(b) right main-stem bronchus
(c) left main-stem bronchus
(d) All of the above.
(e) None of the above.

64. The _______ is lined with cells that produce mucus, which traps particulate matter not already filtered out in the upper airway.
(a) trachea
(b) right main-stem bronchus
(c) left main-stem bronchus
(d) All of the above.
(e) None of the above.

65. The wider and straighter bronchus is the _______.
(a) trachea
(b) right main-stem bronchus
(c) left main-stem bronchus
(d) All of the above.
(e) None of the above.

66. The more angled bronchus is the _________.
(a) trachea
(b) right main-stem bronchus
(c) left main-stem bronchus
(d) All of the above.
(e) None of the above.

67. The mainstream bronchi divide into the _______, which ultimately divide into the _______ (the small airways).
(a) bronchioles/alveoli
(b) bronchioles/lung parenchyma
(c) secondary bronchi/bronchioles
(d) secondary bronchi/alveolar ducts
(e) alveolar ducts/alveoli

68. The most important functional unit of the respiratory system is the
(a) alveoli.
(b) epiglottis.
(c) pulmonary arteries.
(d) trachea.
(e) bronchi.

69. Oxygen and carbon dioxide exchange primarily occurs in the
(a) alveoli.
(b) epiglottis.
(c) pulmonary arteries.
(d) trachea.
(e) bronchi.

70. Tidal volume, the volume of gas inhaled or exhaled during a single respiratory cycle, is normally about _____ in the adult.

(a) 150 cc
(b) 250 cc
(c) 350 cc
(d) 500 cc
(e) 600 cc

71. The volume of alveolar air, the air reaching the alveoli for gas exchange, is approximately _____ in the adult.

(a) 150 cc
(b) 250 cc
(c) 350 cc
(d) 500 cc
(e) 600 cc

72. The volume of dead space air, the air remaining in passageways and unavailable for gas exchange, is approximately _____ in the adult.

(a) 150 cc
(b) 250 cc
(c) 350 cc
(d) 500 cc
(e) 600 cc

73. The expansion of the lungs is partially dependent upon the adherence of two layers of pleura. This adherence is created by a small amount of lubricating fluid between the layers, combined with an airtight seal. The pleural layer that lines the interior of the thoracic cavity (chest wall) is called the

(a) partial pleura.
(b) visceral pleura.
(c) parietal pleura.
(d) vital pleura.
(e) thoracic pleura.

74. The pleural layer that lines the exterior of the lungs is called the

(a) partial pleura.
(b) visceral pleura.

(c) parietal pleura.
(d) vital pleura.
(e) thoracic pleura.

75. The main respiratory center lies in the medulla, which is located in the
(a) mediastinum.
(b) diaphragm.
(c) brain stem
(d) medullar cavity, near the vagus nerve.
(e) frontal lobe of the brain.

76. Which of the following statements regarding the act of respiration is true?
(a) Breathing occurs as a result of pressure changes in the lungs.
(b) Inspiration begins by contraction of the diaphragm and the intercostal muscles.
(c) Expiration occurs when the inspiratory muscles relax.
(d) All of the above are true.
(e) None of the above are true.

77. The normal resting respiratory rate for an adult is
(a) 5 to 12 respirations per minute.
(b) 12 respirations per minute.
(c) 12 to 20 respirations per minute.
(d) 18 to 24 respirations per minute.
(e) 40 to 60 respirations per minute.

78. The normal resting respiratory rate for a child is
(a) 5 to 12 respirations per minute.
(b) 12 respirations per minute.
(c) 12 to 20 respirations per minute.
(d) 18 to 24 respirations per minute.
(e) 40 to 60 respirations per minute.

79. The normal resting respiratory rate for an infant is
(a) 5 to 12 respirations per minute.
(b) 12 respirations per minute.
(c) 12 to 20 respirations per minute.
(d) 18 to 24 respirations per minute.
(e) 40 to 60 respirations per minute.

80. The approximate percentage of oxygen found in room air is
(a) 11 percent.
(b) 50 percent.
(c) 21 percent.
(d) 30 percent.
(e) 31 percent.

81. Which of the following statements regarding the exchange of oxygen and carbon dioxide is true?
(a) Gas diffuses from areas of higher partial pressure to areas of lower partial pressure.
(b) Oxygen will diffuse from the pulmonary capillaries into the alveolar spaces.
(c) Carbon dioxide will diffuse from the alveolar air into the pulmonary circulation.
(d) All of the above are true.
(e) None of the above are true.

82. Oxygen diffuses into blood plasma and combines with
(a) fibrinogen.
(b) hemoglobin.
(c) leukocytes.
(d) All of the above.
(e) None of the above.

83. Room air contains 79 percent nitrogen, which, when inspired,
(a) serves no metabolic function.
(b) is necessary for maintaining inflation of body cavities.
(c) replaces the oxygen carried by plasma.
(d) Answers (a) and (b) only.
(e) Answers (a), (b), and (c).

84. Normal arterial PO_2 (at sea level) is
(a) 8.0–10.0 mmHg (torr).
(b) 7.35–7.40 mmHg (torr).
(c) 35–40 mmHg (torr).
(d) 50–60 mmHg (torr).
(e) 80–100 mmHg (torr).

85. Normal arterial PCO_2 (at sea level) is
(a) 8.0–10.0 mmHg (torr).
(b) 7.35–7.40 mmHg (torr).

(c) 35–40 mmHg (torr).
(d) 50–60 mmHg (torr).
(e) 80–100 mmHg (torr).

86. The most common cause of upper airway obstruction is
(a) aspiration of unchewed food (especially hot dogs or steak).
(b) occlusion of the posterior pharynx by the tongue.
(c) consumption of alcohol during meals.
(d) food consumption while engaged in physical activity.
(e) laryngeal spasm secondary to foreign body stimulation of the larynx.

87. Foreign body obstruction of the upper airway frequently occurs secondary to
(a) children placing toys in their mouths during play.
(b) adults consuming alcohol with meals.
(c) trauma producing loose teeth, clotted blood, or vomitus.
(d) All of the above.
(e) None of the above.

88. Aspiration of vomitus may cause
(a) increased interstitial fluid and pulmonary edema.
(b) alveoli damage.
(c) bronchial obstruction.
(d) All of the above.
(e) None of the above.

89. Which of the following statements regarding saliva is true?
(a) Saliva contains digestive enzymes for starches.
(b) Saliva contains protein-dissolving enzymes.
(c) Saliva contains hydrochloric acid.
(d) All of the above are true.
(e) None of the above are true.

90. Which of the following statements regarding vomitus is false?
(a) Vomitus contains digestive enzymes for starches.
(b) Vomitus contains protein-dissolving enzymes.
(c) Vomitus contains hydrochloric acid.
(d) All of the above are false.
(e) None of the above are false.

91. Laryngeal spasms may occur secondary to
(a) edema of the glottis.
(b) bronchospasm.
(c) spinal trauma.
(d) All of the above.
(e) None of the above.

92. Assessment of patient's airway/ventilatory status includes observation of the
(a) rise and fall of the chest.
(b) color of the skin.
(c) flaring of the nares.
(d) Answers (a) and (c) only.
(e) Answers (a), (b), and (c).

93. The preferred sites for auscultating air-movement quality (when the patient's condition permits) are the
(a) superior and inferior epigastrium sites.
(b) bilateral supraclavicular sites.
(c) six posterior chest auscultation sites.
(d) bilateral axillary sites.
(e) suprasternal and xiphoid sites.

94. Respiratory difficulty may also be noted by observation of the patient's use of accessory muscles (the diaphragm or neck muscles) accompanied by retractions of
(a) intracostal and subcostal spaces.
(b) suprasternal notch and supraclavicular fossa.
(c) submandibular and submucosal fossa.
(d) Answers (a) and (b) only.
(e) Answers (a), (b), and (c).

95. While ventilating an intubated patient with the bag-valve-mask device you notice that it becomes more and more difficult to compress the bag. This may indicate
(a) decreasing lung compliance.
(b) an occlusion of the endotracheal tube.
(c) endobronchial intubation.
(d) All of the above.
(e) None of the above.

96. Which of the following manual airway maneuvers is the preferred method for all but spine-injured patients?

(a) head-tilt with chin-lift
(b) chin-lift alone
(c) head-tilt with jaw-thrust
(d) jaw-thrust alone
(e) head-tilt with jaw-lift

97. Which of the following manual airway maneuvers is the preferred method for spine-injured patients?

(a) head-tilt with chin-lift
(b) chin-lift alone
(c) head-tilt with jaw-thrust
(d) jaw-thrust alone
(e) head-tilt with jaw-lift

98. Which of the following statements regarding an oropharyngeal airway is false?

(a) The oropharyngeal airway will keep the base of the tongue from occluding the posterior oropharynx.
(b) When the patient is intubated, the oropharyngeal airway may be used as a "bite block" to prevent occlusion of the endotracheal tube.
(c) The oropharyngeal airway will isolate the trachea, preventing aspiration of vomitus.
(d) The oropharyngeal airway may obstruct the airway further.
(e) Patients with a gag reflex will not tolerate placement of an oropharyngeal airway.

99. Which of the following statements regarding a nasopharyngeal airway is false?

(a) The nasopharyngeal airway can be used in the presence of oral cavity trauma.
(b) The nasopharyngeal airway may cause severe epistaxis.
(c) The nasopharyngeal airway may be used when the patient's teeth are clenched closed.
(d) The nasopharyngeal airway will bypass the tongue.
(e) Patients with a gag reflex will not tolerate placement of a nasopharyngeal airway.

100. Which of the following statements regarding the esophageal obturator airway (the EOA) is false?

(a) The EOA delivers oxygen at the level of the laryngopharynx.
(b) The EOA is inserted with the patient's head/neck in a neutral or flexed position.
(c) The EOA may cause obstruction of the trachea.
(d) The EOA may be used in persons of all ages and sizes.
(e) The EOA is contraindicated in patients who have ingested caustic poisons.

101. The inflatable cuff at the distal end of the endotracheal tube holds approximately

(a) 0 to 5 ml of air.
(b) 5 to 10 ml of air.
(c) 20 to 25 ml of air.
(d) 25 to 30 ml of air.
(e) 30 to 35 ml of air.

102. The inflatable cuff at the distal end of the EOA holds approximately

(a) 0 to 5 ml of air.
(b) 5 to 10 ml of air.
(c) 20 to 25 ml of air.
(d) 25 to 30 ml of air.
(e) 30 to 35 ml of air.

103. The inflatable cuff at the distal end of the esophageal gastric tube airway (EGTA) holds approximately

(a) 0 to 5 ml of air.
(b) 5 to 10 ml of air.
(c) 20 to 25 ml of air.
(d) 25 to 30 ml of air.
(e) 30 to 35 ml of air.

104. Which of the following statements regarding the EGTA is true?

(a) The EGTA increases the opportunity for regurgitation, but allows direction of emesis away from the airway.
(b) The EGTA permits the passage of a nasogastric tube through the ventilation port for decompression of the stomach.

(c) Unlike the EOA, the EGTA has no contraindications.
(d) All of the above are true.
(e) None of the above is true.

105. The preferred technique for airway management of an unconscious patient who does not have a gag reflex is

(a) nasopharyngeal airway and assisted ventilations with a bag-valve-mask device.
(b) oropharyngeal airway and assisted ventilations with a bag-valve-mask device.
(c) EOA- and BVM-assisted ventilations (in the event of spinal trauma).
(d) nasotracheal intubation (especially in the event of epiglottitis) with BVM-assisted ventilations.
(e) endotracheal intubation (except in the event of epiglottitis) with BVM-assisted ventilations.

106. Which of the following medications may be administered via the endotracheal tube?

(1) naloxone
(2) epinephrine 1:10,000
(3) atropine sulphate
(4) lidocaine
(5) 50% dextrose in water
(6) epinephrine 1:1000

(a) 1, 2, 3, 4, 5
(b) 1, 2, 3, 4, 6
(c) 1, 2, 3, 4
(d) 1, 2, 3, 4, 5, 6
(e) None of the above.

107. The straight laryngoscope blade is also called a ______ blade.

(a) MacIntosh
(b) MacBurnie
(c) Miller
(d) Monroe
(e) Magill

108. The curved laryngoscope blade is also called a ______ blade.

(a) MacIntosh
(b) MacBurnie
(c) Miller
(d) Monroe
(e) Magill

109. The forceps usually used to remove visualized obstructive matter are called ______ forceps.

(a) MacIntosh
(b) MacBurnie
(c) Miller
(d) Monroe
(e) Magill

110. The straight laryngoscope blade is used to visualize the vocal cords by

(a) elevating the larynx.
(b) elevating the epiglottis.
(c) indirectly elevating the epiglottis.
(d) indirectly elevating the larynx.
(e) elevating only the posterior tongue.

111. The curved laryngoscope blade is used to visualize the vocal cords by

(a) elevating the larynx.
(b) elevating the epiglottis.
(c) indirectly elevating the epiglottis.
(d) indirectly elevating the larynx.
(e) elevating only the posterior tongue.

112. The curved laryngoscope blade is designed to be placed

(a) under the vallecula.
(b) under the epiglottis.
(c) in the larynx.
(d) in the vallecula.
(e) in the right hand.

113. The straight laryngoscope blade is designed to be placed

(a) under the vallecula.
(b) under the epiglottis.
(c) in the larynx.
(d) in the vallecula.
(e) in the right hand.

114. Children younger than 8 years old

(a) should not be intubated in the field.
(b) should be intubated with uncuffed pediatric endotracheal tubes only.

(c) should receive sedation to allow for successful intubation.
(d) Both answers (b) and (c).
(e) None of the above.

115. When a flexible/malleable stylet is used to curve the endotracheal tube, the stylet tip must be recessed at least _______ from the distal end of the tube.
(a) 6 cm or 2¼ in.
(b) 5 cm or 2 in.
(c) 4 cm or 1½ in.
(d) 2 cm or ½–¾ in.
(e) 1 cm or ⅓ in.

116. Hyperventilation with 100 percent oxygen should
(a) precede any intubation attempt (or repeat attempt) for at least 5 to 10 seconds.
(b) not be utilized on the COPD patient; a 50 percent delivery (8 LPM with BVM) should be employed with these patients.
(c) be utilized only with head-injured patients.
(d) Both answers (a) and (b).
(e) Both answers (b) and (c).

117. The maximum allowable time for an intubation attempt is
(a) 5 seconds.
(b) 10 seconds.
(c) 15 seconds.
(d) 30 seconds.
(e) 1 minute.

118. Which of the following statements regarding intubation is false?
(a) Esophageal intubation may result in severe hypoxia, brain damage, and/or brain death.
(b) Trauma to the vallecula may cause subcutaneous emphysema.
(c) Trauma to the pyriform sinus may cause subcutaneous emphysema.
(d) All of the above are false.
(e) None of the above is false.

119. Endobronchial intubation is evidenced by

(a) increasing gastric distention.
(b) bilaterally diminished or absent breath sounds.
(c) unilaterally diminished or absent breath sounds.
(d) All of the above.
(e) None of the above.

120. When endobronchial intubation is suspected, the paramedic should

(a) hyperinflate the distal cuff to provide an improved seal.
(b) deflate the distal cuff to allow for passage of air around the tube.
(c) deflate the cuff and extubate the patient; hyperventilate with BVM and 100 percent oxygen; reintubate the patient with a fresh endotracheal tube.
(d) Any of the above.
(e) None of the above.

121. Successful endotracheal intubation is evidenced by

(a) observation of chest rise and fall.
(b) auscultation of equal bilateral breath sounds.
(c) auscultation of the absence of gastric noises on ventilation.
(d) Any of the above.
(e) All of the above, in combination.

122. Which of the following statements regarding nasotracheal intubation is false?

(a) Spinal injury is not a contraindication for nasotracheal intubation.
(b) Blind nasotracheal intubation may be performed on any apneic patient.
(c) Nasotracheal intubation does not require a laryngoscope.
(d) Nasotracheal intubation presents a greater risk of infection.
(e) Nasotracheal intubation may cause additional trauma.

123. Nasotracheal intubation is indicated in all of the following situations, except

(a) the unconscious patient with nasal and basilar skull fractures requiring hyperventilation.
(b) the conscious patient nearing respiratory arrest secondary to asthma, anaphylactic shock, or inhalation injury.
(c) the conscious patient with severe oral trauma and suspected spine injury.
(d) the unconscious patient with a fractured jaw.
(e) the unconscious patient with clenched teeth or profound gag reflex.

124. During blind nasotracheal intubation, advancement of the ET tube into the glottic opening is accomplished

(a) between the patient's respiratory efforts.
(b) during the patient's expirations.
(c) during the patient's inspirations.
(d) as the patient swallows.
(e) None of the above.

125. When the anterior neck is palpated from the top, downwards, the cricoid cartilage is

(a) the first prominent structure felt.
(b) the second prominent structure felt.
(c) the third prominent structure felt.
(d) between the first and second prominent structures.
(e) between the second and third prominent structures.

126. When the anterior neck is palpated from the top, downwards, the thyroid cartilage is

(a) the first prominent structure felt.
(b) the second prominent structure felt.
(c) the third prominent structure felt.
(d) between the first and second prominent structures.
(e) between the second and third prominent structures.

127. When the anterior neck is palpated from the top, downwards, the cricothyroid membrane is

(a) the first prominent structure felt.
(b) the second prominent structure felt.
(c) the third prominent structure felt.
(d) between the first and second prominent structures.
(e) between the second and third prominent structures.

128. Which of the following statements regarding transtracheal jet insufflation is false?

(a) The needle is inserted into the cricoid cartilage.
(b) Upon removal of the needle, the catheter allows for both ventilation and suctioning of secretions.
(c) Attaching the catheter to a 15 mm adapter allows for BVM or demand valve ventilation.
(d) All of the above are false.
(e) None of the above is false.

129. Which of the following statements regarding transtracheal jet insufflation is false?

(a) The placement of a second large-bore catheter next to the first one may be required in cases of total upper airway obstruction to promote more effective expiration.
(b) Suction may be applied to the second catheter to promote more effective expiration.
(c) High-pressure ventilation and air entrapment during transtracheal jet insufflation may cause a pneumothorax.
(d) All of the above are false.
(e) None of the above is false.

130. Which of the following statements regarding cricothyrotomy is false?

(a) An incision is made in the cricoid cartilage.
(b) Attaching the inserted tracheostomy tube to a 15 mm adapter allows for BVM ventilation.
(c) An 18 gauge over-the-needle catheter may be used for small children.
(d) Complications include hemorrhage or subcutaneous emphysema.
(e) None of the above is false.

131. Which of the following statements regarding suction is true?

(a) Whistle-tip suction may be used up to 30 seconds each time, as its smaller bore depletes less oxygen than the Yankauer device.
(b) Tonsil-tip suction removes larger particles but less fluid volume than the whistle-tip device.
(c) A Yankauer device can be used to suction the nares, oropharynx, larynx, and esophagus.
(d) A whistle-tip device can be used to suction the nares, oropharynx, nasopharynx, or endotracheal tube.
(e) Suction is applied as the device is inserted and continued as the device is withdrawn.

132. With a 4- to 6-liter-per-minute oxygen flow rate, a nasal cannula will deliver an oxygen concentration of

(a) 36–44 percent.
(b) 26–54 percent.
(c) 40–60 percent.
(d) 60–80 percent.
(e) 80–100 percent.

133. The liter flow rate range for oxygen administration when a nasal cannula is used is

(a) 1–4 LPM.
(b) 1–6 LPM.
(c) 1–8 LPM.
(d) 4–8 LPM.
(e) 6–10 LPM.

134. With a liter flow rate range of 8 to 12 LPM, a simple face mask can deliver an oxygen concentration of

(a) 36–44 percent.
(b) 26–54 percent.
(c) 40–60 percent.
(d) 60–80 percent.
(e) 80–100 percent.

135. The liter flow rate range for oxygen administration when a simple face mask is used is

(a) 15 LPM only.
(b) 1–6 LPM.
(c) 8–10 LPM.
(d) 6–12 LPM.
(e) 12–15 LPM.

136. With a 10- to 15-liter-per-minute oxygen flow rate, a nonrebreathing mask will deliver an oxygen concentration of

(a) 36–44 percent.
(b) 26–54 percent.
(c) 40–60 percent.
(d) 60–80 percent.
(e) 80–100 percent.

137. The liter flow rate range for oxygen administration when a nonrebreathing mask is used is
(a) 6–8 LPM.
(b) 8–10 LPM.
(c) 6–15 LPM.
(d) 8–15 LPM.
(e) 10–15 LPM.

138. The venturi mask is a special oxygen delivery mask designed primarily for
(a) pediatric patients.
(b) geriatric patients.
(c) COPD patients.
(d) All of the above.
(e) None of the above.

139. The venturi mask delivers four different oxygen concentrations depending upon the liter flow of oxygen. These different concentrations are
(a) 12 percent, 16 percent, 24 percent, or 40 percent.
(b) 16 percent, 24 percent, 28 percent, or 35 percent.
(c) 24 percent, 28 percent, 35 percent, or 40 percent.
(d) 28 percent, 40 percent, 44 percent, or 50 percent.
(e) 40 percent, 44 percent, 50 percent, or 54 percent.

140. Limitations of mouth-to-mouth, mouth-to-nose, and mouth-to-stoma ventilations include all of the following, except
(a) inadequate ventilatory volume.
(b) delivery of only 17 percent oxygen.
(c) lack of protection against disease transmission.
(d) lack of protection against patient aspiration.
(e) inability to hyperventilate the patient without causing rescuer hyperventilation.

141. Which of the following statements regarding mouth-to-mask ventilations is false?
(a) With an oxygen inlet valve system, 10 LPM of oxygen will provide the patient with approximately 50 percent oxygen with mouth-to-mask ventilations.
(b) The tidal volume provided with mouth-to-mask ventilations exceeds that of the average single-operator BVM ventilations.

(c) With a one-way valve device, the risk of contact with patient secretions and expired air is minimized.
(d) Safe hyperventilation is not possible with mouth-to-mask ventilations.
(e) An oropharyngeal airway cannot be used during mouth-to-mask ventilations.

142. Which of the following statements regarding the bag-valve-mask (BVM) device is false?

(a) The BVM device will deliver 21 percent oxygen without oxygen supplementation.
(b) The BVM device will deliver only approximately 60 percent oxygenation when run at 12 LPM without a reservoir.
(c) The BVM device can easily be adequately operated by a single rescuer, delivering a ventilation tidal volume of 800 ml of air.
(d) The BVM device will deliver 90–95 percent oxygen when operated at 10 to 15 LPM with an oxygen reservoir attachment.
(e) The BVM device can be efficiently used by one rescuer to ventilate an intubated patient.

143. Which of the following statements regarding demand valve ventilation is true?

(a) The demand valve connects to a mask, esophageal intubation device, or endotracheal tube.
(b) The demand valve provides 100 percent oxygenation when operated at 40 to 60 LPM.
(c) The demand valve can ventilate past a minor airway obstruction.
(d) All of the above are true.
(e) None of the above is true.

144. Which of the following statements regarding demand valve ventilation is false?

(a) Lung compliance is not detectable with use of demand valve ventilation.
(b) Use of demand valve ventilation is limited to patients 16 years old and older, secondary to the risk of injury from hyperexpansion of the lungs.
(c) Potential pulmonary rupture is increased with use of demand valve ventilation on intubated patients.
(d) Gastric distension is a frequent side effect when demand valve ventilation is used on nonintubated patients.
(e) Because of the relatively low oxygen flow rates required, a demand valve can be portably operated for a prolonged period of time.

145. Gastric distension

(a) occurs when adequate airway positioning is not achieved.
(b) occurs when airway positioning is adequate but ventilatory volumes are excessive, even in the intubated patient.
(c) will significantly increase the risk of aspiration and decrease the ventilatory capacity of the lungs.
(d) Answers (a) and (c) only.
(e) Answers (a), (b), and (c).

146. Electrolytes are defined as substances that

(a) dissociate in water and are capable of conducting electrical current.
(b) generate electrical current when combined with water.
(c) are negatively charged until combined with water.
(d) All of the above.
(e) None of the above.

147. *Cations* are defined as

(a) electrolytes able to pass freely across a semipermeable membrane.
(b) positively charged ions.
(c) negatively charged ions.
(d) All of the above.
(e) None of the above.

148. *Anions* are defined as

(a) electrolytes able to pass freely across a semipermeable membrane.
(b) positively charged ions.
(c) negatively charged ions.
(d) All of the above.
(e) None of the above.

149. The chief extracellular ion is

(a) sodium.
(b) potassium.
(c) calcium.
(d) chloride.
(e) magnesium.

150. The chief intracellular ion is

(a) sodium.
(b) potassium.
(c) calcium.
(d) chloride.
(e) magnesium.

151. The symbol/abbreviation for sodium is

(a) S^+.
(b) So^+.
(c) N^+.
(d) Na^+.
(e) Na^-.

152. The symbol/abbreviation for chloride is

(a) C^-.
(b) C^+.
(c) Cl^-.
(d) Cl^+.
(e) Cl^{++}.

153. The symbol/abbreviation for magnesium is

(a) Ma^+.
(b) Ma^-.
(c) $Mang^+$.
(d) Mg^-.
(e) Mg^{++}.

154. The symbol/abbreviation for bicarbonate is

(a) B^+.
(b) B^-.
(c) HCO^+.
(d) BC^-.
(e) HCO_3^-.

155. The symbol/abbreviation for calcium is

(a) C^{++}.
(b) Ca^{++}.
(c) C^-.
(d) Ca^{--}.
(e) K^+.

156. The symbol/abbreviation for potassium is

(a) PO^+.
(b) p^+.
(c) K^+.
(d) p^-.
(e) PO^-.

157. The symbol/abbreviation for phosphate is

(a) HPO_3^-.
(b) $H_2O_3^+$.
(c) $H_2O_3^-$.
(d) HCO_2^-.
(e) HPO_2^-.

158. The principal buffer of the body is

(a) K^+.
(b) HCO_3^-.
(c) HPO_3^-.
(d) $CaCl_2$.
(e) HPO_2^+.

159. Which of the following electrolytes are positively charged ions?

(a) Sodium, potassium, and chloride.
(b) Chloride, bicarbonate, and phosphate.
(c) Sodium, chloride, and phosphate.
(d) Sodium, potassium, and calcium.
(e) Chloride, bicarbonate, and magnesium.

160. Which of the following electrolytes are negatively charged ions?

(a) Sodium, potassium, and chloride.
(b) Chloride, bicarbonate, and phosphate.
(c) Sodium, chloride, and phosphate.
(d) Sodium, potassium, and calcium.
(e) Chloride, bicarbonate, and magnesium.

161. Fluid with osmotic pressure equal to normal body fluid is called

(a) homeotonic.
(b) hypertonic.
(c) isotonic.
(d) hypotonic.
(e) homeostatic.

162. Fluid with osmotic pressure less than that of normal body fluid is called

(a) homeotonic.
(b) hypertonic.
(c) isotonic.
(d) hypotonic.
(e) homeostatic.

163. Fluid with osmotic pressure greater than that of normal body fluid is called

(a) homeotonic.
(b) hypertonic.
(c) isotonic.
(d) hypotonic.
(e) homeostatic.

164. Water will move across a semipermeable membrane

(a) in the direction of least resistance.
(b) from an isotonic solution to a homeotonic solution.
(c) from a homeostatic solution to an isotonic solution.
(d) from an area of higher solute concentration to that of lower solute concentration.
(e) from an area of lower solute concentration to that of higher solute concentration.

165. Sodium will move across a semipermeable membrane

(a) in the direction of least resistance.
(b) from an isotonic solution to a homeotonic solution.
(c) from a homeostatic solution to an isotonic solution.
(d) from an area of higher solute concentration to that of lower solute concentration.
(e) from an area of lower solute concentration to that of higher solute concentration.

166. Movement of a solute across a semipermeable membrane is called

(a) osmosis.
(b) simple transfer.
(c) diaphoresis.
(d) diffusion.
(e) suffusion.

167. Movement of a solvent across a semipermeable membrane is called

(a) osmosis.
(b) simple transfer.
(c) diaphoresis.
(d) diffusion.
(e) suffusion.

168. Of the following forms of movement across a semipermeable membrane, the fastest is

(a) osmosis.
(b) simple transfer.
(c) diaphoresis.
(d) diffusion.
(e) suffusion.

169. Electrolytes move across a semipermeable membrane using

(a) osmosis.
(b) simple transfer.
(c) diaphoresis.
(d) diffusion.
(e) suffusion.

170. Water will move across a semipermeable membrane using

(a) osmosis.
(b) simple transfer.
(c) diaphoresis.
(d) diffusion.
(e) suffusion.

171. Which of the following statements regarding active transport is false?

(a) Energy is required to accomplish active transport.
(b) Active transport is slower than diffusion or osmosis.
(c) Larger molecules can be moved across semipermeable membranes with active transport.
(d) Molecules can move toward areas of higher concentration with active transport.
(e) Proteins are moved across semipermeable membranes with active transport.

172. Facilitated diffusion

(a) employs "helper proteins" to cross the cell membrane.
(b) requires energy to occur.
(c) is a selective process, occurring only with certain molecules.
(d) All of the above.
(e) None of the above.

173. An iron-based compound called _______ is contained within the red blood cells and is responsible for the transportation of oxygen to the cells.

(a) hemoglobin
(b) erythroglobin
(c) thromboglobin
(d) leukoglobin
(e) plasmaglobin

174. Red blood cells (RBCs) are also known as

(a) hemocytes.
(b) erythrocytes.
(c) thrombocytes.
(d) leukocytes.
(e) plasma.

175. White blood cells (WBCs) are also known as
(a) hemocytes.
(b) erythrocytes.
(c) thrombocytes.
(d) leukocytes.
(e) plasma.

176. Approximately 54 percent of the total body blood volume consists of
(a) hemocytes.
(b) erythrocytes.
(c) thrombocytes.
(d) leukocytes.
(e) plasma.

177. Approximately 45 percent of the total body blood volume consists of
(a) hemocytes.
(b) erythrocytes.
(c) thrombocytes.
(d) leukocytes.
(e) plasma.

178. The _______ are/is responsible for the clotting factor of blood.
(a) hemocytes
(b) erythrocytes
(c) thrombocytes
(d) leukocytes
(e) plasma

179. Immunity and the combatting of infection are the responsibility of the
(a) hemocytes.
(b) erythrocytes.
(c) thrombocytes.
(d) leukocytes.
(e) plasma.

180. Platelets are also known as
(a) hemocytes.
(b) erythrocytes.

(c) thrombocytes.
(d) leukocytes.
(e) plasma.

181. The primary system of blood classification is called the
(a) ABO system.
(b) ABC system.
(c) AB system.
(d) Rh system.
(e) O system.

182. Blood type B contains
(a) antibody A, antibody B, and no antigen.
(b) antigen A, antigen B, and no antibody.
(c) antigen B and antibody A.
(d) antigen A and antibody B.
(e) no antigen and no antibody.

183. Blood type O contains
(a) antibody A, antibody B, and no antigen.
(b) antigen A, antigen B, and no antibody.
(c) antigen B and antibody A.
(d) antigen A and antibody B.
(e) no antigen and no antibody.

184. Blood type A contains
(a) antibody A, antibody B, and no antigen.
(b) antigen A, antigen B, and no antibody.
(c) antigen B and antibody A.
(d) antigen A and antibody B.
(e) no antigen and no antibody.

185. Blood type AB contains
(a) antibody A, antibody B, and no antigen.
(b) antigen A, antigen B, and no antibody.
(c) antigen B and antibody A.
(d) antigen A and antibody B.
(e) no antigen and no antibody.

186. A person who has blood type ___ is called the universal donor.

(a) C
(b) A
(c) B
(d) AB
(e) O

187. A person who has blood type ___ is called the universal recipient.

(a) C
(b) A
(c) B
(d) AB
(e) O

188. The universal donor may only receive blood from persons who have type ___ blood.

(a) C
(b) A
(c) B
(d) AB
(e) O

189. A person with Rh negative blood usually does not have anti-Rh antibodies. Exposure to Rh positive blood will cause

(a) no adverse reaction to any receipt of Rh+ blood.
(b) the development of anti-Rh antibodies after the initial exposure.
(c) severe, perhaps fatal allergic reactions to any subsequent receipt of Rh+ blood.
(d) Answers (a) and (b) only.
(e) Answers (b) and (c) only.

190. Protein containing intravenous fluids are called

(a) isotonic osmolloids.
(b) colloids.
(c) hypertonic osmolloids.
(d) crystalloids.
(e) hypotonic osmolloids.

191. Intravenous fluids not containing proteins are called

(a) isotonic osmolloids.
(b) colloids.

(c) hypertonic osmolloids.
(d) crystalloids.
(e) hypotonic osmolloids.

192. Prehospital intravenous fluid replacement is initiated with
(a) isotonic osmolloid solutions.
(b) colloid solutions.
(c) hypertonic osmolloid solutions.
(d) crystalloid solutions.
(e) hypotonic osmolloid solutions.

193. A greater increase in intravascular fluid can be accomplished sooner with
(a) isotonic osmolloid solutions.
(b) colloid solutions.
(c) hypertonic osmolloid solutions.
(d) crystalloid solutions.
(e) hypotonic osmolloid solutions.

194. Normal saline (NS) and 5% dextrose in water (D_5W) are examples of
(a) isotonic osmolloid solutions.
(b) colloid solutions.
(c) hypertonic osmolloid solutions.
(d) crystalloid solutions.
(e) hypotonic osmolloid solutions.

195. Plasmanate and Hetastarch are examples of
(a) isotonic osmolloid solutions.
(b) colloid solutions.
(c) hypertonic osmolloid solutions.
(d) crystalloid solutions.
(e) hypotonic osmolloid solutions.

196. Lactated Ringers (LR or RL) is an example of
(a) an isotonic osmolloid solution.
(b) a colloid solution.
(c) a hypertonic osmolloid solution.
(d) a crystalloid solution.
(e) a hypotonic osmolloid solution.

197. Solutions containing higher solute concentrations than that within the cell are called

(a) isotonic.
(b) isotonic or hypertonic.
(c) hypotonic.
(d) hypertonic or hypotonic.
(e) hypertonic.

198. Solutions with a similar solute concentration as that within the cell are called

(a) isotonic.
(b) isotonic or hypertonic.
(c) hypotonic.
(d) hypertonic or hypotonic.
(e) hypertonic.

199. If a normally hydrated person receives an infusion of a/an ______ solution, a shift of fluid will occur from the extracellular to the intracellular compartments.

(a) isotonic
(b) isotonic or hypertonic
(c) hypotonic
(d) hypertonic or hypotonic
(e) hypertonic

200. If a normally hydrated person receives an infusion of a/an ______ solution, fluid will shift from the intracellular to the extracellular compartments.

(a) isotonic
(b) isotonic or hypertonic
(c) hypotonic
(d) hypertonic or hypotonic
(e) hypertonic

201. Infusion of solution that is ______ into a normally hydrated person will not cause a significant shift of fluid.

(a) isotonic
(b) isotonic or hypertonic
(c) hypotonic
(d) hypertonic or hypotonic
(e) hypertonic

202. Solutions containing a lesser concentration of solutes than that within the cell are called

(a) isotonic.
(b) isotonic or hypertonic.
(c) hypotonic.
(d) hypertonic or hypotonic.
(e) hypertonic.

203. Acid-base balance refers to the concentration of

(a) the chief extracellular ion of body fluids.
(b) the chief intracellular ion of body fluids.
(c) hydrogen ions in body fluids.
(d) All of the above.
(e) None of the above.

204. Which of the following statements regarding pH is true?

(a) The term "pH" is used to express the hydrogen-ion (H^+) concentration of a fluid.
(b) The pH equals the negative log of the hydrogen-ion (H^+) concentration.
(c) At a pH of 7, a solution is neutral (water has a pH of 7).
(d) All of the above.
(e) None of the above.

205. A fluid with a pH below 7 has

(a) an increased concentration of H^+ and is called "acid."
(b) a decreased concentration of H^+ and is called "alkaline" or basic.
(c) an increased concentration of alkaline ions and is called "alkaline" or basic.
(d) Both answers (a) and (c).
(e) None of the above.

206. A fluid with a pH above 7 has

(a) an increased concentration of H^+ and is called "acid."
(b) a decreased concentration of H^+ and is called "alkaline" or basic.
(c) an increased concentration of alkaline ions and is called "alkaline" or basic.
(d) Both answers (a) and (c).
(e) None of the above.

207. Homeostasis requires a body fluid pH range of

(a) 7.0–7.6.
(b) 7.25–7.35.
(c) 7.35–7.45.
(d) 7.45–7.55.
(e) 7.80–7.95.

208. Acidosis is described as a pH

(a) below that of normal pH.
(b) above that of normal pH.
(c) within that of normal pH range.
(d) All of the above.
(e) None of the above.

209. Alkalosis is described as a pH

(a) below that of normal pH.
(b) above that of normal pH.
(c) within that of normal pH range.
(d) All of the above.
(e) None of the above.

210. The extreme limits of pH compatible with life are approximately _____ on the alkaline side, and _____ on the acid side.

(a) 7.8/6.9
(b) 6.9/7.8
(c) 8.0/6.0
(d) 6.0/8.0
(e) 7.3/7.5

211. The body's normal pH is maintained by

(a) the carbonate buffer system.
(b) the respiratory system.
(c) the renal system.
(d) Answers (a) and (b) only.
(e) Answers (a), (b), and (c).

212. The fastest-acting pH defense mechanism, responding within seconds, is

(a) the carbonate buffer system.
(b) the respiratory system.
(c) the renal system.
(d) Answers (a) and (b) only.
(e) Answers (a), (b), and (c).

213. The slowest-acting pH defense mechanism, requiring hours or days, is

(a) the carbonate buffer system.
(b) the respiratory system.
(c) the renal system.
(d) Answers (a) and (b) only.
(e) Answers (a), (b), and (c).

214. The carbonate system has two components, which are

(a) carbonic acid ($H_3CO_2^-$) and bicarbonate (HCO_2).
(b) carbonic acid (H_2CO_3) and bicarbonate (HCO_3^-).
(c) carbonic acid (HCO_2) and bicarbonate ($H_3CO_2^-$).
(d) Any of the above.
(e) None of the above.

215. Carbonic acid is a weak acid that constantly breaks down into

(a) water (H_20) and carbon dioxide (CO_2).
(b) hydrogen ions (H^+) and bicarbonate ions (HCO^+).
(c) hydrogen ions (H^+) and bicarbonate ions (HCO_3^-).
(d) Both answers (a) and (b).
(e) Both answers (a) and (c).

216. Increased respirations result in

(a) an increased release of CO_2 and a decreased H^+.
(b) a decreased release of CO_2 and an increased H^+.
(c) a decreased release of CO_2 and a decreased H^+.
(d) an increased release of CO_2 and an increased H^+.
(e) None of the above.

217. Decreased respirations result in

(a) an increased release of CO_2 and a decreased H^+.
(b) a decreased release of CO_2 and an increased H^+.
(c) a decreased release of CO_2 and a decreased H^+.
(d) an increased release of CO_2 and an increased H^+.
(e) None of the above.

218. Respiratory acidosis is caused by

(a) an increased release of CO_2 and a decreased H^+.
(b) a decreased release of CO_2 and an increased H^+.
(c) a decreased release of CO_2 and a decreased H^+.
(d) an increased release of CO_2 and an increased H^+.
(e) None of the above.

219. Respiratory alkalosis is caused by

(a) an increased release of CO_2 and a decreased H^+.
(b) a decreased release of CO_2 and an increased H^+.
(c) a decreased release of CO_2 and a decreased H^+.
(d) an increased release of CO_2 and an increased H^+.
(e) None of the above.

220. The kidneys regulate pH by excretion of excess

(a) hydrogen ions.
(b) carbon ions.
(c) bicarbonate ions.
(d) Both answers (a) and (b).
(e) Both answers (a) and (c).

221. If the pH falls, the kidneys eliminate more

(a) CO_2.
(b) H^+.
(c) HCO_3^-.
(d) $H_2CO_3^-$.
(e) None of the above.

222 If the pH rises, the kidneys eliminate more

(a) CO_2.
(b) H^+.
(c) HCO_3^-.
(d) $H_2CO_3^-$.
(e) None of the above.

223. Anaerobic metabolism can cause

(a) metabolic alkalosis.
(b) respiratory acidosis.
(c) respiratory alkalosis.
(d) homeostasis.
(e) metabolic acidosis.

224. Prolonged vomiting and excessive use of diuretics can cause

(a) metabolic alkalosis.
(b) respiratory acidosis.
(c) respiratory alkalosis.
(d) homeostasis.
(e) metabolic acidosis.

225. *Shock* is defined as inadequate

(a) blood pressure (less than 100 systolic).
(b) pulse (less than 80 per minute).
(c) respirations (less than 20 per minute).
(d) hydration.
(e) cellular oxygenation.

226. Adequate oxygenation, according to the Fick principle, is dependent upon

(a) red blood cell oxygenation.
(b) delivery of oxygen to the tissues.
(c) off-loading of the oxygen at the cellular level.
(d) All of the above.
(e) None of the above.

227. Perfusion is dependent upon

(a) the pump (heart).
(b) the fluid (oxygenated blood).
(c) the container (blood vessels).
(d) Answers (a) and (b) only.
(e) Answers (a), (b), and (c).

228. Which of the following statements regarding baroreceptors is false?

(a) Baroreceptors are located in the aortic arch and carotid arteries.
(b) Baroreceptors are composed of nerve tissue and are designed to detect changes in blood pressure.
(c) When baroreceptors are stimulated by a change in blood pressure, the adrenal glands send signals to the brain to alter body functions and return the blood pressure to a normal state.
(d) If the baroreceptors note an increase in blood pressure, the brain responds by decreasing the heart rate, preload, and/or after load.
(e) If the baroreceptors note a decrease in blood pressure, the brain responds by activating the sympathetic nervous system.

229. Activation of the sympathetic nervous system produces all of the following responses, except

(a) increased strength of cardiac contractions.
(b) increased rate of cardiac contractions.
(c) peripheral arterial constriction resulting in an increased container size.
(d) increased peripheral vascular resistance.
(e) stimulation of the adrenal glands to release epinephrine and norepinephrine to further enhance the sympathetic effects.

230. Which of the following statements regarding cellular oxygenation is false?

(a) The arterial blood reaching the alveoli from the right side of the heart has a partial pressure of approximately 16 percent, which causes oxygen to diffuse into the capillary blood flow.
(b) Hemoglobin is the transportation molecule with which oxygen binds to travel to the cells.
(c) The arterial blood reaching the body tissues has a greater PO_2 than the cells because of oxygen depletion from normal metabolic activity.
(d) Oxygen will diffuse into the body cells and the CO_2 will diffuse into the capillaries.
(e) An inadequate number of RBCs will not affect the transportation of oxygen to the cells.

231. Inadequate cellular perfusion

(a) causes anaerobic cellular metabolism, producing acid by-products (pyruvic acid, lactic acid).
(b) causes acidosis, leading to cellular death, tissue death, and organ failure.
(c) prevents cellular utilization of glucose, causing immediate and profound hypoglycemia.
(d) Answers (a) and (b) only.
(e) Answers (a), (b), and (c).

232. If the patient is taking beta-blocker medications, the initial signs and symptoms of ______ will be hindered.

(a) compensated shock
(b) irreversible shock
(c) delayed shock
(d) superficial shock
(e) decompensated shock

233. Which of the following statements regarding cardiac effects during decompensated shock is false?

(a) Decompensated shock is indicated by a decreasing pulse pressure (increasing systolic and decreasing diastolic blood pressures).
(b) Decreasing diastolic blood pressure results in decreased coronary blood flow.
(c) Increased cardiac ischemia results in myocardial infarction, or an increased size of previous infarct, secondary to decreased perfusion.
(d) All of the above are false.
(e) None of the above is false.

234. Which of the following statements regarding peripheral effects during shock is false?

(a) During compensatory shock, increased peripheral vascular resistance is enhanced by pre- and postcapillary sphincter contraction.
(b) As ischemia develops, anaerobic metabolism is changed to aerobic metabolism.
(c) During decompensated shock, precapillary sphincters relax and open, while postcapillary sphincters remain contracted.
(d) During decompensated shock, capillary blood flow becomes stagnant, resulting in increased ischemia, increased H^+ production, and Rouleaux formation of RBCs.
(e) Stacking of RBCs (Rouleaux formation) produces an emboli threat; when irreversible shock ensues, postcapillary sphincters relax and open, releasing H^+ and Rouleaux RBCs into systemic circulation.

235. Hypovolemic shock can be caused by

(a) internal or external hemorrhage, or burns.
(b) vomiting, diarrhea, or excessive sweating.
(c) pancreatitis, peritonitis, or bowel obstruction.
(d) Answers (a) and (b) only.
(e) Answers (a), (b), and (c).

236. Which of the following statements regarding neurogenic shock is false?

(a) Reflex peripheral vasoconstriction is absent when neurogenic shock occurs.
(b) Reflex sympathetic cardiac stimulation is absent when neurogenic shock occurs.
(c) Central nervous system depressant drugs may cause neurogenic shock.
(d) All of the above are false.
(e) None of the above is false.

237. The presence of a radial pulse suggests a systolic blood pressure of at least

(a) 90 mmHg.
(b) 60 mmHg.
(c) 80 mmHg.
(d) 40 mmHg.
(e) 70 mmHg.

238. If the radial pulse is absent, the presence of a femoral pulse suggests a systolic blood pressure of at least

(a) 90 mmHg.
(b) 60 mmHg.
(c) 80 mmHg.
(d) 40 mmHg.
(e) 70 mmHg.

239. If the femoral pulse is absent, the presence of a carotid pulse suggests a systolic blood pressure of at least

(a) 90 mmHg.
(b) 60 mmHg.
(c) 80 mmHg.
(d) 40 mmHg.
(e) 70 mmHg.

240. The best indicator of decreased cerebral perfusion from shock is the presence of

(a) delayed capillary refill.
(b) altered level of consciousness.
(c) hyperventilation.
(d) tachycardia.
(e) cool, pale, clammy skin.

241. Which of the following statements regarding the pneumatic antishock garment (PASG) is true?

(a) PASG is considered beneficial in the treatment of shock, despite the lack of confirmation of its mechanism of action.
(b) PASG has been proved to autotransfuse up to two units of blood from the lower extremities to the central circulation.
(c) PASG has been proved to increase blood pressure secondary to increased peripheral vascular resistance achieved by pneumatic counter pressure.
(d) Both answers (b) and (c).
(e) None of the above.

242. Possible complications of PASG application in the treatment of the hypotensive patient includes all of the following, except

(a) increased intercranial pressure.
(b) decreased diaphragmatic excursion (ventilatory impedance).
(c) increased pulmonary edema.
(d) decreased perfusion to lower body organs.
(e) decreased renal function.

243. Prehospital intravenous access is initiated for all of the following reasons, except

(a) fluid replacement with a crystalloid solution.
(b) fluid replacement with a colloid solution.
(c) to secure a drug administration route.
(d) to obtain predextrose blood samples.
(e) to obtain blood type and cross samples.

244. Which of the following statements regarding IV fluids is true?

(a) All crystalloid solutions contain glucose, which prevents effective fluid replacement; colloid solutions are preferred for prehospital settings.
(b) Crystalloid solutions improve the oxygen transportation capability of the patient's RBCs.
(c) 3 ml of crystalloid solution is required to replace each milliliter of blood loss.
(d) Both answers (a) and (b).
(e) Both answers (b) and (c).

245. Macrodrip IV administration sets have a drops-to-milliliter ratio

(a) of 80 drops to 1 ml.
(b) of 60 drops to 1 ml.
(c) of 15 drops to 1 ml.
(d) of 10 drops to 1 ml.
(e) that may vary, depending upon the manufacturer.

246. Microdrip IV administration sets have a drops-to-milliliter ratio

(a) of 80 drops to 1 ml.
(b) of 60 drops to 1 ml.
(c) of 15 drops to 1 ml.
(d) of 10 drops to 1 ml.
(e) that may vary, depending upon the manufacturer.

247. Which of the following statements regarding catheter size effect upon fluid administration rates is false?

(a) The larger the catheter lumen, the greater the rate of flow.
(b) The longer the catheter length, the faster the rate of flow.
(c) The smaller the catheter gauge number, the greater the catheter lumen.
(d) Both answers (a) and (b) are false.
(e) Both answers (b) and (c) are false.

248. Infiltration of fluid or medication into the tissues surrounding the IV site is called

(a) extravasation.
(b) extravenous shift.
(c) third spacing.
(d) Any of the above.
(e) None of the above.

249. The cause of a pyrogenic reaction to IV cannulation is

(a) irritation and inflammation of the cannulated vein by the IV solution, needle, catheter, or infused medication.
(b) advancing the catheter incompletely and then withdrawing the catheter back over (or through) the needle.
(c) the infusion of air from an incompletely flushed administration set.
(d) the presence of pyrogen proteins within the IV solution or administration set.
(e) an antigen/antibody reaction to the IV solution.

250. The cause of thrombophlebitis from IV therapy is

(a) irritation and inflammation of the cannulated vein by the IV solution, needle, catheter, or infused medication.
(b) advancing the catheter incompletely and then withdrawing the catheter back over (or through) the needle.
(c) the infusion of air from an incompletely flushed administration set.
(d) the presence of pyrogen proteins within the IV solution or administration set.
(e) an antigen/antibody reaction to the IV solution.

251. Signs and symptoms of a pyrogenic reaction to IV cannulation include

(a) complaints of pain along the path of the vein, and eurythema with edema at the puncture site.
(b) onset of complaints within 30 to 60 minutes of cannulation.
(c) complaints of fever, chills, nausea, headache, and/or backache.
(d) Both answers (a) and (b).
(e) Both answers (b) and (c).

252. Signs and symptoms of thrombophlebitis caused by IV cannulation include

(a) complaints of pain along the path of the vein, and eurythema with edema at the puncture site.
(b) onset of complaints within 30 to 60 minutes of cannulation.
(c) complaints of fever, chills, nausea, headache, and/or backache.
(d) Both answers (a) and (b).
(e) Both answers (b) and (c).

253. Using all of the following activities, indicate the correct order of treatment for severely injured trauma patients.

(1) IV cannulation and fluid administration
(2) extrication
(3) initiation of transportation
(4) immediate life threat stabilization

(a) 2, 1, 4, 3
(b) 2, 4, 3, 1
(c) 2, 3, 1, 4
(d) 2, 4, 1, 3
(e) 2, 1, 3, 4

The answer key to Section Two begins on page 370.

3

Test Section Three

Test Section Three covers the following subjects:

* Drug Information and Medication Terminology
* Drug Doses, Administrations and Calculations
* Factors Affecting Drug Actions
* Epinephrine (1:10, 1:1, racemic solutions, and "high-dose" regimen).
  Norepinephrine, Isoproterenol, Dopamine, Dobutamine, Lidocaine, Bretylium Tosylate, Procainamide, Verapamil, Adenosine, Atropine, Sodium Bicarbonate, Morphine Sulfate, Nitronox, Furosemide, Nitroglycerin, Calcium Chloride, Aminophylline, Albuterol, $D_{50}W$, Thiamine, Dexamethasone, Methylprednisolone, Diazepam, Oxytocin, Magnesium Sulfate, Diphenhydramine, Ipecac, Activated Charcoal, and Naloxone

EMT - Paramedic National Standards Review Self Test Third Edition

1. Sodium bicarbonate and calcium chloride are examples of medications originally derived from
- (a) animal sources.
- (b) vegetable (plant) sources.
- (c) mineral sources.
- (d) synthetic sources.
- (e) None of the above.

2. Insulin and epinephrine are examples of medications originally derived from
- (a) animal sources.
- (b) vegetable (plant) sources.
- (c) mineral sources.
- (d) synthetic sources.
- (e) None of the above.

3. Digitalis, morphine sulfate, and atropine are examples of medications originally derived from
- (a) animal sources.
- (b) vegetable (plant) sources.
- (c) mineral sources.
- (d) synthetic sources.
- (e) None of the above.

4. Lidocaine and bretylium tosylate are examples of medications originally derived from
- (a) animal sources.
- (b) vegetable (plant) sources.
- (c) mineral sources.
- (d) synthetic sources.
- (e) None of the above.

5. In 1970 an act was passed to regulate and control the manufacturing, distribution, and dispensing of drugs that have abuse potential. This act is called the
- (a) Federal Food, Drug, and Cosmetic Act.
- (b) Harrison Narcotic Act.
- (c) Controlled Substance Act.
- (d) Narcotic Control Act.
- (e) Pure Food and Drug Act.

6. Medications that contain only small amounts of narcotic and have a limited abuse potential are classed as

(a) schedule I drugs.
(b) schedule II drugs.
(c) schedule III drugs.
(d) schedule IV drugs.
(e) schedule V drugs.

7. Drugs that have a high potential for abuse and a severe dependence liability but also have accepted medical uses are classed as

(a) schedule I drugs.
(b) schedule II drugs.
(c) schedule III drugs.
(d) schedule IV drugs.
(e) schedule V drugs.

8. Drugs that are frequently abused and have no accepted medical uses are classed as

(a) schedule I drugs.
(b) schedule II drugs.
(c) schedule III drugs.
(d) schedule IV drugs.
(e) schedule V drugs.

9. The official name for epinephrine is

(a) epi.
(b) epinephrine hydrochloride.
(c) Adrenalin.
(d) epinephrine hydrochloride, U.S.P.
(e) beta-(3,4-dihydroxyphenl)-a-methylaminoethanol.

10. The chemical name for epinephrine is

(a) epi.
(b) epinephrine hydrochloride.
(c) Adrenalin.
(d) epinephrine hydrochloride, U.S.P.
(e) beta-(3,4-dihydroxyphenl)-a-methylaminoethanol.

11. The generic name for epinephrine is

(a) epi.
(b) epinephrine hydrochloride.
(c) Adrenalin.
(d) epinephrine hydrochloride, U.S.P.
(e) beta-(3,4-dihydroxyphenl)-a-methylaminoethanol.

12. The trade or proprietary name for epinephrine is

(a) epi.
(b) epinephrine hydrochloride.
(c) Adrenalin.
(d) epinephrine hydrochloride, U.S.P.
(e) beta-(3,4-dihydroxyphenl)-a-methylaminoethanol.

13. The slang or abbreviated name for epinephrine is

(a) epi.
(b) epinephrine hydrochloride.
(c) Adrenalin.
(d) epinephrine hydrochloride, U.S.P.
(e) beta-(3,4-dihydroxyphenl)-a-methylaminoethanol.

14. When drugs are dissolved in water (or other solvents) the preparations are called

(a) syrups.
(b) tinctures.
(c) solutions.
(d) suspensions.
(e) spirits.

15. When drugs are chemically extracted with alcohol, the preparations usually contain a small amount of that alcohol, and are called

(a) syrups.
(b) tinctures.
(c) solutions.
(d) suspensions.
(e) spirits.

16. Some liquid drug preparations do not remain dissolved. The solute settles to the bottom and they require shaking before use. These preparations are called

(a) syrups.
(b) tinctures.

(c) mixers.
(d) suspensions.
(e) spirits.

17. Most drugs are prepared in solid form. These drug forms include all of the following, except
(a) ampules.
(b) powders.
(c) pills.
(d) capsules.
(e) tablets.

18. The term ______ refers to the action of two drugs combining to create a much stronger effect than either drug alone.
(a) synergism
(b) potentiation
(c) cumulative action
(d) antagonism
(e) therapeutic action

19. The term ______ refers to an increased drug effect achieved by repeated doses.
(a) synergism
(b) potentiation
(c) cumulative action
(d) antagonism
(e) therapeutic action

20. The term ______ refers to an opposition between two or more drugs.
(a) synergism
(b) potentiation
(c) cumulative action
(d) antagonism
(e) therapeutic action

21. The term ______ refers to one drug's effect being increased by its combination with another drug.
(a) synergism
(b) potentiation
(c) cumulative action
(d) antagonism
(e) therapeutic action

22. The term ______ refers to the desired medical effect of the drug at appropriate doses.

(a) synergism
(b) potentiation
(c) cumulative action
(d) antagonism
(e) therapeutic action

23. The term or phrase that best refers to undesired and unavoidable drug effects that are harmful to the patient is

(a) side effects.
(b) untoward effects.
(c) contraindications.
(d) idiosyncrasy.
(e) hypersensitivity.

24. The term or phrase that best refers to any undesired and unavoidable drug effects is

(a) side effects.
(b) untoward effects.
(c) contraindications.
(d) idiosyncrasy.
(e) hypersensitivity.

25. The term or phrase that best refers to an allergic reaction to a medication is

(a) side effects.
(b) untoward effects.
(c) contraindications.
(d) idiosyncrasy.
(e) hypersensitivity.

26. The term or phrase that best refers to an undesired and unavoidable drug effect occurring in a specific patient that is not seen in other patients who take the same medication is

(a) side effects.
(b) untoward effects.
(c) contraindications.
(d) idiosyncrasy.
(e) hypersensitivity.

27. The term or phrase _______ refers to specific physiological or medical reasons that a patient should not receive a particular medication.

(a) side effects
(b) untoward effects
(c) contraindications
(d) idiosyncrasy
(e) hypersensitivity

28. When a condition or complaint is resistant to normal treatment or stimulation, it is said to be _______ to that treatment or stimulation.

(a) refractory
(b) antagonistic
(c) refractional
(d) synergistic
(e) reciprocal

29. The administration of an IV solution by regulating its flow rate based upon observation of desired or undesired effects is called

(a) estimation.
(b) calculated dosages.
(c) approximation.
(d) fluctuation.
(e) titration.

30. A medication known to increase urine output is called

(a) an antihypertensive drug.
(b) a diuretic.
(c) a vasopressor.
(d) a dilator.
(e) a vasodilator.

31. A medication known to increase blood pressure is called

(a) an antihypertensive drug.
(b) a diuretic.
(c) a vasopressor.
(d) a dilator.
(e) a vasodilator.

32. Medications administered orally, rectally, or vaginally are called

(a) injectable drugs.
(b) noninjectable drugs.
(c) enteral drugs.
(d) parenteral drugs.
(e) transenteral drugs.

33. Medications that are administered by IV, IM, or SQ routes are called

(a) injectable drugs.
(b) noninjectable drugs.
(c) enteral drugs.
(d) parenteral drugs.
(e) transenteral drugs.

34. Medications that are administered by transtracheal, intracardiac, or intralingual routes are called

(a) injectable drugs.
(b) noninjectable drugs.
(c) enteral drugs.
(d) parenteral drugs.
(e) transenteral drugs.

35. Which of the following statements regarding drug administration is false?

(a) The patient's age, weight, and general health condition will influence the action of drugs.
(b) As a rule, enteral drug administration achieves a higher blood level of medication faster than parenteral administration.
(c) The IV and endotracheal (transtracheal) routes provide approximately the same speed of drug absorption.
(d) Sublingual drug absorption is faster than subcutaneous drug absorption.
(e) Hypothermia, shock, and acidosis will inhibit drug absorption and/or action.

36. Routes employed for drug elimination include all of the following, except

(a) excretion by the liver into the bile.
(b) excretion by the kidneys via urination.
(c) excretion by the lungs via expiration.
(d) excretion by the digestive system via feces.
(e) excretion by the oropharynx via salivation.

37. Which of the following statements regarding the pharmacodynamics of drugs is true?

(a) The majority of medications achieve their effect by attaching to specific proteins present on cell membranes called "drug receptors sites."
(b) Drugs that achieve their effect by binding with drug receptors are called "agonists."
(c) Drugs that bind with drug receptors for the purpose of blocking another drug from attachment to that receptor are called "antagonists."
(d) Both answers (a) and (b) are true.
(e) Answers (a), (b), and (c) are true.

38. The chemical mediator acetylcholine is the primary neurotransmitter employed by the

(a) central nervous system.
(b) peripheral nervous system.
(c) parasympathetic nervous system.
(d) sympathetic nervous system.
(e) independent nervous system.

39. The hormones epinephrine and norepinephrine are employed by the

(a) central nervous system.
(b) peripheral nervous system.
(c) parasympathetic nervous system.
(d) sympathetic nervous system.
(e) independent nervous system.

40. Vegetative functions are the responsibility of the

(a) central nervous system.
(b) peripheral nervous system.
(c) parasympathetic nervous system.
(d) sympathetic nervous system.
(e) independent nervous system.

41. Functions during stress are the responsibility of the

(a) central nervous system.
(b) peripheral nervous system.
(c) parasympathetic nervous system.
(d) sympathetic nervous system.
(e) independent nervous system.

42. Constricted pupils, increased salivation, and decreased heart rate are effects that occur with

(a) stimulation of acetylcholine release.
(b) stimulation of alpha 1 receptors.
(c) stimulation of beta 1 receptors.
(d) stimulation of beta 2 receptors.
(e) stimulation of dopaminergic receptors.

43. Bronchodilation and vasodilation are effects that occur with

(a) stimulation of acetylcholine release.
(b) stimulation of alpha 1 receptors.
(c) stimulation of beta 1 receptors.
(d) stimulation of beta 2 receptors.
(e) stimulation of dopaminergic receptors.

44. Renal, cerebral, and coronary arterial dilation is believed to occur with

(a) stimulation of acetylcholine release.
(b) stimulation of alpha 1 receptors.
(c) stimulation of beta 1 receptors.
(d) stimulation of beta 2 receptors.
(e) stimulation of dopaminergic receptors.

45. Increased heart rate is an effect that occurs with

(a) stimulation of acetylcholine release.
(b) stimulation of alpha 1 receptors.
(c) stimulation of beta 1 receptors.
(d) stimulation of beta 2 receptors.
(e) stimulation of dopaminergic receptors.

46. Peripheral vasoconstriction and occasional bronchoconstriction are effects that occur with

(a) stimulation of acetylcholine release.
(b) stimulation of alpha 1 receptors.
(c) stimulation of beta 1 receptors.
(d) stimulation of beta 2 receptors.
(e) stimulation of dopaminergic receptors.

47. Increased cardiac automaticity, conduction, and contractile force are effects that occur in response to

(a) stimulation of acetylcholine release.
(b) stimulation of alpha 1 receptors.

(c) stimulation of beta 1 receptors.
(d) stimulation of beta 2 receptors.
(e) stimulation of dopaminergic receptors.

48. Drugs that are said to be adrenergic
(a) provide sympathetic nervous system stimulation.
(b) block sympathetic nervous system stimulation.
(c) are sympathetic nervous system agonists.
(d) Both answers (a) and (c).
(e) Both answers (b) and (c).

49. Drugs that are said to be antiadrenergic
(a) provide sympathetic nervous system stimulation.
(b) block sympathetic nervous system stimulation.
(c) are sympathetic nervous system agonists.
(d) Both answers (a) and (c).
(e) Both answers (b) and (c).

50. Drugs that are referred to as sympatholytic
(a) provide sympathetic nervous system stimulation.
(b) block sympathetic nervous system stimulation.
(c) are sympathetic nervous system agonists.
(d) Both answers (a) and (c).
(e) Both answers (b) and (c).

51. Drugs that are referred to as sympathomimetic
(a) provide sympathetic nervous system stimulation.
(b) block sympathetic nervous system stimulation.
(c) are sympathetic nervous system agonists.
(d) Both answers (a) and (c).
(e) Both answers (b) and (c).

52. Medications that stimulate the actions of the parasympathetic nervous system are said to be
(a) parasympathomimetic.
(b) cholinergic.
(c) anticholinergic.
(d) Both answers (a) and (b).
(e) Both answers (a) and (c).

53. Medications that block the actions of the parasympathetic nervous system are said to be

(a) parasympathetic blockers.
(b) parasympatholytic.
(c) anticholinergic.
(d) Answers (a) and (b) only.
(e) Answers (a), (b), and (c).

54. A drug that influences the force of cardiac muscular contractility is said to be

(a) inotropic.
(b) phototropic.
(c) photosensitive.
(d) chemotropic.
(e) chronotropic.

55. A drug that influences the rate of the heart is referred to as being

(a) inotropic.
(b) phototropic.
(c) photosensitive.
(d) chemotropic.
(e) chronotropic.

56. Your patient weighs 154 pounds. How many kilograms (kg) does your patient weigh?

(a) 70 kg
(b) 85 kg
(c) 77 kg
(d) 93.5 kg
(e) 74.8 kg

57. Your patient weighs 187 pounds. How many kilograms (kg) does your patient weigh?

(a) 70 kg
(b) 85 kg
(c) 77 kg
(d) 93.5 kg
(e) 74.8 kg

58. Your patient weighs 55 kilograms. How many pounds (lb) does your patient weigh?

(a) 137.5 lb
(b) 190 lb

(c) 209 lb
(d) 110 lb
(e) 121 lb

59. Your patient weighs 95 kilograms. How many pounds (lb) does your patient weigh?

(a) 137.5 lb
(b) 190 lb
(c) 209 lb
(d) 110 lb
(e) 121 lb

The following seven questions are designed to test your mathematics skills and do not reflect common prehospital medication ratios of milligrams-to-milliliters. The answers may not reflect realistic infusion rates. All supplies referred to, however, represent standard EMS equipment (for example, microdrip administration sets).

60. You are presented with an ampule that contains 200 mg of a medication in 40 ml of solution. What is the concentration of medication per milliliter?

(a) 0.5 mg per ml
(b) 5.0 mg per ml
(c) 1.0 mg per ml
(d) 50 mg per ml
(e) 10 mg per ml

61. You are presented with a tubex (preloaded syringe) that contains a medication concentration of 2 mg per cubic centimeter of solution. There are 150 mg of medication in the tubex. The tubex contains _____.

(a) 30 cc
(b) 100 cc
(c) 75 cc
(d) 37.5 cc
(e) 300 cc

62. If 1200 mg of a medication is injected into a 250 cc bag of D_5W, what concentration of micrograms of medication per milliliter is achieved?

(a) 480 µg/ml
(b) 4800 µg/ml
(c) 0.48 mg/ml
(d) 4.80 mg/ml
(e) 300,000 mg/ml

63. You have just spiked a 500 cc bag of normal saline with a macrodrip infusion set that delivers 15 gtt/ml. You are ordered to run the IV at 24 gtt/min. How much time will it require to infuse the 500 cc?

(a) 5 hours and 12 minutes
(b) 3 hours and 6 minutes
(c) 6 hours and 12 minutes
(d) 45 minutes
(e) 34 minutes

64. You are ordered to administer one liter of fluid at a rate of 250 ml/hr. Your infusion set delivers 10 gtt/ml. How many drops per minute must you infuse?

(a) 4 gtt/min
(b) 21 gtt/min
(c) 25 gtt/min
(d) 42 gtt/min
(e) 250 gtt/min

65. You are ordered to administer a medication at a rate of 4 µg/kg/min to a 220 pound patient. You prepare the infusion using 750 mg of the medication, a 250 cc bag of D_5W, and a microdrip infusion set. How many drops per minute do you infuse?

(a) 25 gtt/min
(b) 15 gtt/min
(c) 8 gtt/min
(d) 9 gtt/min
(e) 32 gtt/min

66. You are ordered to administer 9 mg/kg of medication to a 176 pound patient. The medication is packaged as 1 gram in 5 cc of solution. How many cubic centimeters must be administered?

(a) 3.6 cc
(b) 3.9 cc

(c) 72 cc
(d) 7.2 cc
(e) 3600 cc

67. Which of the following statements regarding epinephrine is false?

(a) Epi affects both alpha and beta receptors.
(b) Epi increases myocardial oxygen demand and can increase the size of an infarct.
(c) Epi's alpha effects are stronger than its beta effects.
(d) Epi is a short-acting drug and requires repeat boluses every 5 minutes to maintain a therapeutic blood level of the drug.
(e) Epi is ineffective when administered to an acidotic patient.

68. Epinephrine 1:10,000 is indicated in all of the following situations, except

(a) ventricular fibrillation.
(b) asystole.
(c) pulmonary edema.
(d) pulseless electrical activity (PEA).
(e) anaphylactic shock.

69. Administration of epinephrine 1:10,000 is contraindicated in patients who are

(a) not in need of significant respiratory or cardiopulmonary resuscitation.
(b) suffering from minor asthma.
(c) hypovolemic.
(d) Answers (a) and (c) only.
(e) Answers (a), (b), and (c).

70. Which of the following statements regarding epinephrine 1:10,000 is true?

(a) Epi requires protection from light.
(b) Epi is deactivated when combined with alkaline solutions.
(c) Epi is contraindicated in pediatric patients.
(d) Both answers (a) and (b) are true.
(e) Both answers (b) and (c) are true.

71. For administration in adult cardiac arrest, epinephrine 1:10,000 is administered in doses of
(a) 1.0 to 2.0 mg IVP every 5 to 10 minutes.
(b) 0.5 to 1.0 mg IVP every 3 to 5 minutes.
(c) 1.0 to 2.5 mg ET every 3 to 5 minutes.
(d) Both answers (a) and (c).
(e) Both answers (b) and (c).

72. Doses of epinephrine 1:10,000 may be administered via
(a) IV only.
(b) SQ only.
(c) IO or SQ only.
(d) IV or ET only.
(e) IV, ET, or IO only.

73. Which of the following statements regarding norepinephrine (Levophed) is false?
(a) Levophed, in therapeutic doses, provides vasodilation of the renal and mesenteric vasculature to preserve their functions.
(b) Norepinephrine primarily affects alpha receptors, providing extensive peripheral vasoconstriction.
(c) Levophed is indicated for use in hypotension secondary to cardiogenic or neurogenic shock, and is contraindicated in hypovolemic shock.
(d) Tissue necrosis will result if norepinephrine is infused through an infiltrated IV site.
(e) Levophed is administered only by IV drip, "piggybacked" onto a preexisting IV line.

74. The trade name for isoproterenol is
(a) Inderal.
(b) Intropin.
(c) Isuprel.
(d) Isoptin.
(e) Stadol.

75. Which of the following statements regarding isoproterenol is false?
(a) Isoproterenol is used to increase peripheral vascular resistance.
(b) Isoproterenol acts only on beta receptors.
(c) Isoproterenol increases cardiac output by increasing the heart rate and strength of contractility.

(d) Isoproterenol increases myocardial oxygen demand and can increase the size of an infarct.
(e) Isoproterenol is deactivated when combined with alkaline solutions.

76. When transcutaneous pacing is unavailable, isoproterenol is administered for
(a) symptomatic bradycardias, refractory to atropine.
(b) symptomatic heart blocks, refractory to atropine.
(c) asystole, refractory to epinephrine.
(d) Answers (a) and (b) only.
(e) Answers (a), (b), and (c).

77. Methods of isoproterenol administration include
(a) 2 mg IV injection for asystole refractory to epinephrine.
(b) 2 mg diluted in 250 ml of 5% dextrose in water, infused over 20 to 30 minutes.
(c) 0.5 to 1 mg IV bolus, repeated every 5 minutes.
(d) All of the above.
(e) None of the above.

78. The most commonly used trade name for dopamine is
(a) Inotropine.
(b) Dobutrex.
(c) Intropin.
(d) Inopress.
(e) Dopastat.

79. Which of the following statements regarding dopamine is false?
(a) Dopamine in therapeutic doses maintains blood flow through the renal and mesenteric vasculature to preserve their functions.
(b) Dopamine is a vasopressor and may cause hypertensive crisis in a patient taking monoamine oxidase (MAO) inhibitors for depression.
(c) At 2–10 μg/kg/minute, dopamine primarily affects beta 1 receptors.
(d) Dopamine's special chemistry allows for administration of sodium bicarbonate through a dopamine infusion line.
(e) None of the above is false.

80. Dopamine is indicated for use in cases of
(a) cardiogenic hypotension only.
(b) cardiogenic and neurogenic hypotension (or hypovolemic hypotension after fluid resuscitation only).
(c) hypovolemic hypotension only.
(d) cardiogenic and/or neurogenic hypotension only.
(e) neurogenic hypotension only.

81. Dopamine is administered by a "piggybacked" infusion prepared by adding
(a) 800 mg of dopamine to 1000 ml of normal saline.
(b) 800 mg of dopamine to 500 ml of 5% dextrose in water.
(c) 400 mg of dopamine to 250 ml of 5% dextrose in water.
(d) Both answers (a) and (b).
(e) Both answers (b) and (c).

82. The desired vasopressor effect of dopamine is best achieved when administered in the dose range of
(a) 2 to 5 μg/kg/min.
(b) 5 to 20 μg/kg/min.
(c) over 20 μg/kg/min.
(d) All of the above.
(e) None of the above.

83. Which of the following statements regarding dobutamine is false?
(a) Dobutamine is administered in IV boluses of 0.5 to 1.0 mg, and repeated every 10 minutes.
(b) Dobutamine is used for congestive heart failure patients who are hypotensive.
(c) Dobutamine has a positive inotropic effect with little chronotropic activity.
(d) Dobutamine produces less increase in heart rate than isoproterenol or dopamine.
(e) Dobutamine should be administered at a flow rate based on the patient's response.

84. Which of the following statements regarding lidocaine is false?
(a) Lidocaine lowers the threshold of ventricular fibrillation.
(b) Lidocaine decreases the incidence of PVCs by suppressing ventricular ectopy.

(c) Lidocaine may depress the central nervous system.
(d) Ectopy suppression occurs only when adequate blood levels of lidocaine are maintained by repeated boluses or a maintenance infusion.
(e) Successful defibrillation should be followed by administration of lidocaine.

85. Malignant PVCs include all of the following, except
(a) a single PVC per minute when occasional PACs are present.
(b) more than six unifocal PVCs per minute.
(c) occasional couplets or runs of PVCs.
(d) two to five multifocal PVCs.
(e) any single R-on-T PVC.

86. Lidocaine is indicated in all of the following situations, except
(a) atrial fibrillation with a ventricular rate above 100.
(b) malignant PVCs.
(c) ventricular tachycardia with pulses.
(d) ventricular fibrillation.
(e) wide complex PSVT.

87. Lidocaine is contraindicated in all of the following situations, except
(a) second-degree AV block with more than six PVCs per minute.
(b) third-degree AV block with R-on-T PVCs.
(c) bradycardia with couplet PVCs.
(d) bradycardia refractory to atropine and isuprel, with R-on-T PVCs.
(e) sinus rhythm without ectopy, after defibrillation.

88. Untoward effects of lidocaine administration include all of the following, except
(a) agitation or irritability.
(b) euphoria or elation.
(c) drowsiness or altered level of consciousness.
(d) muscle twitching or seizures.
(e) unresponsiveness and death.

89. The standard IV bolus dosage of lidocaine is

(a) 1 mg/kg IV bolus, repeated every 8 to 10 minutes only if ectopy continues.
(b) 1 mg/kg IV bolus, repeated every 5 minutes until a lidocaine infusion is prepared and administered.
(c) 0.5-0.75 mg/kg IV bolus followed by 1.0-1.5 mg/kg repeat boluses every 5 to 10 minutes until a lidocaine infusion is prepared and administered, or maximum bolus dose of 3 mg/kg is reached.
(d) 1.0-1.5 mg/kg IV bolus, repeated every 8 to 10 minutes until a lidocaine infusion is prepared and administered.
(e) 1.0-1.5 mg/kg IV bolus followed by 0.5-0.75 mg/kg repeat boluses every 5 to 10 minutes until a lidocaine infusion is prepared and administered, or maximum bolus dose of 3 mg/kg is reached.

90. The initial bolus of lidocaine should be

(a) administered by infusion, at a TKO rate, if the patient is complaining of chest pain and has malignant PVCs.
(b) reduced by half if the patient is 70 years old or older.
(c) doubled if the patient is in cardiopulmonary arrest.
(d) Answers (a) and (b) only.
(e) Answers (a), (b), and (c).

91. Administration routes for lidocaine include

(a) IV only.
(b) ET only.
(c) IO only.
(d) IV or ET.
(e) IV, ET, or IO.

92. Which of the following statements regarding bretylium tosylate (Bretylol) is false?

(a) Bretylium's effects are seen within one minute of administration.
(b) Bretylium increases the threshold of ventricular fibrillation.
(c) When used for indicated situations, there are no contraindications for bretylium.
(d) Bretylium is only used when lidocaine has proved unsuccessful (or when a patient is hypersensitive to lidocaine).
(e) Postural hypotension will occur in at least 50 percent of all conscious patients receiving bretylium.

93. Bretylium is indicated for

(a) malignant PVCs or ventricular dysrhythmias that are refractory to lidocaine.
(b) atrial fibrillation that is refractory to lidocaine.
(c) malignant PVCs, ventricular dysrhythmias, or atrial fibrillation in patients allergic to lidocaine.
(d) Answers (a) and (c) only.
(e) Answers (a), (b), and (c).

94. Bretylium is administered in which of the following manners?

(a) 1.0 to 1.5 mg/kg IV bolus every 5 minutes, until dysrhythmia is discontinued.
(b) 1.0 to 1.5 mg/kg IV bolus. If dysrhythmia persists, 0.5 to 0.75 mg/kg boluses may be given every 5 minutes, not to exceed a total dose of 30 mg/kg.
(c) 5 mg/kg initial IV bolus. If dysrhythmia persists, 10 mg/kg boluses may be given every 5 minutes, not to exceed a total dose of 30 mg/kg.
(d) 10 mg/kg initial IV bolus. If dysrhythmia persists, 5 mg/kg boluses may be given every 5 minutes, not to exceed a total dose of 30 mg/kg.
(e) IV infusion only; prepared by adding 500 mg of bretylium to 100 ml 5% dextrose in water, infused over 5 to 50 minutes.

95. Administration routes for bretylium include

(a) IV only.
(b) ET only.
(c) IO only.
(d) IV or ET.
(e) IV, ET, or IO.

96. Which of the following statements regarding procainamide (Pronestyl) is false?

(a) Procainamide is indicated for ventricular dysrhythmias only.
(b) 3 mg/kg of lidocaine must be found unsuccessful prior to initiation of procainamide therapy.
(c) Procainamide is particularly effective for treatment of PVCs in the presence of bradycardias or heart blocks.
(d) The presence of heart blocks contraindicates procainamide administration.
(e) The effect-duration of procainamide is shorter than that of lidocaine.

97. Which of the following statements regarding procainamide administration is false?

(a) Discontinue procainamide administration if the patient becomes hypotensive.
(b) Discontinue administration when 20 mg/kg of procainamide have been delivered.
(c) Procainamide is never administered via the ET tube.
(d) Discontinue procainamide administration if the original QRS width has widened by 50 percent or more.
(e) Discontinue procainamide administration when the dysrhythmia is suppressed.

98. Verapamil is available in capsule form under the trade name Verelan. In tablet form, verapamil is available under the trade names of

(a) Isoptin and Cardene.
(b) Cardene and Calan.
(c) Isoptin and Calan.
(d) Isoxsuprine and Isoptin.
(e) Calan and Cardizem.

99. Verapamil

(a) is used to treat narrow-complex PSVTs.
(b) is used to treat symptomatic atrial flutter or fibrillation with too rapid a ventricular response.
(c) causes coronary vasodilation and reduces myocardial oxygen demand.
(d) Answers (a) and (b) only.
(e) Answers (a), (b), and (c).

100. Verapamil

(a) causes peripheral vasodilation and is contraindicated in hypotensive patients.
(b) is contraindicated for patients on beta-blocking medications or patients with a history of WPW syndrome.
(c) may be administered IV, ET, or IO.
(d) Answers (a) and (b) only.
(e) Answers (a), (b), and (c).

101. Which of the following statements regarding verapamil administration is true?

(a) The initial dose is 5 to 10 slow IVP, followed by a repeat bolus (if the dysrhythmia persists) of 2.5 to 5 mg after 15 to 30 minutes, not to exceed a total dose of 30 mg in 30 minutes.
(b) The initial dose is 2.5 to 5 mg slow IVP, followed by a repeat bolus (if the dysrhythmia persists) of 5 to 10 mg after 15 to 30 minutes, not to exceed a total dose of 30 mg in 30 minutes.
(c) The initial dose is 5 to 10 mg slow IVP, followed by a repeat bolus (if the dysrhythmia persists) of 20 to 30 mg after 15 to 30 minutes, not to exceed a total dose of 60 mg in 30 minutes.
(d) Any of the above are true.
(e) None of the above is true.

102. Which of the following statements is false?

(a) Adenosine is a natural substance present in all body cells.
(b) Another name for adenosine is Tonocard.
(c) Adenosine decreases AV conduction.
(d) The half-life of adenosine is less than 5 seconds.
(e) None of the above is false.

103. Indications for adenosine administration include

(a) narrow complex PSVT refractory to vagal maneuvers.
(b) Wolff-Parkinson-White (WPW) tachycardias, refractory to vagal maneuvers.
(c) wide complex tachycardias, refractory to vagal maneuvers.
(d) Answers (a) and (b) only.
(e) Answers (a) and (c) only.

104. Contraindications for adenosine administration include

(a) second- or third-degree heart blocks.
(b) hypersensitivity to adenosine.
(c) sick sinus syndrome.
(d) Answers (a) and (b) only.
(e) Answers (a), (b), and (c).

105. Following cardioversion, adenosine may cause all of the following, except
(a) PVCs or PACs.
(b) prolonged (greater than 10 minutes) nausea.
(c) sinus bradycardia or AV blocks.
(d) sinus tachycardia.
(e) transient asystole.

106. Which of the following statements is true?
(a) Adenosine administration may precipitate bronchospasm in patients with asthma.
(b) Chest pain, shortness of breath, dizziness, and nausea are common side effects of adenosine.
(c) Adenosine side effects are usually self-limited because of its short half-life.
(d) Answers (b) and (c) are true.
(e) All of the above are true.

107. The initial dose of adenosine is
(a) 6 mg rapid IVP (within 1 to 2 seconds).
(b) 6 mg slow IVP (over 30 seconds to 1 minute).
(c) 3 mg rapid IVP (within 1 to 2 seconds).
(d) 3 mg slow IVP (over 30 seconds to 1 minute).
(e) 15 mg rapid IVP (over 15 to 30 seconds).

108. If conversion of the tachycardia is not achieved within 1 to 2 minutes, the repeat dosage of adenosine is
(a) 12 mg rapid IVP (within 1 to 2 seconds).
(b) 12 mg slow IVP (over 30 seconds to 1 minute).
(c) 6 mg rapid IVP (within 1 to 2 seconds).
(d) 6 mg slow IVP (over 30 seconds to 1 minute).
(e) 30 mg rapid IVP (over 15 to 30 seconds).

109. Which of the following statements regarding adenosine administration is false?
(a) Adenosine may be administered IV or ET.
(b) Adenosine should be administered only by rapid IVP directly into the vein or into the medication administration port closest to the patient.
(c) The second dose of adenosine may be repeated once, at the same dose.
(d) Single bolus doses greater than 12 mg of adenosine are contraindicated.
(e) Each bolus of adenosine should be followed by a rapid saline flush.

110. Which of the following statements regarding atropine is false?

(a) Atropine is a parasympathetic blocker.
(b) Atropine blocks organophosphate insecticide effects.
(c) Atropine has positive inotropic effects with little to no chronotropic effects.
(d) Five or more mg of atropine may be required to reverse organophosphate poisoning.
(e) In the presence of atropine's indications, there are no contraindications for its use.

111. Atropine is indicated for all of the following situations, except

(a) symptomatic bradycardias with or without ectopy.
(b) symptomatic second- and third-degree AV blocks.
(c) symptomatic PSVT.
(d) asystole.
(e) bradycardic PEA (EMD).

112. In a cardiac arrest situation, atropine is administered in doses of

(a) 0.5 mg every 3 to 5 minutes, not to exceed 0.04 mg/kg or 3 mg.
(b) 1.0 mg every 3 to 5 minutes, not to exceed 0.4 mg/kg or 3 mg.
(c) 1.0 mg every 3 to 5 minutes, not to exceed 0.5 mg/kg or 4 mg.
(d) 1.0 mg every 3 to 5 minutes, not to exceed 0.04 mg/kg or 3 mg.
(e) 0.1 mg/kg every 3 to 5 minutes, not to exceed 0.05 mg/kg or 4 mg.

113. In a symptomatic patient with a pulse (unrelated to organophosphate exposure situations), atropine is administered in doses of

(a) 0.5 mg every 3 to 5 minutes, not to exceed 0.04 mg/kg or 3 mg.
(b) 1.0 mg every 3 to 5 minutes, not to exceed 0.4 mg/kg or 3 mg.
(c) 1.0 mg every 3 to 5 minutes, not to exceed 0.5 mg/kg or 4 mg.
(d) 1.0 mg every 3 to 5 minutes, not to exceed 0.04 mg/kg or 3 mg.
(e) 0.1 mg/kg every 3 to 5 minutes, not to exceed 0.05 mg/kg or 4 mg.

114. Administration routes for atropine include

(a) IV only.
(b) ET only.
(c) IO only.
(d) IV or ET.
(e) IV, ET, or IO.

115. Which of the following statements regarding sodium bicarbonate is false?

(a) Sodium bicarb is administered immediately after intubation during a cardiac arrest with a known "down time" of more than 10 minutes.
(b) Sodium bicarb can cause metabolic acidosis if administered in large doses.
(c) The administration of sodium bicarb has not been proven effective in the early treatment of cardiac arrest.
(d) Both answers (a) and (b) are false.
(e) None of the above is false.

116. Sodium bicarbonate is used in the treatment of

(a) tricyclic antidepressant overdose.
(b) cardiac arrest refractory to intubation, hyperventilation, and first-line ACLS medications.
(c) cardiac arrest prior to administration of lidocaine.
(d) Answers (a) and (b) only.
(e) Answers (a), (b), and (c).

117. The correct administration of sodium bicarbonate is

(a) a 0.5 mEq/kg bolus, repeated (if needed) after 10 minutes with a 1.0 mEq/kg bolus.
(b) a 1.0 mEq/kg bolus, repeated (if needed) after 10 minutes with a 0.5 mEq/kg bolus.
(c) a 0.5 mEq/kg bolus, repeated (if needed) after 10 minutes.
(d) a 1.0 mEq/kg bolus, repeated (if needed) after 10 minutes.
(e) None of the above.

118. Administration routes for sodium bicarbonate include

(a) IV only.
(b) ET only.
(c) IO only.
(d) IV or ET.
(e) IV, ET, or IO.

119. Which of the following statements regarding morphine sulfate is false?

(a) Morphine is classified as a schedule V drug.
(b) Morphine is a narcotic analgesic derived from opium.
(c) Morphine reduces systemic vascular resistance and increases venous capacitance, thus relieving pulmonary edema.
(d) Morphine decreases venous return and diminishes myocardial oxygen demand.
(e) Morphine alleviates anxiety that might otherwise result in increased infarct size.

120. Morphine sulfate is indicated for the treatment of all of the following, except

(a) chest pain in the setting of an AMI.
(b) symptomatic pulmonary edema with associated chest pain.
(c) symptomatic pulmonary edema without chest pain.
(d) chest pain in the setting of blunt trauma to the chest.
(e) severe pain from isolated extremity trauma or kidney stones.

121. Morphine sulfate is contraindicated for administration to patients who are

(a) hypotensive or hypovolemic.
(b) head injured or complaining of abdominal pain.
(c) allergic to barbiturates.
(d) Answers (a) and (b) only.
(e) Answers (a), (b), and (c).

122. Administration of morphine sulfate may cause

(a) life-threatening allergic reactions.
(b) respiratory depression or arrest.
(c) hypotension in a normotensive patient.
(d) Answers (a) and (b) only.
(e) Answers (a), (b), and (c).

123. Morphine sulfate is administered in

(a) 10 mg boluses every few minutes until total pain relief is achieved.
(b) an initial dose of 5 mg, followed by 4 mg every few minutes until pain relief is achieved or respiratory depression occurs.
(c) an initial dose of 2 to 10 mg, followed by 2 mg every few minutes until pain tolerance is achieved or respiratory depression occurs.
(d) 2 mg injections only, repeated only as necessary to achieve total pain relief.
(e) None of the above.

124. Administration routes for morphine sulfate include

(a) IV only.
(b) IV or IM only.
(c) IV or ET only.
(d) IV, ET, or IM.
(e) IV, ET, IM, or IO.

125. Which of the following statements regarding Nitronox is false?

(a) Nitronox is an inhaled analgesic, administered instead of oxygen to achieve pain relief without the dangerous side effects of morphine.
(b) Nitronox is administered via a standard nonrebreathing mask.
(c) Nitronox may be administered to patients with altered levels of consciousness when pain is clearly a complaint.
(d) All of the above are false.
(e) None of the above is false.

126. Nitronox is indicated in all of the following situations, except

(a) burns.
(b) severe anxiety reactions secondary to alcohol or drug abuse.
(c) musculoskeletal pain.
(d) cardiac chest pain.
(e) musculoskeletal trauma.

127. Nitronox is particularly useful in cases of

(a) severe head injury with altered level of consciousness.
(b) chest trauma with suspected pneumothorax.
(c) bowel obstruction.
(d) injured COPD patients who should not receive high concentrations of oxygen.
(e) None of the above.

128. Which of the following statements regarding furosemide (Lasix) is false?

(a) Lasix is prescribed to patients who suffer from hypertension.
(b) Lasix inhibits reabsorption of sodium in the kidneys.
(c) Lasix is a potent diuretic, but its diuresis effects are not observed until 30 or more minutes after administration.
(d) Lasix inhibits reabsorption of chloride in the kidneys.
(e) Lasix is a vasodilator and must be protected from light to maintain potency.

129. Furosemide is indicated for the treatment of

(a) congestive heart failure with pulmonary edema.
(b) congestive heart failure with hypotension.
(c) anaphylaxis with pulmonary edema.
(d) Answers (a) and (b) only.
(e) Answers (a), (b), and (c).

130. Furosemide may cause

(a) fetal abnormalities if administered to a pregnant woman.
(b) dehydration from fluid loss.
(c) electrolyte depletion from fluid loss.
(d) Answers (b) and (c) only.
(e) Answers (a), (b), and (c).

131. Furosemide may be administered in a dose of

(a) 20 mg when the patient is already on diuretic therapy.
(b) 40 mg when the patient is not on diuretic therapy.
(c) as high as 80 mg when severe threat is indicated.
(d) All of the above.
(e) None of the above.

132. Administration routes for furosemide include
(a) IV only.
(b) ET only.
(c) IO only.
(d) IV or ET.
(e) IV, ET, or IO.

133. Which of the following statements regarding nitroglycerin (NTG) is false?
(a) NTG is a vascular smooth-muscle relaxant.
(b) NTG will dilate cerebral arteries and decrease intercranial pressure.
(c) NTG will dilate coronary arteries and improve myocardial perfusion.
(d) NTG commonly causes headaches and may cause hypotension.
(e) NTG deteriorates when exposed to air and light.

134. Nitroglycerin is indicated for all of the following situations, except
(a) angina.
(b) chest pain associated with AMI.
(c) pulmonary edema.
(d) cerebral edema.
(e) chest pain associated with congestive heart failure.

135. Administration of 0.4 mg nitroglycerin tablets or sprays is done sublingualy, in a frequency of
(a) one tablet/spray every 5 to 10 minutes, until pain relief is achieved or headache ensues.
(b) one tablet/spray every 3 to 5 minutes, until pain is achieved, hypotension ensues, or a total of 3 doses has been taken.
(c) one to three tablets/sprays every 5 to 10 minutes as needed to achieve pain relief.
(d) Any of the above.
(e) None of the above.

136. Which of the following statements regarding calcium chloride is false?
(a) Calcium is used in the treatment of asystole and PEA (EMD).
(b) Calcium may cause digitalis toxicity in patients on digitalis medications.

(c) When combined with sodium bicarb, calcium will precipitate as carbonates.
(d) Calcium is administered in doses of 250–500 mg and may be repeated every 10 minutes as ordered.
(e) Calcium ions increase the strength of myocardial contraction.

137. Which of the following statements regarding epinephrine 1:1000 is false?

(a) Epinephrine 1:1000 will increase myocardial effort and oxygen demand, potentially causing angina or AMI.
(b) Pulmonary edema or emboli may produce asthmalike wheezing, but use of epinephrine 1:1000 is contraindicated.
(c) In an emergency, epinephrine 1:1000 may be diluted with 100 cc of 5% dextrose in water, and used as epinephrine 1:10,000 (IV or ET).
(d) Epinephrine 1:1000 may be administered only subcutaneously.
(e) Frequent side effects of epinephrine 1:1000 are anxiety, complaints of "palpitations," headache, tremors, and tachycardia.

138. Epinephrine 1:1000 is indicated for the treatment of

(a) severe, localized allergic reactions.
(b) anaphylactic shock.
(c) asthma (in patients less than 40 years old only).
(d) Both answers (a) and (b).
(e) Both answers (a) and (c).

139. Epinephrine 1:1000 is administered to adults in doses of

(a) 0.1 to 0.3 mg.
(b) 0.3 to 0.5 mg.
(c) 0.5 to 1.0 mg.
(d) 1.0 to 2.0 mg.
(e) Both answers (b) and (c).

140. Which of the following statements regarding aminophylline is false?

(a) Patients already taking medications containing theophylline must receive a reduced dose of aminophylline.
(b) Aminophylline may cause ventricular ectopy, tachycardias, and hypotension.
(c) Aminophylline continues to be preferred over nebulized beta agonists.
(d) All of the above are false.
(e) None of the above is false.

141. Aminophylline may be indicated in all of the following situations, except

(a) congestive heart failure with pulmonary edema.
(b) asthma.
(c) chronic bronchitis with bronchospasm.
(d) allergic reactions.
(e) emphysema with bronchospasm.

142. Aminophylline is administered by "piggybacked" IV infusion, prepared by adding

(a) 250–500 mg to 80 or 90 ml of 5% dextrose in water, to be infused over 20 to 30 minutes.
(b) 250–500 mg to 20 ml of 5% dextrose in water, to be infused over 20 to 30 minutes.
(c) 250–500 mg to 500 ml of 5% dextrose in water, to be infused over 20 to 30 minutes.
(d) Answers (a) or (b) only.
(e) Answers (a), (b), or (c).

143. All of the following statements regarding racemic epinephrine are true, except

(a) Racemic epinephrine is administered by nebulized inhalation or diluted IV infusion.
(b) Racemic epinephrine is used to treat croup.
(c) Racemic epinephrine is contraindicated for epiglottis.
(d) Racemic epinephrine may cause tachycardia or dysrhythmias.
(e) The standard racemic epinephrine dose is 0.25 to 0.75 ml diluted with 2 ml of normal (or respiratory) saline.

144. Which of the following statements regarding albuterol (Proventil, Ventolin) is false?

(a) Albuterol is a sympatholytic medication.
(b) Albuterol produces bronchodilation without the frequency of tachycardia seen with epi or aminophylline.
(c) Albuterol is used to treat asthma and bronchospasm.
(d) The standard adult albuterol dose is 2.5 mg (0.5 ml of a 0.5% solution diluted with 2.5 ml of saline).
(e) In the prehospital setting, albuterol is administered by inhalation only.

145. Indications for administration of 50% dextrose in water ($D_{50}W$) include

(a) altered level of consciousness, suspected to be from acute alcohol abuse.
(b) altered level of consciousness, suspected to be from acute drug overdose.
(c) altered level of consciousness, despite suspicion of hyperglycemia, when reagent strip confirmation is unavailable.
(d) Answers (a) and (b) only.
(e) Answers (a), (b), and (c).

146. When hypoglycemia is confirmed by a reagent strip, $D_{50}W$ is contraindicated in cases of

(a) altered level of consciousness, suspected to be secondary to a seizure disorder.
(b) altered level of consciousness, suspected to be secondary to a cerebrovascular accident.
(c) altered level of consciousness, suspected to be secondary to hyperglycemia.
(d) All of the above.
(e) None of the above.

147. $D_{50}W$ is administered in a dosage of

(a) 25 grams (25 ml of a 50 percent solution).
(b) 2500 grams (50 ml of a 50 percent solution).
(c) 25 mg (50 ml of a 50 percent solution).
(d) 250 mg (25 m of a 50 percent solution).
(e) 25 grams (50 ml of a 50 percent solution).

148. Administration routes for $D_{50}W$ include
(a) IV only.
(b) ET only.
(c) IM only.
(d) IV or IM.
(e) IV, IM, or IO.

149. Thiamine deficiency may
(a) alter the cell's ability to metabolize glucose.
(b) be responsible for neurologic deficits observed in alcoholics who have been on a "binge."
(c) contribute to altered levels of consciousness.
(d) All of the above.
(e) None of the above.

150. Administration of thiamine is indicated in situations involving
(a) altered level of consciousness with a history of brittle diabetes.
(b) altered level of consciousness with a history of alcoholism.
(c) delerium tremens.
(d) Answers (a) and (b) only.
(e) Answers (b) and (c) only.

151. Potential indications for the administration of steroid preparations include all of the following, except
(a) asthma.
(b) anaphylaxis.
(c) acute pulmonary edema or congestive heart failure.
(d) spinal cord injury.
(e) neurogenic shock.

152. Which of the following statements regarding methylprednisolone (Solu-Medrol) is true?
(a) Solu-Medrol is a synthetic steroid with potent anti-inflammatory properties.
(b) For acute histamine reactions, 80 to 125 mg of Solu-Medrol are administered to an adult, IV.
(c) For adult spinal cord injury, 30 mg/kg of Solu-Medrol is infused IV, over 15 minutes (followed 45 minutes later by a 5.4 mg/kg/hr maintenance infusion).
(d) Both answers (a) and (c) are true.
(e) Answers (a), (b), and (c) are true.

153. For management of seizures, the standard adult diazepam dosage is
- (a) 1 to 3 mg.
- (b) 2 to 5 mg.
- (c) 5 to 10 mg.
- (d) 5 to 15 mg.
- (e) 15 to 20 mg.

154. For treatment of acute anxiety reactions, the standard adult diazepam dosage is
- (a) 1 to 3 mg.
- (b) 2 to 5 mg.
- (c) 5 to 10 mg.
- (d) 5 to 15 mg.
- (e) 15 to 20 mg.

155. For sedation prior to cardioversion, the standard adult diazepam dosage is
- (a) 1 to 3 mg.
- (b) 2 to 5 mg.
- (c) 5 to 10 mg.
- (d) 5 to 15 mg.
- (e) 15 to 20 mg.

156. Administration routes for diazepam include
- (a) IV and IM.
- (b) ET and IO.
- (c) rectally.
- (d) Both answers (a) and (b).
- (e) Both answers (a) and (c).

157. In the prehospital setting, oxytocin is administered
- (a) to induce labor during a long transport.
- (b) to enhance delivery of a second fetus in a multiple birth situation with prolonged labor.
- (c) to assist in control of severe postpartum hemorrhage.
- (d) to delay labor during a long transport.
- (e) None of the above.

158. Which of the following statements regarding magnesium sulfate is false?

(a) Magnesium sulfate is a CNS stimulant and may cause tachycardia, tachypnea, and hypertension.
(b) Magnesium sulfate is indicated for the treatment of seizures secondary to toxemia of pregnancy (eclampsia).
(c) Calcium chloride antagonizes magnesium sulfate effects.
(d) All of the above are false.
(e) None of the above are false.

159. Administration of 1 to 2 grams of magnesium sulfate, diluted in 100 ml 5% dextrose in water, over 1 to 2 minutes, is recommended for

(a) ventricular tachycardia.
(b) ventricular fibrillation.
(c) Torsade de pointes.
(d) Both answers (a) and (b).
(e) Both answers (a) and (c).

160. The most familiar trade name for diphenhydramine is

(a) Valium.
(b) Benadryl.
(c) Demerol.
(d) Vistaril.
(e) Haldol.

161. Which of the following statements regarding diphenhydramine is false?

(a) Diphenhydramine is indicated for the treatment of allergic reactions or anaphylactic shock, but only after initial administration of epinephrine.
(b) Diphenhydramine is indicated for the treatment of asthma, but only after the initial administration of epinephrine.
(c) Administration routes for diphenhydramine include IV and IM only.
(d) Diphenhydramine acts as a sedative.
(e) None of the above is false.

162. The normal adult dosage of diphenhydramine is

(a) 1 mg/kg every 2–3 minutes.
(b) 10–20 mg diluted in 50 ml D_5W and infused over 15 minutes.

(c) a 25-50 mg bolus.
(d) Both answers (a) and (c).
(e) Both answers (b) and (c).

163. Which of the following statements regarding syrup of ipecac is true?

(a) Ipecac is a particularly effective emetic when used for treatment of thorazine overdose.
(b) If the overdose patient is obtunded, ipecac may be administered sublingually to effect gastric evacuation without aspiration.
(c) After the administration of 15 to 30 ml of oral (or sublingual) ipecac, the patient should remain NPO until arrival at the emergency room.
(d) All of the above are true.
(e) None of the above is true.

164. Which of the following statements regarding activated charcoal is true?

(a) Oral administration of activated charcoal is contraindicated in the patient with a depressed level of consciousness.
(b) If ipecac has been administered, activated charcoal is contraindicated until after emesis.
(c) Activated charcoal is indicated for treatment of ingestions that contraindicate administration of ipecac.
(d) All of the above are true.
(e) None of the above is true.

165. Which of the following statements regarding naloxone (Narcan) is false?

(a) Naloxone antagonizes the effects of narcotics.
(b) Administration of naloxone may precipitate withdrawal signs and symptoms in the addicted patient.
(c) Naloxone is not effective in the treatment of synthetic narcotic overdose.
(d) Naloxone is not effective in the treatment of barbiturate overdose, but may still be administered to rule out narcotic ingestion.
(e) Naloxone is indicated for treatment of unconsciousness of unknown etiology, especially when alcohol ingestion is suspected.

166. Naloxone is administered in doses of

(a) 1–2 mg, repeated every 5 minutes as needed.
(b) 2–5 mg if Darvon overdose is suspected.
(c) 2–5 mg if alcohol overdose is suspected.
(d) Answers (a) and (c) only.
(e) Answers (a), (b), and (c).

167. Administration routes for naloxone include

(a) IV only.
(b) IV or ET only.
(c) IM or SQ only.
(d) IV, ET, IM, or SQ
(e) IV, ET, IM, SQ, or IO.

168. There are three regimens suggested by ACLS guidelines for high dose epinephrine administration (in a protracted arrest). The "intermediate" regimen consists of IV epinephrine;

(a) 2 mg to 5 mg, every 3 to 5 minutes.
(b) 1 mg, then 3 mg, then 5 mg boluses, 3 minutes apart.
(c) 0.1 mg/kg every 3 to 5 minutes.
(d) 1 mg every 3 to 5 minutes.
(e) 5 mg every 10 minutes.

169. The "escalating" regimen consists of IV epinephrine;

(a) 2 mg to 5 mg, every 3 to 5 minutes.
(b) 1 mg, then 3 mg, then 5 mg boluses, 3 minutes apart.
(c) 0.1 mg/kg every 3 to 5 minutes.
(d) 1 mg every 3 to 5 minutes.
(e) 5 mg every 10 minutes.

170. The "high" regimen consists of IV epinephrine;

(a) 2 mg to 5 mg, every 3 to 5 minutes.
(b) 1 mg, then 3 mg, then 5 mg boluses, 3 minutes apart.
(c) 0.1 mg/kg every 3 to 5 minutes.
(d) 1 mg every 3 to 5 minutes.
(e) 5 mg every 10 minutes.

The answer key to Section Three begins on page 376.

4

Test Section Four

Test Section Four covers the following subjects:

* Trauma
* Burns (thermal and electrical)

EMT - Paramedic National Standards Review Self Test Third Edition

1. Which of the following procedures are required prior to transporting the seriously injured trauma victim?
(a) a primary examination
(b) a secondary examination and initiation of IV fluid replacement
(c) obtaining a medical history and immobilizing extremity fractures
(d) Answers (a) and (b) only.
(e) Answers (a), (b), and (c).

2. Newton's first law of motion states that
(a) a body in motion will remain in motion unless acted upon by an outside force.
(b) force equals mass multiplied by acceleration or deceleration.
(c) a body at resta will remain at rest unless acted upon by an outside force.
(d) Both answers (a) and (b).
(e) Both answers (a) and (c).

3. Newton's second law of motion states that
(a) a body in motion will remain in motion unless acted upon by an outside force.
(b) force equals mass multiplied by acceleration or deceleration.
(c) a body at resta will remain at rest unless acted upon by an outside force.
(d) Both answers (a) and (b).
(e) Both answers (a) and (c).

4. An automobile traveling approximately 30 mph is driven head-on into a cement bridge abutment, coming to a sudden stop. Which of the following statements regarding the energy of this accident is false?
(a) A destructive energy was created the instant the auto struck the abutment.
(b) The auto absorbed energy at impact, manifested by vehicle deformity and bending of the frame.

(c) The restrained driver of this auto also absorbed energy on impact.
(d) Three impacts occurred: the auto impacted the abutment, the driver impacted the seatbelts, the driver's internal organs impacted various internal structures and/or restraining ligaments.
(e) The transfer of energy may have resulted in tissue injuries despite the presence of restraints.

5. The branch of physics that deals with force and its effects in producing or modifying motion in bodies is called

(a) kinematics.
(b) kinetosis.
(c) kinesthesia.
(d) kinetics.
(e) kinesiology.

6. Energy manifested in the form of motion is called

(a) potential energy.
(b) thermal energy.
(c) kinetic energy.
(d) radiant energy.
(e) latent energy.

7. Deceleration forces may result in life-threatening hemorrhage of the _____ secondary to laceration by the ligamentum teres.

(a) aorta
(b) liver
(c) kidneys
(d) spleen
(e) bladder

8. Deceleration forces may result in life-threatening hemorrhage of the _____ secondary to laceration by the ligamentum arteriosum.

(a) aorta
(b) liver
(c) kidneys
(d) spleen
(e) bladder

9. A frontal impact accident, involving "down and under" travel of the driver, has caused the driver's knees to impact with the dashboard. From this impact alone, you would anticipate all of the following, except

(a) knee injuries.
(b) femoral shaft injuries.
(c) distal tibial injuries.
(d) posterior hip fractures.
(e) posterior hip dislocations.

10. A frontal impact accident, involving "up and over" travel of the driver, has produced a star fracture of the windshield, high above the steering wheel. From this sign you would anticipate

(a) skull or facial fractures.
(b) compression, hyperflexion, or hyperextension injuries of the cervical spine.
(c) femoral fractures.
(d) Answers (a) and (b) only.
(e) Answers (a), (b), and (c).

11. Which of the following statements regarding ejection injuries is false?

(a) Ejection produces multiple impacts.
(b) Ejection occurs most frequently in frontal impacts.
(c) Ejection frequently saves the victim from more serious injuries.
(d) All of the above are false.
(e) None of the above is false.

12. In an accident involving lateral impact on the driver's side, with a restrained driver, you would anticipate injuries

(a) anywhere along the right side of the patient and the patient's spine.
(b) anywhere along the left side of the patient and the patient's spine.
(c) anywhere about either side of the patient and the patient's spine.
(d) involving the upper extremities, chest, and spine only.
(e) involving the lower extremities, pelvis, and spine only.

13. When a vehicle is impacted from the rear, the occupant's body is propelled forward to impact the restraint device (a restrained occupant) or the vehicle's interior (an unrestrained occupant). Meanwhile, the occupant's head

(a) has moved forward, preceding the body (hyperflexion) and then snapped backward (hyperextension), producing a "whiplash" injury.
(b) has first snapped backwards (hyperflexion) and then snapped forward (hyperextension), producing a "whiplash" injury.
(c) remains stationary, causing hyperflexion of the cervical spine with a more rapid rebound hyperextension, producing a "whiplash" injury.
(d) remains stationary, causing hyperextension of the cervical spine with a more rapid rebound hyperflexion, producing a "whiplash" injury.
(e) None of the above.

14. In an accident involving off-center impact, causing rotation of the auto around the point of impact, you would anticipate

(a) lateral injury patterns.
(b) frontal injury patterns.
(c) the occupant nearest to the point of impact to suffer significant injury.
(d) the occupant furthest from the auto's center mass to suffer significant injury.
(e) All of the above.

15. Which of the following statements regarding restraint systems is false?

(a) Even properly positioned lap and shoulder belts may produce significant injuries during an accident.
(b) Airbags are effective only in frontal collisions.
(c) Survival rates are greatly improved by the use of restraint systems.
(d) All of the above are false.
(e) None of the above is false.

16. Suspected alcohol intoxication or substance abuse is an important factor in consideration of the mechanism of injury because

(a) the relaxation produced by drugs or alcohol lessens the severity of injury.
(b) blood testing is required by law in many states.
(c) altered ability to perceive signs and symptoms of injury will frequently result in missed injuries.
(d) All of the above.
(e) None of the above.

17. Motorcycle accidents with frontal impact frequently cause

(a) abdominal injury from handlebar contact.
(b) pelvic injury from handlebar contact.
(c) bilateral femoral fractures from handlebar contact.
(d) Answers (b) and (c) only.
(e) Answers (a), (b), and (c)

18. When involved in a motorcycle accident, if the victim was wearing a helmet

(a) the likelihood of head injury is greatly diminished.
(b) the likelihood of spinal injury is greatly diminished.
(c) the likelihood of deceleration injury is greatly diminished.
(d) Answers (a) and (b) only.
(e) Answers (a), (b), and (c).

19. The most commonly fractured bone in the body is the

(a) mandible.
(b) clavicle.
(c) humerus.
(d) femur.
(e) tibia.

20. Your patient jumped from a second-story window to escape a person threatening assault. Her chief complaint is bilateral ankle and heel pain from first landing erect, on both feet. Given this mechanism of injury, you must also suspect

(a) spinal fracture.
(b) internal injuries.
(c) upper extremity injuries.
(d) Answers (a) and (b) only.
(e) Answers (a), (b), and (c).

21. Which of the following statements regarding bullet wounds is false?

(a) Bullets designed to expand on entry create an entrance wound that will be larger than the exit wound.
(b) A higher-velocity bullet will produce less cavitation injury because the bullet will be less likely to tumble or fragment as it travels through the body.
(c) All bullets create greater entrance wounds than exit wounds because the velocity is slowed by the body tissue prior to exit.
(d) All of the above are false.
(e) None of the above is false.

22. Facial bones include all of the following, except

(a) the zygoma.
(b) the maxilla.
(c) the mandible.
(d) the sphenoid bone.
(e) the palantine bone.

23. The cranium consists of the frontal, temporal,

(a) occipital, maxillary, and nasal concha bones.
(b) sphenoid, occipital, and parietal bones.
(c) ethmoid, sphenoid, and zygomatic bones.
(d) parietal, zygomatic, and sphenoid bones.
(e) occipital, parietal, and zygomatic bones.

24. The cervical spine consists of

(a) 3 to 5 vertebrae.
(b) 5 vertebrae.
(c) 7 vertebrae.
(d) 12 vertebrae.
(e) 14 vertebrae.

25. The thoracic spine consists of

(a) 3 to 5 vertebrae.
(b) 5 vertebrae.
(c) 7 vertebrae.
(d) 12 vertebrae.
(e) 14 vertebrae.

26. The lumbar spine consists of
(a) 3 to 5 vertebrae.
(b) 5 vertebrae.
(c) 7 vertebrae.
(d) 12 vertebrae.
(e) 14 vertebrae.

27. The sacrum consists of _____ that are fused together.
(a) 3 to 5 vertebrae
(b) 5 vertebrae
(c) 7 vertebrae
(d) 12 vertebrae
(e) 14 vertebrae

28. The coccyx consists of _____ that are fused together.
(a) 3 to 5 vertebrae
(b) 5 vertebrae
(c) 7 vertebrae
(d) 12 vertebrae
(e) 14 vertebrae

29. Each vertebra consists of a vertebral body, a spinal foramen, and three processes:
(a) two transverse and one spinous.
(b) two spinous and one posterior.
(c) two spinal and one transverse.
(d) two peripheral and one anterior.
(e) two posterior and one spinous.

30. The outermost meningeal layer lines the interior surface of the skull and vertebral column, and is called the
(a) arachnoid membrane.
(b) erytnoid membrane.
(c) dura mater.
(d) pia mater.
(e) tia mater.

31. The tissue between the two mater layers is called the
(a) arachnoid membrane.
(b) erytnoid membrane.
(c) dura mater.
(d) pia mater.
(e) tia mater.

32. The innermost meningeal layer lines the brain and spinal cord, and is called the

(a) arachnoid membrane.
(b) erytnoid membrane.
(c) dura mater.
(d) pia mater.
(e) tia mater.

33. Which of the following statements regarding central (CNS) and peripheral nervous system (PNS) neurons is true?

(a) CNS neurons are covered with a thin, fibrous sheath that prevents their regeneration in the event of injury or disease.
(b) PNS neurons are covered with a thin, fibrous sheath that prevents their regeneration in the event of injury or disease.
(c) PNS neurons are covered with a thin, fibrous sheath that allows regeneration in the event of injury or disease.
(d) CNS neurons are covered with a thin, fibrous sheath that allows regeneration in the event of injury or disease.
(e) None of the above is true.

34. Deceleration forces may cause the brain to slide across the _______, which has sharp and irregular bony prominences that may produce abrasions, lacerations, or contusions of the brain tissue.

(a) zygomatic plate
(b) basilar plate
(c) tentorial plate
(d) cribriform plate
(e) sphenoidal plate

35. *Dermatomes* are defined as specific body surface areas of sensation innervated by

(a) peripheral sensory nerves.
(b) visceral somatic nerves.
(c) cranial sensory nerves.
(d) All of the above.
(e) None of the above.

36. The body surface region of the clavicles is sensed by the

(a) first cranial nerve.
(b) third cervical nerve.
(c) fourth thoracic nerve.
(d) tenth cranial nerve.
(e) tenth thoracic nerve.

37. The nipple line body surface region is sensed by the

(a) fourth cranial nerve.
(b) third cervical nerve.
(c) fourth thoracic nerve.
(d) fifth lumbar nerve.
(e) tenth thoracic nerve.

38. The umbilical body surface region is sensed by the

(a) fourth cranial nerve.
(b) third cervical nerve.
(c) fourth thoracic nerve.
(d) fifth lumbar nerve.
(e) tenth thoracic nerve.

39. Blood is carried to the brain by the carotid artery system and vertebral artery system. These systems connect at the base of the brain in the _______, from which arteries supply blood to the brain itself.

(a) sinus of Wallace
(b) sinus of Willis
(c) circle of Wallace
(d) circle of Willis
(e) Haversian canals

40. The orbits of the eye (eye sockets) are formed by

(a) the frontal bones.
(b) the zygomatic bones.
(c) the maxillary and nasal bones.
(d) Both answers (a) and (b).
(e) Answers (a), (b), and (c).

41. The anterior chamber of the eye is filled with a clear liquid called

(a) visceral humor.
(b) aqueous humor.
(c) vitreous humor.
(d) aquial gel.
(e) globular gel.

42. The posterior chamber of the eye is filled with a clear gel-like substance called

(a) visceral humor.
(b) aqueous humor.
(c) vitreous humor.
(d) aquial gel.
(e) globular gel.

43. The anterior and posterior chambers of the eye are separated by the

(a) retina.
(b) cornea.
(c) lacrimal ducts.
(d) pupil.
(e) lens.

44. The colored area of the eye is called the

(a) iris.
(b) cornea.
(c) sclera.
(d) conjunctiva.
(e) lens.

45. The white area of the eye is called the

(a) iris.
(b) cornea.
(c) sclera.
(d) conjunctiva.
(e) lens.

46. The protective clear covering of the anterior eye assists in focusing light and is called the

(a) iris.
(b) cornea.
(c) sclera.
(d) conjunctiva.
(e) lens.

47. The light- and color-sensing tissue of the eye is called the

(a) retina.
(b) cornea.
(c) lacrimal ducts.
(d) pupil.
(e) lens.

48. The membrane that lines the undersurfaces of the eyelids is called the

(a) iris.
(b) cornea.
(c) sclera.
(d) conjunctiva.
(e) lens.

49. The opening through which light travels is called the

(a) retina.
(b) cornea.
(c) lacrimal ducts.
(d) pupil.
(e) lens.

50. A structure of the eye that does not have its own vasculature and is dependant upon lacrimal fluid for nutrients and oxygen is called the

(a) iris.
(b) cornea.
(c) sclera.
(d) conjunctiva.
(e) lens.

51. The inner ear is susceptible to injuries from

(a) blast trauma.
(b) diving trauma.
(c) basilar skull fracture.
(d) Answers (a) and (b) only.
(e) Answers (a), (b), and (c).

52. The ear is a sensory structure, functioning to provide

(a) hearing.
(b) positional sense (physical sense of balance).
(c) intercranial pressure release.
(d) Answers (a) and (b) only.
(e) Answers (a), (b), and (c).

53. The ______ the actual organ(s) of hearing; vibrations here stimulate the auditory nerve, which sends the sound signal to the brain.

(a) tympanic membrane is
(b) cochlea is

(c) semicircular canals are
(d) ossicles are
(e) auditory nerve is

54. The _______ provide(s) the sense of motion, position, or balance.

(a) tympanic membrane
(b) cochlea
(c) semicircular canals
(d) ossicles
(e) auditory nerve

55. The first structure(s) to perceive sound wave vibrations is/are the _______.

(a) tympanic membrane
(b) cochlea
(c) semicircular canals
(d) ossicles
(e) auditory nerve

56. Disturbances of the sense of balance can be caused by injury or illness and produce a variety of symptoms and complaints. The complaint, "The room is spinning!" would be referred to as

(a) vitiligo.
(b) dizziness.
(c) "the vapors."
(d) syncope.
(e) vertigo.

57. Even superficial scalp and facial injuries bleed profusely because of the relatively large amount of vasculature present. This concentration of vessels is present to prevent

(a) hypoxia of the brain during injury.
(b) facial disfiguration from hypoxia.
(c) heat loss in cold climates.
(d) Answers (a) and (b) only.
(e) Answers (a), (b), and (c).

58. Open head injuries may involve the presence of
- (a) linear skull fractures.
- (b) depressed skull fractures.
- (c) comminuted skull fractures.
- (d) Answers (b) and (c) only.
- (e) Answers (a), (b), and (c).

59. Closed head injuries may involve the presence of
- (a) linear skull fractures.
- (b) depressed skull fractures.
- (c) comminuted skull fractures.
- (d) Answers (a) and (b) only.
- (e) Answers (a), (b), and (c).

60. Which of the following statements regarding basilar skull fractures is true?
- (a) Bilateral periorbital ecchymosis (raccoon eyes) occurs within 10 minutes of a basilar skull fracture.
- (b) Mastoid ecchymosis (Battle's sign) indicates that basilar skull fracture has just occurred.
- (c) Leakage of blood and/or cerebral spinal fluid (CSF) may assist in slowing a rise in intracranial pressure (ICP).
- (d) All of the above are true.
- (e) None of the above is true.

61. Compression fractures of the spine and/or intervertebral disc ruptures generally result from
- (a) axial loading forces.
- (b) hyperextension injuries.
- (c) hyperflexion injuries.
- (d) Any of the above.
- (e) None of the above.

62. Spinal fracture occurs most commonly in the
- (a) cervical and thoracic spine.
- (b) thoracic and sacral spine.
- (c) cervical and lumbar spine.
- (d) lumbar and sacral spine.
- (e) thoracic and lumbar spine.

63. Limited or painful ocular range of motion may indicate a fracture of the
- (a) basilar skull.
- (b) zygomatic bone.

(c) mandible.
(d) Any of the above.
(e) None of the above.

64. The contrecoup brain injury is produced by
(a) direct brain injury from penetrating trauma.
(b) deceleration trauma with rebound injury.
(c) direct blunt trauma with rebound injury.
(d) Both answers (a) and (b).
(e) Both answers (b) and (c).

65. A transient period of dysfunction or loss of consciousness may accompany the brain injury called
(a) contusion.
(b) concussion.
(c) confusion.
(d) Both answers (a) and (b).
(e) Both answers (b) and (c).

66. Actual physical injury of the brain tissue does not accompany a
(a) contusion.
(b) concussion.
(c) confusion.
(d) Both answers (a) and (b).
(e) Both answers (b) and (c).

67. Cellular damage, edema, and a potential for increased intracranial pressure accompanies
(a) contusion.
(b) concussion.
(c) confusion.
(d) Both answers (a) and (b).
(e) Both answers (b) and (c).

68. Rupture of the middle meningeal artery will result in
(a) an epidural hematoma.
(b) a subdural hematoma.
(c) intracerebral hemorrhage.
(d) Any of the above.
(e) None of the above.

69. Immediate onset of strokelike signs and symptoms most frequently accompanies

(a) an epidural hematoma.
(b) a subdural hematoma.
(c) intracerebral hemorrhage.
(d) Any of the above.
(e) None of the above.

70. A transient loss of consciousness, followed by a lucid interval, and then a diminishing level of consciousness is the classic presentation of

(a) an epidural hematoma.
(b) a subdural hematoma.
(c) intracerebral hemorrhage.
(d) Any of the above.
(e) None of the above.

71. Bleeding within the meninges produces

(a) an epidural hematoma.
(b) a subdural hematoma.
(c) intracerebral hemorrhage.
(d) Any of the above.
(e) None of the above.

72. Several hours or days may pass before the manifestation of signs and symptoms when head trauma results in

(a) an epidural hematoma.
(b) a subdural hematoma.
(c) intracerebral hemorrhage.
(d) Any of the above.
(e) None of the above.

73. Which of the following statements regarding increased intracranial pressure (ICP) is false?

(a) ICP is increased by edema or hemorrhage within the cranium.
(b) Increased ICP compresses cerebral vasculature, resulting in diminished cerebral blood flow.
(c) Hypercarbia results from diminished cerebral blood flow and causes cerebral vascular constriction.
(d) Systolic blood pressure will rise in an attempt to improve cerebral circulation.
(e) Rising systolic blood pressure increases ICP.

74. Increasing ICP will displace the brain away from the site of insult, which will result in compression of

(a) the third cranial nerve, causing pupillary dilation of one or both pupils.
(b) the reticular activating system, producing diminishing level of consciousness.
(c) the medulla oblongata, producing nausea and/or projectile vomiting.
(d) Answers (a) and (b) only.
(e) Answers (a), (b), and (c).

75. Compression of the medula oblongata into the foramen magnum produces vital sign changes associated with increased ICP. These changes are called

(a) Cushing's reflex (or triad).
(b) Kussmaul's reflex (or triad).
(c) Biot's reflex (or triad).
(d) Beck's reflex (or triad).
(e) Stoke's reflex (or triad).

76. Which of the following best describes the vital signs produced by increased ICP?

(a) Increased systolic blood pressure, bradycardia, and apneustic respirations.
(b) Decreased systolic blood pressure, tachycardia, and hyperpnea.
(c) Decreased systolic blood pressure, bradycardia, and tachypnea.
(d) Increased systolic blood pressure, bradycardia, and deep or erratic respirations.
(e) Decreased systolic blood pressure, tachycardia, and deep or erratic respirations.

77. Which of the following statements regarding spinal cord injury is false?

(a) Spinal cord injury may present with paresthesia, parasthenia, anesthesia, or paralysis of bilateral lower extremities only.
(b) Spinal cord injury may present with paresthesia, parasthenia, anesthesia, or paralysis of all extremities.
(c) Spinal cord injury may present with paresthesia, parasthenia, anesthesia, or localized pain at the site of insult.
(d) Spinal cord injury may present with paresthesia, parasthenia, anesthesia, or paralysis of the arm and leg on one side of the body.
(e) Spinal cord injury may present without any symptoms whatsoever.

78. Paralysis of the lower extremities only is called

(a) paraplegia.
(b) quadriplegia.
(c) hemiplegia.
(d) paresis.
(e) paresthesia.

79. Weakness or partial paralysis is referred to as

(a) paraplegia.
(b) quadriplegia.
(c) hemiplegia.
(d) paresis.
(e) paresthesia.

80. The sensation of numbness, prickling, and/or tingling is called

(a) paraplegia.
(b) quadriplegia.
(c) hemiplegia.
(d) paresis.
(e) paresthesia.

81. Paralysis of one side of the body is called

(a) paraplegia.
(b) quadriplegia.
(c) hemiplegia.
(d) paresis.
(e) paresthesia.

82. Paralysis of the upper and lower extremities is called
(a) paraplegia.
(b) quadriplegia.
(c) hemiplegia.
(d) paresis.
(e) paresthesia.

83. The most reliable indication of the presence of spinal injury is
(a) the presence of significant head injury.
(b) lack of sensation below the site of the injury.
(c) lack of movement below the site of the injury.
(d) complaint of pain at the site of the injury.
(e) the mechanism of injury.

84. Neurogenic shock is caused by
(a) systemic vasodilation.
(b) systemic vasoconstriction.
(c) autonomic nervous system interruption.
(d) Both answers (a) and (c).
(e) Both answers (b) and (c).

85. Cervical spine injury may result in
(a) apnea or diaphragmatic breathing.
(b) priapism.
(c) "hold-up" positioning of the arms above the head.
(d) Answers (a) and (b) only.
(e) Answers (a), (b), and (c).

86. The patient's complaint of a "dark curtain" suddenly obstructing a portion of her/his field of vision accompanies an ocular emergency called
(a) hyphema.
(b) acute retinal artery occlusion.
(c) retinal detachment.
(d) conjunctival hemorrhage.
(e) corneal abrasions.

87. Extreme pain involving the anterior eye (often with the sensation that "Something is in my eye!") accompanies
(a) hyphema.
(b) acute retinal artery occlusion.
(c) retinal detachment.
(d) conjunctival hemorrhage.
(e) corneal abrasions.

88. Sudden, painless vision loss involving one eye is caused by
- (a) hyphema.
- (b) acute retinal artery occlusion.
- (c) retinal detachment.
- (d) conjunctival hemorrhage.
- (e) corneal abrasions.

89. The appearance of blood within the iris and pupil is called
- (a) hyphema.
- (b) acute retinal artery occlusion.
- (c) retinal detachment.
- (d) conjunctival hemorrhage.
- (e) corneal abrasions.

90. The appearance of blood within the sclera of the eye is produced by
- (a) hyphema.
- (b) acute retinal artery occlusion.
- (c) retinal detachment.
- (d) conjunctival hemorrhage.
- (e) corneal abrasions.

91. Bleeding that occurs within the anterior chamber of the eye produces
- (a) hyphema.
- (b) acute retinal artery occlusion.
- (c) retinal detachment.
- (d) conjunctival hemorrhage.
- (e) corneal abrasions.

92. Vitreous humor loss
- (a) may regenerate, producing only transient loss of sight.
- (b) can be replaced with synthetic humor to preserve esthetic appearance and function of sight.
- (c) will produce blindness in the affected eye.
- (d) will produce loss of depth perception only.
- (e) will produce loss of color perception only.

93. Which of the following statements regarding cervical spine immobilization is true?
- (a) A rigid cervical collar alone will not maintain immobilization of the C-spine.
- (b) In-line traction of the cervical spine is contraindicated in trauma patients.

(c) Even when oral intubation is required, cervical spinal immobilization is not discontinued.
(d) All of the above are true.
(e) None of the above is true.

94. Endotracheal intubation may stimulate a parasympathetic response, which could result in
(a) increased ICP.
(b) bradycardias and/or dysrhythmias.
(c) tachycardias and/or dysrhythmias.
(d) Both answers (a) and (b).
(e) Both answers (a) and (c).

95. Increased ICP may produce a respiratory pattern characterized by several short breaths followed by irregular periods of apnea. This pattern is called
(a) apneustic respirations.
(b) ataxic respirations.
(c) Biot's breathing/respirations.
(d) Cheyne–Stokes respirations.
(e) central neurogenic hyperventilation.

96. Increased ICP may produce a respiratory pattern characterized by hyperpnea and tachypnea. This pattern is called
(a) apneustic respirations.
(b) ataxic respirations.
(c) Biot's breathing/respirations.
(d) Cheyne–Stokes respirations.
(e) central neurogenic hyperventilation.

97. Increased ICP may produce a respiratory pattern characterized by cycles of increasing, then decreasing respiratory effort, separated by periods of apnea. This pattern is called
(a) apneustic respirations.
(b) ataxic respirations.
(c) Biot's breathing/respirations.
(d) Cheyne–Stokes respirations.
(e) central neurogenic hyperventilation.

98. Increased ICP may produce a respiratory presentation characterized by irregular and ineffective respiratory efforts. This presentation is called

(a) apneustic respirations.
(b) ataxic respirations.
(c) Biot's breathing/respirations.
(d) Cheyne–Stokes respirations.
(e) central neurogenic hyperventilation.

99. Increased ICP may produce a respiratory presentation characterized by prolonged inspiratory effort without coordinated expiratory patterns. This presentation is called

(a) apneustic respirations.
(b) ataxic respirations.
(c) Biot's breathing/respirations.
(d) Cheyne–Stokes respirations.
(e) central neurogenic hyperventilation.

100. The first priority in care for the spinally immobilized, head-injured patient with increasing ICP is

(a) establishment of IV access with 5% dextrose in water to facilitate administration of ICP-reducing medications without overhydration.
(b) establishment of IV access with NS or LR to facilitate administration of ICP-reducing medications.
(c) hyperventilation with high flow oxygen and intubation.
(d) Both answers (a) and (c).
(e) None of the above.

101. All of the following medications may be indicated for administration in the setting of increased ICP and/or spinal trauma, except

(a) morphine sulfate.
(b) oxygen.
(c) diazepam.
(d) furosemide.
(e) methylprednisolone.

102. Which of the following statements regarding hyperventilation of a head-injured patient is false?

(a) Hyperventilation of the nonintubated patient frequently produces gastric distention, increasing the risk of emesis and subsequent aspiration.
(b) Hyperventilation with high-flow oxygen increases cerebral hypercarbia, providing a decrease in ICP via cerebral vascular dilation.

(c) Hyperventilation with high-flow oxygen decreases cerebral carbon dioxide levels and increases cerebral oxygenation, thereby reducing development of cerebral edema and increased ICP.
(d) All of the above are false.
(e) None of the above is false.

103. The presence of tachycardia and hypotension in the patient with a closed head injury
(a) confirms the absence of increased ICP.
(b) confirms the presence of increased ICP.
(c) suggests the presence of hidden injuries elsewhere, but does not rule out the presence of increased ICP.
(d) suggests the presence of hidden injuries elsewhere, and rules out the presence of increased ICP.
(e) confirms the presence of increased ICP, but does not rule out the presence of hidden injuries elsewhere.

104. The Glasgow Coma Scale evaluates a patient's level of consciousness based on
(a) respiratory rate, respiratory expansion, systolic blood pressure, and capillary refill.
(b) the history mnemonic, "AMPLE."
(c) alertness, pulse, respirations, blood pressure, and motor responses.
(d) eye opening responses, verbal responses, and motor responses.
(e) the patient's ability to localize and identify sound stimulus.

105. When provided with a pain stimulus, your patient attempts to interfere with the stimulus application by grabbing at the source or pushing the source away. This response characterizes
(a) a withdrawal response to pain.
(b) decerebrate posturing.
(c) the ability to localize pain and coordinate a response.
(d) decorticate posturing.
(e) Either answer (b) or (d).

106. When provided with a pain stimulus, your patient flexes and/or retracts the stimulated area to avoid or escape the stimulus. This response characterizes

(a) a withdrawal response to pain.
(b) decerebrate posturing.
(c) the ability to localize pain and coordinate a response.
(d) decorticate posturing.
(e) Either answer (b) or (d).

107. When provided with a pain stimulus, your patient flexes and adducts both arms. This response characterizes

(a) a withdrawal response to pain.
(b) decerebrate posturing.
(c) the ability to localize pain and coordinate a response.
(d) decorticate posturing.
(e) Either answer (b) or (d).

108. When provided with a pain stimulus, your patient extends and abducts both arms. This response characterizes

(a) a withdrawal response to pain.
(b) decerebrate posturing.
(c) the ability to localize pain and coordinate a response.
(d) decorticate posturing.
(e) Either answer (b) or (d).

109. Decorticate or decerebrate posturing indicates the presence of

(a) coordinated and localizing responses to stimulation.
(b) irreversible brain damage.
(c) a high (C-1 to C-3) spinal cord lesion, resulting in reflexive muscle movement of the extremities.
(d) a significant brain injury that is life-threatening.
(e) Either answer (c) or (d).

110. Open wounds to the neck should

(a) increase suspicion of spinal injury.
(b) increase anticipation of airway compromise.
(c) be sealed with an occlusive dressing and the patient placed in a slightly Trendelenburg position to reduce the potential for an air embolus.
(d) Answers (a) and (b) only.
(e) Answers (a), (b), and (c).

111. Which of the following statements regarding PASG application in the head-injured patient is true?

(a) Presence of a head-injury contraindicates the application of PASG.
(b) PASG should be applied, but not inflated until the patient displays signs and symptoms of increased ICP.
(c) PASG may be applied but not inflated until the patient displays signs and symptoms of neurogenic or hypovolemic shock.
(d) Application and inflation of PASG is recommended to assist in obtaining IV access in the presence of increased ICP.
(e) Both answers (b) and (d).

112. The IV fluid of choice in the presence of severe head injury is

(a) D_5W only, because fluid administration is contraindicated in the presence of increased ICP.
(b) D_5W only, because seizures can be anticipated and dilantin is incompatible with NS or LR.
(c) any colloid solution.
(d) NS or LR only, because additional (hidden) injuries must be anticipated.
(e) any crystalloid solution.

113. The presence of unequal pupils is called

(a) dysconjugate gaze.
(b) doll's eye response.
(c) sympathetic ophthalmia.
(d) nystagmus.
(e) aniscoria.

114. Failure of the pupils to move together (looking in different directions simultaneously) is called

(a) dysconjugate gaze.
(b) doll's eye response.
(c) sympathetic ophthalmia.
(d) nystagmus.
(e) aniscoria.

115. Which of the following statements regarding unequal pupils is false?

(a) A lesion on, or increased ICP of, the third cranial nerve will produce pupillary size changes.
(b) A single dilated and nonreactive pupil may indicate CNS injury or cerebral hypoxia.
(c) Delays in pupillary response to light can be indicative of drugs that depress the CNS, injury, or cerebrovascular compromise.
(d) Unilateral dilation of a pupil always indicates hemisphere compromise of the opposite side (right hemisphere injury produces left pupillary dilation and vice versa).
(e) None of the above is false.

116. Bilateral pupillary dilation with very delayed or absent response to light (fixed) may be attributed to all of the following, except

(a) brain stem injury.
(b) bilateral hemisphere injury or hypoxia.
(c) herniation of the brain stem through the foramen magnum.
(d) substance abuse.
(e) a congenital defect common to a small percentage of the population.

117. Your patient ambulates without difficulty but complains of neck pain as he twists and rubs his neck. Indicate the correct sequence of treatment for this patient:

(1) Place the patient supine.
(2) Apply a cervical collar.
(3) Immobilize the patient to a long back board.
(4) Manually immobilize the patient's cervical spine.

(a) 2, 4, 1, 3
(b) 4, 2, 3, 1
(c) 1, 4, 2, 3
(d) 4, 2, 1, 3
(e) None of the above (this patient does not require spinal immobilization).

118. Which of the following statements regarding mannitol is false?

(a) Mannitol is a rapid-acting, potent diuretic used to decrease cerebral edema.
(b) Mannitol decreases pulmonary edema.

(c) Mannitol may increase intracranial hemorrhage.
(d) Mannitol can cause dehydration or electrolyte disturbances.
(e) None of the above is false.

119. Which of the following statements regarding furosemide (Lasix) is false?

(a) Lasix is a systemic diuretic that may be used for treatment of cerebral edema.
(b) The vasodilation effect of Lasix can lower the patient's blood pressure.
(c) Use of Lasix may diminish cerebral blood flow, thus increasing cerebral hypoxia and hypercarbia.
(d) All of the above are false.
(e) None of the above is false.

120. Which of the following statements regarding methylprednisolone (Solu-Medrol) is true?

(a) Methylprednisolone is an anti-inflammatory steroid believed to reduce the tissue irritation that produces edema.
(b) Even in short-term treatment of cerebral edema, high doses of Solu-Medrol can produce serious steroidal side effects.
(c) Solu-Medrol is administered by IV infusion only.
(d) Both answers (a) and (b).
(e) Both answers (a) and (c).

121. Which of the following statements regarding seizures and ICP is true?

(a) Increased ICP frequently produces seizure activity.
(b) Seizure activity frequently produces increased ICP.
(c) Because administration of diazepam is contraindicated in the presence of head injury, trauma-related seizures are treated with hyperventilation and fluid administration only.
(d) Both answers (a) and (b).
(e) Both answers (a) and (c).

122. The muscles responsible for chest expansion (inspiration) are the
(a) diaphragm, sternocleidomastoid, and intercostal muscles.
(b) trapezius, pectoralis major, and rectus abdominus muscles.
(c) diaphragm, trapezius, and lattisimus dorsi muscles.
(d) Both answers (a) and (b).
(e) Both answers (b) and (c).

123. The area between the parietal and visceral pleura is called a "potential" space because
(a) there is no space there until trauma occurs to create one.
(b) the area is sealed and the space filled by a small amount of lubricating fluid, providing adhesion of the two layers.
(c) the terms "parietal" and "visceral" refer to opposite surfaces of a single pleura and do not involve spaces.
(d) Any of the above.
(e) None of the above.

124. Atelectasis is defined as
(a) failure of respiratory effort.
(b) absence of respiratory effort.
(c) collapsed bronchioles.
(d) collapsed alveoli.
(e) absence of chest expansion.

125. Lung tissue contusion may result in
(a) localized atelectasis.
(b) localized pulmonary edema.
(c) increased $PaCO_2$.
(d) Answers (a) and (b) only.
(e) Answers (a), (b), and (c).

126. Blunt fracture of a single rib, with associated contusion and pain, may result in
(a) atelectasis.
(b) reduced thoracic expansion.
(c) pneumo- and/or hemothoraces.
(d) Answers (b) and (c) only.
(e) Answers (a), (b), and (c).

127. A flail chest segment is best defined as
(a) one rib, fractured in two or more places.
(b) two or more adjacent ribs, each fractured in two or more places.

(c) three or more adjacent ribs, each fractured in a single (common) location.
(d) All of the above.
(e) None of the above.

128. Paradoxical movement of a flail segment is infrequently observed because of the "splinting" effect of localized muscle spasm. However, if it occurs, you would anticipate observing

(a) an absence of movement (expansion) on the affected side.
(b) the flail segment bulging out during inspiration and sinking into the chest during expiration.
(c) the flail segment sinking into the chest during inspiration and bulging out during expiration.
(d) Either answer (a) or (b).
(e) None of the above.

129. Traumatic asphyxia is defined as a

(a) severe crushing injury of the upper or lower airways, producing sudden and complete airway obstruction, resulting in asphyxia.
(b) severe crushing injury of the chest, producing absent or grossly inadequate chest excursion and prevention of venous return to the heart, resulting in asphyxia.
(c) total airway obstruction, produced by any foreign body, resulting in asphyxia.
(d) Either answer (a) or (b).
(e) None of the above.

130. A closed pneumothorax may be caused by any of the following, except

(a) penetrating trauma immediately below the xyphoid process.
(b) ruptured alveoli secondary to the forceful coughing of a COPD patient.
(c) blunt trauma, producing a bronchial or bronchiolar rupture.
(d) a ruptured congenital lung defect, secondary to physical exertion.
(e) intermittent positive pressure ventilation with bag–valve–mask or demand valve devices.

131. A "paper bag syndrome" injury may rupture bronchioles and/or alveoli, resulting in a pneumothorax. This type of injury is produced by

(a) penetrating trauma to a hyperinflated thorax.
(b) blunt trauma to a hyperinflated thorax.
(c) compression trauma to a hyperinflated thorax.
(d) Both answers (a) and (b).
(e) Both answers (a) and (c).

132. A patient presenting with a bulging and cyanotic tongue, jugular venous distention, bloodshot eyes, and generalized upper body cyanosis is classically indicative of

(a) traumatic asphyxia.
(b) paper bag syndrome.
(c) tension pneumothorax.
(d) simple pneumothorax.
(e) hemopneumothorax.

133. All of the following statements regarding tension pneumothorax are true, except

(a) tension pneumothorax is created by the ingress of air into the pleural space without egress.
(b) tension pneumothorax may result in a complete collapse of lung on the injured side.
(c) in the presence of tension pneumothorax, increased intrathoracic pressure impedes cardiac output, but allows for venous return of blood to the heart, thus increasing cardiac dysfunction.
(d) the mediastinal shift produced by a tension pneumothorax reduces the ventilatory potential of the uninjured lung.
(e) tension pneumothorax will reduce cardiac output and may cause occlusion of great vessels.

134. An early sign of tension pneumothorax is

(a) shifting of the trachea toward the affected side.
(b) shifting of the trachea away from the affected side.
(c) severe dyspnea.
(d) Both answers (a) and (c).
(e) Answers (a) or (b), and (c).

135. A late sign of tension pneumothorax is

(a) shifting of the trachea toward the affected side.
(b) shifting of the trachea away from the affected side.

(c) severe dyspnea.
(d) Both answers (a) and (c).
(e) Answers (a) or (b), and (c).

136. Possible signs and symptoms of a simple hemothorax include all of the following, except

(a) diminished or absent breath sounds on the injured side.
(b) jugular venous distention.
(c) tachypnea, tachycardia, and hypotension.
(d) cyanosis.
(e) diaphoresis.

137. Possible signs and symptoms of a tension pneumothorax include all of the following, except

(a) agitation, increasing dyspnea, increasing resistance to ventilation.
(b) subcutaneous emphysema.
(c) jugular venous distention.
(d) hyperresonant breath sounds on the injured side.
(e) tachypnea, tachycardia, and hypotension.

138. Which of the following statements regarding myocardial contusion is false?

(a) Contused myocardium will present with signs and symptoms similar to acute myocardial infarction.
(b) Myocardial contusion presents with chest pain, but rarely is accompanied by dysrhythmias or ectopy.
(c) The prognosis for myocardial contusion is better than that of an acute myocardial infarction.
(d) All of the above are false.
(e) None of the above is false.

139. The heart is surrounded by a protective sac composed of two layers. The inner layer that is contiguous with the cardiac muscle is called the

(a) epicardium.
(b) parietal pericardium.
(c) myocardium.
(d) visceral pericardium.
(e) endocardium.

140. The outer layer of the sac surrounding the heart is called the

(a) epicardium.
(b) parietal pericardium.
(c) myocardium.
(d) visceral pericardium.
(e) endocardium.

141. The area between the layers of the heart's protective sac

(a) is filled with a cushion of air.
(b) is filled with a cushion of cartilage.
(c) is filled with a small amount of lubricating fluid.
(d) varies in amount of space, depending upon the amount of friction generated.
(e) is empty unless penetrated by trauma (and therefore called a "potential space").

142. Pericardial tamponade may be caused by

(a) blunt trauma.
(b) deceleration trauma.
(c) penetrating trauma.
(d) Answers (a) and (b) only.
(e) Answers (a), (b), and (c).

143. Possible signs and symptoms of pericardial tamponade include all of the following, except

(a) tachycardia and tachypnea.
(b) distended jugular veins.
(c) diminished breath sounds in the left thorax.
(d) increasing diastolic blood pressure.
(e) narrowing pulse pressure.

144. Possible signs and symptoms of a dissecting thoracic aneurysm include all of the following, except

(a) a tearing or burning pain in the periumbilical area.
(b) chest pain with numbness or tingling of the left arm.
(c) an absent or diminished radial pulse.
(d) tachycardia and tachypnea.
(e) nausea, vomiting, and/or diaphoresis.

145. Your patient received an isolated blunt injury to the midaxillary line of her right chest. Position for transport should be

(a) left laterally recumbent.
(b) right laterally recumbent.
(c) semi-Fowler's position.
(d) supine and secured to a long backboard.
(e) prone and secured to a long backboard.

146. Your patient has a small penetrating wound in the upper right anterior chest without an exit wound. You should dress the wound with an occlusive dressing

(a) only if sucking sounds are noted.
(b) only if frothy bleeding is noted.
(c) only if the patient complains of dyspnea.
(d) only if the patient is tachypneic.
(e) despite the absence of dyspnea, sucking, or frothing signs.

147. An occlusive dressing on the thorax should be sealed on three sides only, so that

(a) increased thoracic pressure may escape.
(b) thoracic hemorrhage may escape.
(c) atmospheric air may enter and supplement inspired air.
(d) Both answers (a) and (c).
(e) Both answers (b) and (c).

148. Needle decompression of a tension pneumothorax occurs

(a) at the midclavicular line of the affected side, in the fifth or sixth intercostal space (ICS).
(b) at the midaxillary line of the affected side, in the second or third intercostal space (ICS).
(c) either in the fifth or sixth ICS of the midclavicular line or the second or third ICS of the midaxillary line of the affected side.
(d) either in the second or third ICS of the midclavicular line or the fifth or sixth ICS of the midaxillary line of the affected side.
(e) None of the above.

149. Which of the following statements regarding needle decompression of a tension pneumothorax is false?

(a) If a pneumothorax is not present, needle decompression will create one.
(b) Insertion site of the needle is along the lower margin of the upper rib of the appropriate ICS.
(c) Repeat decompression may be necessary proximal to the original site.
(d) Both answers (b) and (c) are false.
(e) None of the above is false.

150. Treatment of the patient with a cardiac contusion

(a) includes administration of standard ACLS medications according to ACLS protocols.
(b) is limited to supportive therapy, with oxygen being the only medication administered to the trauma patient.
(c) includes needle decompression to diminish chest pain secondary to increased mediastinal pressure.
(d) Both answers (b) and (c).
(e) None of the above.

151. Management of the patient with pericardial tamponade includes

(a) high-flow oxygen.
(b) rapid transportation to the ER.
(c) pericardiocentesis, utilizing the subxyphoid approach.
(d) Answers (a) and (b) only.
(e) Answers (a), (b), and (c).

152. Management of an object impaled in the chest includes

(a) occlusive dressing about the entrance (and exit) site of the object.
(b) stabilization with bulky dressings.
(c) removal of the object if its location interferes with required performance of CPR.
(d) Answers (a) and (b) only.
(e) Answers (a), (b), and (c).

153. Organs located within the retroperitoneal space include

(a) the kidneys.
(b) portions of the duodenum, and the pancreas.
(c) the thoracic aorta and the superior vena cava.
(d) Answers (a) and (b) only.
(e) Answers (a), (b), and (c).

154. Signs and symptoms of shock following injury to the lateral left upper abdominal quadrant should cause suspicion of injury to the

(a) liver and left kidney.
(b) spleen and left kidney.
(c) ascending colon, spleen, and left kidney.
(d) descending colon, liver, and left kidney.
(e) None of the above.

155. Signs and symptoms of shock following injury to the lateral right upper abdominal quadrant should cause suspicion of injury to the

(a) liver and right kidney.
(b) spleen and right kidney.
(c) ascending colon, spleen, and right kidney.
(d) descending colon, liver, and right kidney.
(e) None of the above.

156. Complaints of acute pain originating in the right lower abdominal quadrant should cause suspicion of injury or illness involving the

(a) appendix, right ovary or fallopian tube, and descending colon.
(b) appendix, right ovary or fallopian tube, and spleen.
(c) appendix, right ovary or fallopian tube, and ascending colon.
(d) appendix, right ovary or fallopian tube, and gall bladder.
(e) None of the above.

157. Complaints of acute pain originating in the left lower abdominal quadrant should cause suspicion of injury or illness involving the

(a) appendix, left ovary or fallopian tube, and descending colon.
(b) appendix, left ovary or fallopian tube, and spleen.
(c) appendix, left ovary or fallopian tube, and ascending colon.
(d) appendix, left ovary or fallopian tube, and gall bladder.
(e) None of the above.

158. The abdominal aorta bifurcates at its distal end, becoming the left and right

(a) intra-abdominal arteries.
(b) femoral arteries.
(c) common iliac arteries.
(d) great saphenous arteries.
(e) popliteal arteries.

159. Your patient has a gunshot entrance wound in the RLQ of his abdomen. There is no exit wound. You suspect potential injury to the

(a) abdomen only.
(b) abdominopelvic areas only.
(c) abdomen and thorax only.
(d) abdominopelvic and thoracic cavities.
(e) abdomen and retroperitoneal area only.

160. Peritonitis is produced by

(a) free blood in the abdomen.
(b) bowel contents spilled within the abdomen.
(c) digestive fluids loose within the abdomen.
(d) Answers (a) and (c) only.
(e) Answers (a), (b), and (c).

161. *Rebound tenderness* is defined as

(a) the patient's complaint of pain upon initiation of abdominal palpation.
(b) the patient's complaint of pain upon release of abdominal palpation.
(c) the patient's complaint of pain upon palpation of the flanks.
(d) muscle spasm or contraction upon initiation of abdominal palpation.
(e) muscle spasm or contraction upon release of abdominal palpation.

162. *Abdominal guarding* is defined as

(a) the patient's complaint of pain upon initiation of abdominal palpation.
(b) the patient's complaint of pain upon release of abdominal palpation.

(c) the patient's complaint of pain upon palpation of the flanks.
(d) muscle spasm or contraction upon initiation of abdominal palpation.
(e) muscle spasm or contraction upon release of abdominal palpation.

163. Treatment of closed abdominal injuries includes
(a) high-flow oxygen and two or more peripheral crystalloid IVs.
(b) moist dressings covered with occlusive dressings.
(c) application of PASG and inflation of the abdominal section only (in absence of hypotension).
(d) Answers (a) and (b) only.
(e) Answers (a), (b), and (c).

164. Treatment of an eviscerated abdomen includes
(a) high-flow oxygen and two or more peripheral crystalloid IVs.
(b) moist dressings covered with occlusive dressings.
(c) application of PASG and inflation of the abdominal section only (in absence of hypotension).
(d) Answers (a) and (b) only.
(e) Answers (a), (b), and (c).

165. Management of an object impaled in the LUQ of the abdomen includes
(a) occlusive dressing about the entrance (and exit) site of the object, and stabilization with bulky dressings.
(b) high-flow oxygen and two or more peripheral crystalloid IVs.
(c) removal of the object if life-threatening hypotension requires inflation of the abdominal section of the PASG.
(d) Answers (a) and (b) only.
(e) Answers (a), (b), and (c).

166. The human skeleton normally has _____ bones.
(a) 106
(b) 156
(c) 206
(d) 215
(e) 226

167. The area of a long bone where growth occurs is called the
(a) epiphysis.
(b) metaphysis.
(c) diaphysis.
(d) apophysis.
(e) aponeurosis.

168. Each of the widened ends of a long bone is called the
(a) epiphysis.
(b) metaphysis.
(c) diaphysis.
(d) apophysis.
(e) aponeurosis.

169. The cylinderlike shaft of a long bone is called the
(a) epiphysis.
(b) metaphysis.
(c) diaphysis.
(d) apophysis.
(e) aponeurosis.

170. The axial skeleton includes all of the following, except
(a) the skull.
(b) the ribs and sternum.
(c) the clavicles and scapulae.
(d) the sacrum and coccyx.
(e) the spinal column.

171. The appendicular skeleton includes all of the following, except
(a) the clavicles and scapulae.
(b) the arms, forearms, wrists, hands, and fingers.
(c) the legs, ankles, feet, and toes.
(d) the sacrum and coccyx.
(e) the pelvic bones.

172. Red blood cell production occurs in the
(a) yellow bone marrow.
(b) aponeurosis.
(c) red bone marrow.
(d) metaphysis.
(e) periosteum.

173. Fat storage is accomplished within the

(a) yellow bone marrow.
(b) aponeurosis.
(c) red bone marrow.
(d) metaphysis.
(e) periosteum.

174. Initiation of new bone formation is accomplished by the

(a) yellow bone marrow.
(b) aponeurosis.
(c) red bone marrow.
(d) metaphysis.
(e) periosteum.

175. Connective tissue called _______ covers the articular surfaces of bones.

(a) cancellous tissue
(b) a ligament
(c) cartilage
(d) a tendon
(e) articulate tissue

176. Fibrous connective tissue called _______ has the ability to stretch, and connects bone to bone.

(a) cancellous tissue
(b) a ligament
(c) cartilage
(d) a tendon
(e) articulate tissue

177. Fibrous connective tissue called _______ attaches muscle to bone.

(a) cancellous tissue
(b) a ligament
(c) cartilage
(d) a tendon
(e) articulate tissue

178. The medial malleolus is formed by the distal end of the

(a) ulna.
(b) radius.
(c) tibia.
(d) fibula.
(e) femur.

179. The lateral malleolus is formed by the distal end of the

(a) ulna.
(b) radius.
(c) tibia.
(d) fibula.
(e) femur.

180. The acetabulum is

(a) supported by the femoral neck.
(b) located between the femoral neck and head.
(c) the hollow pelvic depression that articulates with the femoral head.
(d) located between the femoral shaft and neck.
(e) None of the above.

181. Muscle tissue is responsible for the body's mobility and

(a) elimination of waste products.
(b) sweating mechanisms.
(c) production of heat.
(d) energy production.
(e) None of the above.

182. In layman's terms, the _____ is called the shoulder blade.

(a) ulna
(b) clavicle
(c) humerus
(d) radius
(e) scapula

183. The bone located on the lateral side of the forearm is the _______.

(a) ulna
(b) clavicle
(c) humerus
(d) radius
(e) scapula

184. The _______ is the bone of the upper arm.

(a) ulna
(b) clavicle
(c) humerus
(d) radius
(e) scapula

185. In layman's terms, the ______ is called the collar bone.

(a) ulna
(b) clavicle
(c) humerus
(d) radius
(e) scapula

186. The bone located on the medial side of the forearm is the ______.

(a) ulna
(b) clavicle
(c) humerus
(d) radius
(e) scapula

187. The ______ are the bones of the wrist.

(a) tarsals
(b) metatarsals
(c) carpals
(d) metacarpals
(e) phalanges

188. The ______ are the bones of the hand.

(a) tarsals
(b) metatarsals
(c) carpals
(d) metacarpals
(e) phalanges

189. The ______ are the bones of the fingers.

(a) tarsals
(b) metatarsals
(c) carpals
(d) metacarpals
(e) phalanges

190. The ______ are the bones of the ankle.

(a) tarsals
(b) metatarsals
(c) carpals
(d) metacarpals
(e) phalanges

191. The _______ are the bones of the foot.

(a) tarsals
(b) metatarsals
(c) carpals
(d) metacarpals
(e) phalanges

192. The _______ are the bones of the toes.

(a) tarsals
(b) metatarsals
(c) carpals
(d) metacarpals
(e) phalanges

193. Each of the large winglike bones of the pelvis is called

(a) an ischium.
(b) an ilium.
(c) an iliac crest.
(d) a sacrum.
(e) a pubic bone.

194. The most superior aspect of either side of the pelvis is the

(a) ischium.
(b) ilium.
(c) iliac crest.
(d) sacrum.
(e) pubic bone.

195. The most inferior portion of either side of the pelvis is the

(a) ischium.
(b) ilium.
(c) iliac crest.
(d) sacrum.
(e) pubic bone.

196. The posterior segment of the pelvis consists of the

(a) ischium.
(b) ilium.
(c) iliac crest.
(d) sacrum.
(e) pubic bone.

197. The anterior portion of the pelvis is formed by the meeting of two

(a) ischiums.
(b) iliums.
(c) iliac crests.
(d) sacrums.
(e) pubic bones.

198. In layman's terms, the _______ is called the kneecap.

(a) patella
(b) tibia
(c) femur
(d) fibula
(e) malleolus

199. The _______ is the bone in the anterior lower leg.

(a) patella
(b) tibia
(c) femur
(d) fibula
(e) malleolus

200. The _______ is the bone in the posterior lower leg.

(a) patella
(b) tibia
(c) femur
(d) fibula
(e) malleolus

201. A _______ is a stretching injury of a muscle.

(a) fracture
(b) strain
(c) dislocation
(d) sprain
(e) cramp

202. A _______ is a partial tear injury of a ligament.

(a) fracture
(b) strain
(c) dislocation
(d) sprain
(e) cramp

203. When a bone is partially fractured, but not severed, it is called

(a) a transverse fracture.
(b) a comminuted fracture.
(c) a greenstick fracture.
(d) an impacted fracture.
(e) an oblique fracture.

204. When a break travels straight across the bone it is called

(a) a transverse fracture.
(b) a comminuted fracture.
(c) a greenstick fracture.
(d) an impacted fracture.
(e) an oblique fracture.

205. A break that travels across the bone at an angle is called

(a) a transverse fracture.
(b) a comminuted fracture.
(c) a greenstick fracture.
(d) an impacted fracture.
(e) an oblique fracture.

206. When an injury causes the bone ends of the broken bone to jam together, it is called

(a) a transverse fracture.
(b) a comminuted fracture.
(c) a greenstick fracture.
(d) an impacted fracture.
(e) an oblique fracture.

207. When sections of bone are crushed and fragmented, it is called

(a) a transverse fracture.
(b) a comminuted fracture.
(c) a greenstick fracture.
(d) an impacted fracture.
(e) an oblique fracture.

208. Which of the following statements regarding fractures is false?

(a) A fracture does not require trauma to occur.
(b) A fracture does not always produce significant pain.

(c) A fracture may be present without a change in the extremity's range of motion, sensation, circulation, or appearance.
(d) All of the above are false.
(e) None of the above is false.

209. Complications of an extremity fracture include all of the following, except

(a) pinched or severed nerves.
(b) soft tissue damage.
(c) pinched or lacerated blood vessels.
(d) contaminated wounds.
(e) hemiparesis.

210. ______ may cause the leg to appear shortened.

(a) An anterior hip dislocation
(b) A posterior hip dislocation
(c) A hip fracture
(d) Either answer (a) or (c).
(e) Answer (a), (b), or (c).

211. ______ usually causes an anterior bulge and lateral rotation of the leg.

(a) An anterior hip dislocation
(b) A posterior hip dislocation
(c) A hip fracture
(d) Either answer (a) or (c).
(e) Answer (a), (b), or (c).

212. ______ usually causes an anterior bulge and lateral rotation of the leg.

(a) An anterior hip dislocation
(b) A posterior hip dislocation
(c) A hip fracture
(d) Either answer (a) or (c).
(e) Answer (a), (b), or (c).

213. Your patient presents with her arm extended above her head and she is unable to move it at the shoulder. She most likely has _______ of her shoulder.

(a) an anterior dislocation
(b) a posterior dislocation
(c) an inferior dislocation
(d) a superior dislocation
(e) None of the above.

214. Your patient presents with a hollow shoulder, his arm hanging forward and away from his body, and his elbow medially rotated. He most likely has _______ of his shoulder.

(a) an anterior dislocation
(b) a posterior dislocation
(c) an inferior dislocation
(d) a superior dislocation
(e) None of the above.

215. Your patient holds his arm close to his anterior chest and appears to have a bulge on his anterior shoulder. He most likely has _______ of his shoulder.

(a) an anterior dislocation
(b) a posterior dislocation
(c) an inferior dislocation
(d) a superior dislocation
(e) None of the above.

216. The pulse that can be found at the posterior area of the knee is called the _______ pulse.

(a) popliteal
(b) dorsalis pedis
(c) posterior fibial
(d) medial tibial
(e) posterior tibial

217. The pulse that can be found on the medial side of the ankle, just posterior and inferior to the "ankle bone," is called the _______ pulse.

(a) popliteal
(b) dorsalis pedis
(c) posterior fibial
(d) medial tibial
(e) posterior tibial

218. The pulse that is found on the top of the foot is called the _______ pulse.

(a) popliteal
(b) dorsalis pedis
(c) posterior fibial
(d) medial tibial
(e) posterior tibial

219. The pulse that sometimes is called "the pedal pulse" is actually the _______ pulse.

(a) popliteal
(b) dorsalis pedis
(c) ulnar
(d) medial tibial
(e) radial

220. Which of the following statements regarding fracture splinting is false?

(a) Use gentle traction to straighten an angulated fracture before splinting, unless significant pain or resistance to correction is encountered.
(b) Splint dislocations or deformities near a joint in the position they are found (manipulate them once only in attempt to restore a distal pulse).
(c) Immobilization of the fracture is not accomplished unless it is immobilized from the joint above to the joint below the fracture.
(d) A fractured long bone may be gently tractioned until protruding bone ends are drawn back into place.
(e) Pad all rigid splints to prevent pressure and discomfort.

221. General rules for the splinting of any fracture include all of the following, except

(a) expose the injury and check for a distal pulse.
(b) remove all jewelry from the injured limb and secure it on the patient's person.
(c) gently move the deformed section to check for the presence of crepitus.
(d) dress the wound and pad all rigid splints before splinting.
(e) always leave fingers and toes exposed unless they are injured and require dressing.

222. Which of the following statements regarding pelvic fracture is false?

(a) Pelvic fractures present a potential life-threat.
(b) Apply and inflate the abdominal section of the PASG to stabilize a pelvic fracture (the legs should remain uninflated unless associated extremity injury or hypotension is present).
(c) Two or more large-bore crystalloid IVs should be initiated regardless of apparently "stable" vital signs in the presence of suspected pelvic fracture.
(d) Consider rapid transport to a trauma facility.
(e) None of the above is false.

223. If the patient is hemodynamically stable, a traction splint is appropriate for stabilization of

(a) a mid-shaft femoral fracture.
(b) a mid-shaft tibial fracture.
(c) a patellar fracture.
(d) Answers (a) and (b) only.
(e) Answers (a), (b), and (c).

224. Which of the following statements regarding upper extremity immobilization is false?

(a) A sling and swathe are required on all upper extremity injuries (except for isolated digital injury).
(b) Place a roll of bandage in the hand of the injured extremity to preserve the position of function.
(c) Use of airsplints requires monitoring to assess for leaks or pressure changes.
(d) Rigid splints may be used in conjunction with a sling and swathe.
(e) An injured finger may be splinted with a tongue depressor or taped to the adjacent finger.

225. Which of the following statements regarding functions of the skin is false?

(a) Skin functions to protect the body from penetration of bacteria and germs.
(b) Skin assists the regulation of body temperature: sweat evaporation warms the skin surface in cold weather and the skin's blood vessels constrict to provide cooling in hot weather.

(c) Skin prevents loss of body fluids and protects underlying structures from minor trauma.
(d) The skin is an organ of sensation.
(e) None of the above is false.

226. The _______ is the outermost layer of skin, consisting of dead cells constantly being rubbed off and replaced.

(a) subcutaneous tissue
(b) subdermis
(c) epidermis
(d) dermis
(e) None of the above.

227. The _______ contains sweat glands, sebaceous glands, and hair follicles.

(a) subcutaneous tissue
(b) subdermis
(c) epidermis
(d) dermis
(e) None of the above.

228. The skin layer composed largely of fat is the _______.

(a) subcutaneous tissue
(b) subdermis
(c) epidermis
(d) dermis
(e) None of the above.

229. The _______ serves as the body's layer of insulation.

(a) subcutaneous tissue
(b) subdermis
(c) epidermis
(d) dermis
(e) None of the above.

230. Sebum

(a) assists the skin to prevent fluid from entering or leaving the body.
(b) moisturizes the epidermis, keeping it soft and pliant.
(c) is a fatty secretion of the epidermal glands.
(d) Answers (a) and (b) only.
(e) Answers (a), (b), and (c).

231. The dark color of extravasated, deoxygenated blood is called

(a) erythema.
(b) ecchymosis.
(c) jaundice.
(d) mottling.
(e) pallor.

232. Inflammation of the skin produces

(a) erythema.
(b) ecchymosis.
(c) jaundice.
(d) mottling.
(e) pallor.

233. An injury in which the skin is not broken, but results in superficial edema and crushing damage to small blood vessels, is called

(a) a hematoma.
(b) a laceration.
(c) an abrasion.
(d) an incision.
(e) a contusion.

234. A pocket of blood that forms within the tissue beneath an injury site is called

(a) a hematoma.
(b) a laceration.
(c) an abrasion.
(d) an incision.
(e) a contusion.

235. Removal of layers of the epidermis or dermis produces an injury that is called

(a) a hematoma.
(b) a laceration.
(c) an abrasion.
(d) an incision.
(e) a contusion.

236. An open soft tissue wound with smooth, linear edges is called

(a) a hematoma.
(b) a laceration.
(c) an abrasion.
(d) an incision.
(e) a contusion.

237. "Goose egg" is a term commonly used by nonmedical personnel to indicate the presence of

(a) a hematoma.
(b) a laceration.
(c) an abrasion.
(d) an incision.
(e) a contusion.

238. An open soft tissue wound with ragged, uneven edges is called

(a) a hematoma.
(b) a laceration.
(c) an abrasion.
(d) an incision.
(e) a contusion.

239. The term *eschar* refers to

(a) the scabs that form on healing wounds or burns.
(b) inelastic, necrotic tissue formation in the area of a full-thickness burn.
(c) formation of blisters associated with a third-degree burn.
(d) Any of the above.
(e) None of the above.

240. An *avulsion* is defined as

(a) the complete removal of a section of tissue from the body.
(b) the incomplete removal of a section of tissue from the body.
(c) a "degloving" injury, where the skin is pulled away from its underlying structure.
(d) Answers (a) and (b) only.
(e) Answers (a), (b), and (c).

241. An *amputation* is defined as

(a) the complete removal of a digit, portion of a limb, or entire limb from the body.
(b) the incomplete removal of a digit, portion of a limb, or entire limb from the body.
(c) a "degloving" injury; where a digit, limb portion, or limb is forcibly pulled from the body.
(d) Answers (a) and (b) only.
(e) Answers (a), (b), and (c).

242. A crushing amputation frequently will result in

(a) profuse bleeding that is difficult to control.
(b) minimal bleeding because of rapid clotting.
(c) minimal bleeding because of vascular muscle contraction, producing tamponade.
(d) profuse bleeding because of muscular contraction.
(e) None of the above.

243. An amputation made with a single blow from a very sharp instrument or object frequently will result in

(a) profuse bleeding that is difficult to control.
(b) minimal bleeding because of rapid clotting.
(c) minimal bleeding because of vascular muscle contraction, producing tamponade.
(d) profuse bleeding because of muscular contraction.
(e) None of the above.

244. The phrase *partial-thickness burn* may be applied to

(a) a first-degree burn.
(b) a second-degree burn.
(c) a third-degree burn.
(d) Both answers (a) and (b).
(e) Both answers (b) and (c).

245. Blisters are associated with

(a) a first-degree burn.
(b) a second-degree burn.
(c) a third-degree burn.
(d) Both answers (a) and (b).
(e) Both answers (b) and (c).

246. Loss of sensation due to destruction of nerve endings is associated with

(a) a first-degree burn.
(b) a second-degree burn.

(c) a third-degree burn.
(d) Both answers (a) and (b).
(e) Both answers (b) and (c).

247. Severe pain may be associated with all areas of
(a) a first-degree burn.
(b) a second-degree burn.
(c) a third-degree burn.
(d) Both answers (a) and (b).
(e) Both answers (b) and (c).

248. The phrase *full-thickness burn* may be applied to
(a) a first-degree burn.
(b) a second-degree burn.
(c) a third-degree burn.
(d) Both answers (a) and (b).
(e) Both answers (b) and (c).

249. Charred black and/or dried white areas are associated with
(a) a first-degree burn.
(b) a second-degree burn.
(c) a third-degree burn.
(d) Both answers (a) and (b).
(e) Both answers (b) and (c).

250. If the rule of nines is used to calculate the percentage of body surface area burned, the head of an adult represents
(a) 1 percent body surface area burned.
(b) 4.5 percent body surface area burned.
(c) 9 percent body surface area burned.
(d) 11 percent body surface area burned.
(e) 18 percent body surface area burned.

251. If the rule of nines is used to calculate the percentage of body surface area burned, the back and buttocks of an adult represents
(a) 1 percent body surface area burned.
(b) 4.5 percent body surface area burned.
(c) 9 percent body surface area burned.
(d) 11 percent body surface area burned.
(e) 18 percent body surface area burned.

252. If the rule of nines is used to calculate the percentage of body surface area burned, the anterior of an adult's leg represents

(a) 1 percent body surface area burned.
(b) 4.5 percent body surface area burned.
(c) 9 percent body surface area burned.
(d) 11 percent body surface area burned.
(e) 18 percent body surface area burned.

253. If the rule of nines is used to calculate the percentage of body surface area burned, the anterior of an adult's arm represents

(a) 1 percent body surface area burned.
(b) 4.5 percent body surface area burned.
(c) 9 percent body surface area burned.
(d) 11 percent body surface area burned.
(e) 18 percent body surface area burned.

254. If the rule of nines is used to calculate the percentage of body surface area burned, the genitalia of an adult represent

(a) 1 percent body surface area burned.
(b) 4.5 percent body surface area burned.
(c) 9 percent body surface area burned.
(d) 11 percent body surface area burned.
(e) 18 percent body surface area burned.

255. If the rule of nines is used to calculate the percentage of body surface area burned, the anterior of an adult's chest represents

(a) 1 percent body surface area burned.
(b) 4.5 percent body surface area burned.
(c) 9 percent body surface area burned.
(d) 11 percent body surface area burned.
(e) 18 percent body surface area burned.

256. If the rule of nines is used to calculate the percentage of body surface area burned, the anterior of an adult's abdomen represents

(a) 1 percent body surface area burned.
(b) 4.5 percent body surface area burned.
(c) 9 percent body surface area burned.
(d) 11 percent body surface area burned.
(e) 18 percent body surface area burned.

257. An area equivalent to the palmar surface of the patient's hand is approximately

(a) 1 percent of the patient's body surface area.
(b) 2 percent of the patient's body surface area.
(c) 4.5 percent of the patient's body surface area.
(d) 5 percent of the patient's body surface area.
(e) None of the above.

258. If the pediatric rule of nines is used to calculate the percentage of body surface area burned, the head represents

(a) 1 percent body surface area burned.
(b) 4.5 percent body surface area burned.
(c) 9 percent body surface area burned.
(d) 7 percent body surface area burned.
(e) 18 percent body surface area burned.

259. If the pediatric rule of nines is used to calculate the percentage of body surface area burned, the anterior leg represents

(a) 1 percent body surface area burned.
(b) 4.5 percent body surface area burned.
(c) 9 percent body surface area burned.
(d) 7 percent body surface area burned.
(e) 18 percent body surface area burned.

260. If the pediatric rule of nines is used to calculate the percentage of body surface area burned, the anterior chest and abdomen represent

(a) 1 percent body surface area burned.
(b) 4.5 percent body surface area burned.
(c) 9 percent body surface area burned.
(d) 7 percent body surface area burned.
(e) 18 percent body surface area burned.

261. If the pediatric rule of nines is used to calculate the percentage of body surface area burned, the posterior chest and abdomen represent

(a) 1 percent body surface area burned.
(b) 4.5 percent body surface area burned.
(c) 9 percent body surface area burned.
(d) 7 percent body surface area burned.
(e) 18 percent body surface area burned.

262. If the pediatric rule of nines is used to calculate the percentage of body surface area burned, the posterior leg represents
- (a) 1 percent body surface area burned.
- (b) 4.5 percent body surface area burned.
- (c) 9 percent body surface area burned.
- (d) 7 percent body surface area burned.
- (e) 18 percent body surface area burned.

263. If the pediatric rule of nines is used to calculate the percentage of body surface area burned, the anterior and posterior arm represent
- (a) 1 percent body surface area burned.
- (b) 4.5 percent body surface area burned.
- (c) 9 percent body surface area burned.
- (d) 7 percent body surface area burned.
- (e) 18 percent body surface area burned.

264. If the pediatric rule of nines is used to calculate the percentage of body surface area burned, the genitalia represent
- (a) 1 percent body surface area burned.
- (b) 4.5 percent body surface area burned.
- (c) 9 percent body surface area burned.
- (d) 7 percent body surface area burned.
- (e) 18 percent body surface area burned.

265. Systemic complications associated with serious burns include all of the following, except
- (a) hyperthermia.
- (b) hypothermia.
- (c) hypovolemia.
- (d) plasma protein loss.
- (e) infection.

266. Circumferential third-degree burns of an extremity are especially dangerous because
- (a) the extremity will require amputation.
- (b) the patient will be more susceptible to hyperthermia.
- (c) the patient will be more susceptible to systemic infection.
- (d) the extremity will heal but will lose all functional ability.
- (e) blood flow may be completely restricted to the underlying and distal tissues.

267. Inhalation injuries are frequently associated with

(a) burns occurring within a closed space.
(b) any period of unconsciousness within a smoky environment.
(c) toxic fumes from combustion of construction materials.
(d) toxic fumes from combustion of petroleum products.
(e) Any of the above.

268. Actual thermal injury to the lungs

(a) is more likely with superheated steam exposure than with dry heat exposure.
(b) is more likely with dry heat exposure than with superheated steam exposure.
(c) occurs with inhalation of toxic fumes.
(d) Both answers (a) and (c).
(e) Both answers (b) and (c).

269. Which of the following statements regarding electrical burn injury is false?

(a) Electrical current causes thermal burns within the body tissues, extending from contact sites to exit sites.
(b) Electrical current may immobilize respiratory muscles and produce respiratory arrest.
(c) Electrical current may produce flash burns or ignite articles of clothing and produce exterior thermal burns as well as internal electrical burns.
(d) Electrical current may produce ventricular fibrillation.
(e) Electrical current follows pathways of muscle or bone fibers more readily than pathways of blood vessels or nerves.

270. Partial-thickness burns of the anterior surface of a lower extremity, occurring in pediatric or geriatric patients, would be categorized as

(a) isolated burns.
(b) moderate burns.
(c) minor burns.
(d) critical burns.
(e) None of the above.

271. Partial-thickness burns of the anterior surface of a lower extremity, occurring in patients with additional physical or medical injuries or disorders, would be categorized as

(a) isolated burns.
(b) moderate burns.
(c) minor burns.
(d) critical burns.
(e) None of the above.

272. Painful sunburns of the face, hands, feet, or genitalia would be categorized as

(a) isolated burns.
(b) moderate burns.
(c) minor burns.
(d) critical burns.
(e) None of the above.

273. The Parkland or Brooke fluid resuscitation formulas use body weight and body surface area burned to determine the patient's total fluid needs for the first 24 hours postburn. Fifty percent of this 24-hour total fluid volume must be infused within the first ______ postburn.

(a) 4 hours
(b) 6 hours
(c) 8 hours
(d) 10 hours
(e) 12 hours

The answer key to Section Four begins on page 380.

5

Test Section Five

Test Section Five covers the following subjects:

* Respiratory Emergencies
* Endocrine Emergencies
* The Nervous System
* The Acute Abdomen

EMT - Paramedic National Standards Review Self Test Third Edition

1. The right lung consists of
(a) a single large lobe of lung tissue.
(b) two lobes of lung tissue: the parietal and visceral lobes.
(c) three lobes of lung tissue: the parietal, visceral, and mediastinal lobes.
(d) two lobes of lung tissue: the upper and lower lobes.
(e) three lobes of lung tissue: the upper, middle, and lower lobes.

2. The left lung consists of
(a) a single large lobe of lung tissue.
(b) two lobes of lung tissue: the parietal and visceral lobes.
(c) three lobes of lung tissue: the parietal, visceral, and mediastinal lobes.
(d) two lobes of lung tissue: the upper and lower lobes.
(e) three lobes of lung tissue: the upper, middle, and lower lobes.

3. Carbon dioxide is ______ times more soluble in water than oxygen.
(a) 10
(b) 12
(c) 21
(d) 120
(e) 210

4. Causes of decreased oxygen levels in the blood include all of the following except
(a) fluid accumulation in alveolar or interstitial spaces.
(b) pneumothorax.
(c) atelectasis.
(d) hyperventilation.
(e) pulmonary embolism.

5. Causes of elevated arterial pressure of carbon dioxide (PCO_2) include increased CO_2 production. All of the following are causes of increased CO_2 production, except
(a) fever.
(b) hypoventilation.

(c) muscular exertion.
(d) shivering.
(e) metabolic processes resulting in formation of acids.

6. PCO_2 is also elevated by decreased CO_2 elimination. All of the following are causes of decreased CO_2 elimination, except

(a) chronic obstructive pulmonary diseases.
(b) hyperventilation.
(c) respiratory depression by drugs.
(d) airway obstruction.
(e) mechanical dysfunction (muscle injury/impairment).

7. Elevated arterial carbon dioxide levels can be decreased by

(a) hyperventilation with supplemental oxygen.
(b) breathing into a paper bag.
(c) correcting the causes of PCO_2 elevation.
(d) Both answers (a) and (c).
(e) Both answers (b) and (c).

8. Which of the following statements regarding regulation of respiratory function is false?

(a) Regulation of respirations is both voluntary and involuntary.
(b) The respiratory center is located in the brain stem.
(c) Inspiration is initiated by nerve impulses sent to the diaphragm and intercostal muscles from the brain stem.
(d) Expiration is initiated by nerve impulses sent to the diaphragm and intercostal muscles from the medulla.
(e) None of the above is false.

9. Microscopic stretch receptors, located in the lung and pleura, prevent overexpansion and stop inspiration. Which of the following statements regarding these receptors is false?

(a) The expansion of inspiration activates the stretch receptors, which then send impulses to the brain stem.
(b) The medulla responds by sending expiration impulses to the diaphragm and intercostal muscles.
(c) Elastic recoil of the lungs occurs as the diaphragm and intercostal muscles relax.
(d) Stretch receptors cease to send impulses to the brain stem and the cycle begins again.
(e) None of the above is false.

10. Chemoreceptors assist in respiratory regulation. These chemoreceptors are located in the

(a) medulla.
(b) aortic arch.
(c) carotid bodies.
(d) Answers (a) and (b) only.
(e) Answers (a), (b), and (c).

11. Respiratory regulating chemoreceptors are stimulated by

(a) decreased arterial oxygen levels.
(b) increased arterial carbon dioxide levels.
(c) decreased pH.
(d) Answers (a) and (b) only.
(e) Answers (a), (b), and (c).

12. In the normal individual, carbon dioxide concentration in the blood affects respiratory activity. Which of the following statements regarding this is true?

(a) Low CO_2 concentration increases respiratory activity.
(b) High CO_2 concentration decreases respiratory activity.
(c) Hypoxemia is the most profound stimulus for respiration.
(d) Both answers (a) and (b) are true.
(e) Answers (a), (b), and (c) are true.

13. Individuals with chronic respiratory disease have a decreased ability to eliminate CO_2, and their respiratory activity is regulated by hypoxic drive. Which of the following statements regarding this is true?

(a) The respiratory center develops accommodation of high PCO_2 levels.
(b) Dominant control of respiration is changes in PO_2.
(c) Respiratory rate and depth increase as PO_2 levels reach 80 to 100 torr.
(d) Both answers (a) and (b) are true.
(e) Answers (a), (b), and (c) are true.

14. A modified form of respiration useful for the reexpansion of areas of atelectasis is

(a) coughing.
(b) sneezing.
(c) hiccoughing.
(d) sighing.
(e) grunting.

15. A modified form of respiration that has no known physiologic purpose is

(a) coughing.
(b) sneezing.
(c) hiccoughing.
(d) sighing.
(e) grunting.

16. A modified form of respiration that serves to protect the airway from obstruction is

(a) coughing.
(b) sneezing.
(c) hiccoughing.
(d) sighing.
(e) grunting.

17. Fever

(a) increases the respiratory rate.
(b) decreases the respiratory rate.
(c) has no effect upon the respiratory rate.
(d) Any of the above.
(e) None of the above.

18. Anxiety

(a) increases the respiratory rate.
(b) decreases the respiratory rate.
(c) has no effect upon the respiratory rate.
(d) Any of the above.
(e) None of the above.

19. Sleep

(a) increases the respiratory rate.
(b) decreases the respiratory rate.
(c) has no effect upon the respiratory rate.
(d) Any of the above.
(e) None of the above.

20. Insufficient oxygenation

(a) increases the respiratory rate.
(b) decreases the respiratory rate.
(c) has no effect upon the respiratory rate.
(d) Any of the above.
(e) None of the above.

21. Ingestion of CNS depressant drugs
- (a) increases the respiratory rate.
- (b) decreases the respiratory rate.
- (c) has no effect upon the respiratory rate.
- (d) Any of the above.
- (e) None of the above.

22. Difficult or labored breathing that occurs whenever the patient is supine or not sitting or standing erect is called
- (a) orthopnea.
- (b) paroxysmal nocturnal dyspnea.
- (c) hyperpnea.
- (d) orthostatic apnea.
- (e) postural respirations.

23. Palpation of the chest will assist in assessing all of the following, except
- (a) presence of subcutaneous emphysema.
- (b) symmetry of excursion.
- (c) presence of a pneumothorax.
- (d) bilateral equality of tactile fremitus.
- (e) structural instability.

24. Friction rub is associated with pleural disease or pleural inflammation and sounds like
- (a) pieces of dried leather rubbing together.
- (b) several strands of hair, rubbed between your thumb and forefinger.
- (c) the rasping noise produced by rubbing your palms together.
- (d) Any of the above.
- (e) None of the above.

25. When the chest is auscultated, the rattling noises produced by the presence of thick mucous or other secretions in the throat or bronchi are called
- (a) snoring.
- (b) stridor.
- (c) wheezing.
- (d) rhonchi.
- (e) rales.

26. When the chest is auscultated, the whistling, musical sound produced by airway narrowing due to bronchoconstriction, edema, or foreign materials is called

(a) snoring.
(b) stridor.
(c) wheezing.
(d) rhonchi.
(e) rales.

27. A harsh, high-pitched sound heard on inspiration, characteristic of upper airway obstruction, is called

(a) snoring.
(b) stridor.
(c) wheezing.
(d) rhonchi.
(e) rales.

28. When the chest is auscultated, fine, moist sounds associated with fluid in the smaller airways are called

(a) snoring.
(b) stridor.
(c) wheezing.
(d) rhonchi.
(e) rales.

29. All of the following observations may indicate the presence of hypoxia or dyspnea, except

(a) anxiety.
(b) confusion.
(c) rapid speech and lengthy sentences.
(d) obesity.
(e) refusal to recline.

The following four questions concern the same event.

30. You are eating at a restaurant with your family when a nearby commotion catches your attention. You notice a conscious middle-aged man sitting in a chair and exhibiting the universal sign of choking. He is cyanotic and quiet with a very distressed expression. You should immediately

(a) perform 6 to 10 back blows.
(b) perform 6 to 10 abdominal thrusts.
(c) ask, "Are you choking? Can you speak? Can you cough?"
(d) solicit information from bystanders.
(e) perform finger sweeps and call for help.

31. After the tenth abdominal thrust, the patient's airway status remains unchanged. However, he no longer has his eyes open and does not appear to be making as much effort to breathe as before. You should

(a) perform 6 to 10 back blows.
(b) perform 6 to 10 abdominal thrusts.
(c) ask, "Are you choking? Can you speak? Can you cough?"
(d) solicit information from bystanders.
(e) perform finger sweeps and call for help.

32. As you continue with your treatment, the patient becomes unconscious and begins to slide from his chair. You support him to prevent trauma and guide him onto the floor into a supine position. You should immediately

(a) call for help, turn his head to the side, open the airway, and perform finger sweeps.
(b) assume the appropriate position and perform 6 to 10 abdominal thrusts, then finger sweeps.
(c) roll him to his side and perform 6 to 10 back blows, then finger sweeps.
(d) call for help, assume the appropriate position, and perform 6 to 10 abdominal thrusts, then finger sweeps.
(e) call for help, roll him to his side, and perform 6 to 10 back blows, then finger sweeps.

33. Your finger sweeps draw some mucous and saliva from the oropharynx, but no large pieces of matter. You should immediately

(a) call for help, turn his head, and repeat the finger sweeps.
(b) call for help, position yourself, and repeat the abdominal thrusts, then the finger sweeps.
(c) call for help, roll the patient to his side, and repeat the back blows, then the finger sweeps.
(d) open the airway and attempt to visualize the matter lodged in the patient's airway.
(e) open the airway and attempt ventilation.

The following three questions concern the same event.

34. A week later, at the same restaurant, you are eating with your family when you hear a commotion coming from the ladies' room. A woman runs out screaming, "She choked to death! She choked to death!" As you enter the bathroom you see an obese woman lying on the floor. There is no apparent trauma, but she is very cyanotic. You should immediately

(a) check for a pulse and call for help.
(b) check for breathing and call for help.
(c) check for level of consciousness and call for help.
(d) perform 6 to 10 chest thrusts and call for help.
(e) perform 6 to 10 back blows and call for help.

35. Your first attempt to ventilate the patient is unsuccessful. You should immediately

(a) perform 6 to 10 chest thrusts, finger sweeps, and attempt to ventilate again.
(b) perform 6 to 10 abdominal thrusts, finger sweeps, and attempt to ventilate again.
(c) reposition the airway and attempt to ventilate again.
(d) check for a pulse and attempt to ventilate again.
(e) perform finger sweeps and attempt to ventilate again.

36. Your next attempt to ventilate this obese patient is also unsuccessful. Using the following, indicate the correct sequence of activities you should perform until ventilation is achieved. (Not all the activities need to be used.)

(1) If ventilation is unsuccessful, reposition the airway and attempt to ventilate again.
(2) Check the patient's pulse.
(3) Perform 6 to 10 abdominal thrusts.
(4) Perform finger sweeps.
(5) Perform 6 to 10 chest thrusts.
(6) Perform 4 slow, firm back blows.
(7) Open the airway and attempt to ventilate the patient.
(8) Perform 4 slow, distinct chest thrusts.

(a) 8, 4, 7, 1, and repeat until ventilation is achieved.
(b) 1, 3, 4, 7, 2, and repeat until ventilation is achieved.
(c) 2, 3, 4, 7, 1, and repeat until ventilation is achieved.
(d) 8, 6, 4, 7, 1, and repeat until ventilation is achieved.
(e) 6, 3, 4, 7, 2, and repeat until ventilation is achieved.

37. An increased number of mucous-secreting cells in the respiratory epithelium, producing large amounts of sputum, is characteristic of

(a) emphysema.
(b) chronic bronchitis.
(c) asthma.
(d) Answers (a) and (b) only.
(e) Answers (a), (b), and (c).

38. Although the alveoli are not seriously affected by the disease process of _______, alveolar hypoventilation occurs and diminishes gas exchange.

(a) emphysema
(b) chronic bronchitis
(c) asthma
(d) Both answers (b) and (c).
(e) Answers (a), (b), and (c).

39. Pursed-lip breathing is indicative of _______ and serves to assist expiration.

(a) emphysema
(b) chronic bronchitis
(c) asthma
(d) Both answers (b) and (c).
(e) Answers (a), (b), and (c).

40. Wheezes may accompany

(a) emphysema.
(b) chronic bronchitis.
(c) asthma.
(d) Both answers (b) and (c).
(e) Answers (a), (b), and (c).

41. Rhonchi may accompany

(a) emphysema.
(b) chronic bronchitis.
(c) asthma.
(d) Both answers (a) and (c).
(e) Both answers (b) and (c).

42. A barrel chest, pink skin coloring, and recent weight loss is characteristic of a patient suffering from

(a) emphysema.
(b) chronic bronchitis.
(c) asthma.
(d) Both answers (b) and (c).
(e) Answers (a), (b), and (c).

43. An overweight patient with cyanosis and peripheral edema is characteristic of a patient suffering from

(a) emphysema.
(b) chronic bronchitis.
(c) asthma.
(d) Both answers (b) and (c).
(e) Answers (a), (b), and (c).

44. Widespread reversible narrowing of the airways by thick, tenacious sputum is characteristic of a patient suffering from

(a) emphysema.
(b) chronic bronchitis.
(c) asthma.
(d) Both answers (a) and (c).
(e) Both answers (b) and (c).

45. Bronchospasm secondary to mucus production is characteristic of a patient suffering from
- (a) emphysema.
- (b) chronic bronchitis.
- (c) asthma.
- (d) Both answers (a) and (c).
- (e) Both answers (b) and (c).

46. Cardiac dysrhythmias may accompany a patient suffering from
- (a) emphysema.
- (b) chronic bronchitis.
- (c) asthma.
- (d) Answers (a) and (b) only.
- (e) Answers (a), (b), and (c).

47. Inhaled irritants, respiratory infection, or emotional stress may precipitate an acute episode of
- (a) emphysema.
- (b) chronic bronchitis.
- (c) asthma.
- (d) Answers (a) and (b) only.
- (e) Answers (a), (b), and (c).

48. Aminophylline may be ordered for the treatment of
- (a) emphysema.
- (b) chronic bronchitis.
- (c) asthma.
- (d) Answers (a) and (b) only.
- (e) Answers (a), (b), and (c).

49. Pneumonia is a respiratory disease caused by a
- (a) bacterial infection.
- (b) virus.
- (c) fungus.
- (d) Answers (a) and (b) only.
- (e) Answers (a), (b), and (c).

50. Signs and symptoms of pneumonia include all of the following, except
- (a) fever, chills, and weakness.
- (b) a productive cough.
- (c) chest pain and tachycardia.
- (d) barrel chest and diminished excursion.
- (e) wheezes and upper abdominal pain.

51. Management of pneumonia includes all of the following, except

(a) supplemental oxygen.
(b) nebulized bronchodilators.
(c) cardiac monitoring.
(d) generalized cooling (in the presence of high fever).
(e) upright positioning for comfort.

52. Carbon monoxide inhalation poisoning is a common

(a) method of suicide, utilizing automobile exhaust.
(b) accidental cause of death from poorly ventilated home heating devices.
(c) hazard for firefighters.
(d) Answers (a) and (b) only.
(e) Answers (a), (b), and (c).

53. Carbon monoxide inhalation causes cellular hypoxia because carbon monoxide

(a) binds to hemoglobin more strongly than oxygen.
(b) prevents the inhalation of oxygen.
(c) prevents oxygen from crossing cell membranes.
(d) Answers (a) and (b) only.
(e) Answers (a), (b), and (c).

54. Signs and symptoms of carbon monoxide poisoning include all of the following, except

(a) headache, irritability, and agitation.
(b) euphoria and elation.
(c) chest pain, nausea, and vomiting.
(d) confusion, loss of coordination, and errors in judgment.
(e) seizures.

55. Cherry red skin coloration has long been associated with carbon monoxide poisoning and occurs

(a) soon after onset of euphoria, but before loss of consciousness.
(b) soon after onset of headache, but before loss of consciousness.
(c) just prior to altered mentation.
(d) very late in the course of poisoning, usually well after loss of consciousness.
(e) only after death, when lividity is noted.

56. Which of the following statements regarding management of carbon monoxide (CO) poisoning is false?

(a) The CO victim should be transported to a hospital that offers hyperbaric oxygen treatment.
(b) In the presence of CO poisoning, nebulized bronchodilators will greatly assist in increasing oxygen transport to the cells.
(c) Oxygen should be administered in as high a concentration as possible, even if the CO-poisoned patient has a history of COPD.
(d) All of the above are false.
(e) None of the above is false.

57. A pulmonary embolism is described as the obstruction of a pulmonary artery by

(a) an air embolus.
(b) a fat embolus or blood clot.
(c) an amniotic fluid embolus.
(d) Answers (a) or (b) only.
(e) Answers (a), (b), or (c).

58. Factors favoring the development of a pulmonary embolus include all of the following, except

(a) atrial fibrillation.
(b) thrombophlebitis.
(c) aspiration of vomitus.
(d) use of oral contraceptives.
(e) prolonged immobilization.

59. Although signs and symptoms will vary depending upon the size of the area obstructed, a patient with a pulmonary embolism may

(a) complain of sudden onset of severe dyspnea, with or without chest pain.
(b) exhibit tachycardia, tachypnea, and hypotension.
(c) have signs and symptoms of right heart failure (distended jugular veins).
(d) Answers (a) and (b) only.
(e) Answers (a), (b), and (c).

60. Hyperventilation may be caused by many medical emergencies. However, when hyperventilation is produced by anxiety alone, it will cause

(a) a fall in arterial CO_2, causing respiratory alkalosis.
(b) a fall in arterial O_2, causing respiratory alkalosis.

(c) a fall in arterial O_2, causing respiratory acidosis.
(d) a fall in arterial CO_2, causing respiratory acidosis.
(e) a fall in arterial CO_2, causing metabolic alkalosis.

61. Signs and symptoms of anxiety-induced hyperventilation syndrome include all of the following, except

(a) onset of seizures in patients with a seizure disorder.
(b) complaints of chest pain.
(c) complaints of dizziness.
(d) focal-motor seizures of the hands and feet.
(e) complaints of numbness and tingling around the mouth and in the hands and feet.

62. Exocrine glands are

(a) glands that secrete hormones.
(b) glands whose secretions reach a target via ducts.
(c) ductless glands that communicate with epithelial surfaces.
(d) glands whose secretions reach a target without the use of ducts.
(e) None of the above.

63. Endocrine glands are

(a) glands that secrete hormones.
(b) glands whose secretions reach a target via ducts.
(c) ductless glands that communicate with epithelial surfaces.
(d) glands whose secretions reach a target without the use of ducts.
(e) None of the above.

64. Which of the following glands (or pairs of glands) has both exocrine and endocrine functions?

(a) the ovaries
(b) the testes
(c) the adrenal glands
(d) the pancreas
(e) the thyroid gland

65. The "master gland" of the body, which regulates the function of most other endocrine glands by hormone secretion, is the

(a) brain.
(b) adrenal gland.
(c) pituitary gland.
(d) thyroid gland.
(e) parathyroid gland.

66. Epinephrine and norepinephrine are secreted by the

(a) ovaries.
(b) adrenal glands.
(c) testes.
(d) thyroid gland.
(e) kidneys.

67. The metabolism of calcium and phosphorus is controlled by the pea-shaped

(a) ovaries or testes.
(b) adrenal glands.
(c) pituitary gland.
(d) thyroid gland.
(e) parathyroid glands.

68. The body's general metabolic rate is regulated by hormones secreted by the

(a) ovaries or testes.
(b) adrenal glands.
(c) pituitary gland.
(d) thyroid gland.
(e) parathyroid glands.

69. Estrogen and progesterone are the hormones secreted by the

(a) ovaries.
(b) adrenal glands.
(c) testes.
(d) thyroid gland.
(e) kidneys.

70. Testosterone is a hormone secreted by the

(a) ovaries.
(b) adrenal glands.

(c) testes.
(d) thyroid gland.
(e) kidneys.

71. Cushing's disease is caused by a hypersecretion of glucocorticoids by the
(a) ovaries or testes.
(b) adrenal glands.
(c) pituitary gland.
(d) thyroid gland.
(e) parathyroid glands.

72. Antidiuretic hormone (ADH) is secreted by the
(a) kidneys.
(b) adrenal glands.
(c) pituitary gland.
(d) thyroid gland.
(e) parathyroid glands.

73. Follicle-stimulating hormone (FSH) controls the function of the
(a) ovaries or testes.
(b) adrenal glands.
(c) pituitary gland.
(d) thyroid gland.
(e) parathyroid glands.

74. Oxytocin is a hormone that stimulates uterine contraction and mammary gland release of milk. Oxytocin is secreted by the
(a) ovaries or testes.
(b) adrenal glands.
(c) pituitary gland.
(d) thyroid gland.
(e) parathyroid glands.

75. Luteinizing hormone (LH) affects the activity of the
(a) ovaries or testes.
(b) adrenal glands.
(c) pituitary gland.
(d) thyroid gland.
(e) parathyroid glands.

76. The adrenal glands are located
(a) at the base of the brain.
(b) in the center of the brain.
(c) anterior and lateral to the trachea.
(d) on top of each kidney.
(e) below each kidney.

77. The pituitary gland is located
(a) at the base of the brain.
(b) in the center of the brain.
(c) anterior and lateral to the trachea.
(d) on top of each kidney.
(e) below each kidney.

78. The thyroid gland is located
(a) at the base of the brain.
(b) in the center of the brain.
(c) anterior and lateral to the trachea.
(d) on top of each kidney.
(e) below each kidney.

79. Alpha cells within the islets of Langerhans secrete
(a) glucagon.
(b) glycogen.
(c) glucose.
(d) insulin.
(e) None of the above.

80. Beta cells within the islets of Langerhans secrete
(a) glucagon.
(b) glycogen.
(c) glucose.
(d) insulin.
(e) None of the above.

81. The form in which carbohydrate (sugar) is stored within the liver is called
(a) glucagon.
(b) glycogen.
(c) glucose.
(d) insulin.
(e) None of the above.

82. The free form of carbohydrate (sugar) in the blood is called
(a) glucagon.
(b) glycogen.
(c) glucose.
(d) insulin.
(e) None of the above.

83. The hormone produced in the pancreas that stimulates an increase in blood sugar is called
(a) glucagon.
(b) glycogen.
(c) glucose.
(d) insulin.
(e) None of the above.

84. The hormone responsible for allowing blood sugar to enter the cell is called
(a) glucagon.
(b) glycogen.
(c) glucose.
(d) insulin.
(e) None of the above.

85. _______ develops when the blood sugar level becomes too high.
(a) Hyperglycemia or diabetic ketoacidosis
(b) Hypoglycemia or diabetic ketoacidosis
(c) Hyperglycemia
(d) Hypoglycemia
(e) Hyperglycemia, hypoglycemia, or diabetic ketoacidosis

86. _______ develops when the blood sugar level becomes too low.
(a) Hyperglycemia or diabetic ketoacidosis
(b) Hypoglycemia or diabetic ketoacidosis
(c) Hyperglycemia
(d) Hypoglycemia
(e) Hyperglycemia, hypoglycemia, or diabetic ketoacidosis

87. _______ develops when there is not enough insulin available in the blood.

(a) Hyperglycemia or diabetic ketoacidosis
(b) Hypoglycemia or diabetic ketoacidosis
(c) Hyperglycemia
(d) Hypoglycemia
(e) Hyperglycemia, hypoglycemia, or diabetic ketoacidosis

88. _______ develops when there is too much insulin available in the blood.

(a) Hyperglycemia or diabetic ketoacidosis
(b) Hypoglycemia or diabetic ketoacidosis
(c) Hyperglycemia
(d) Hypoglycemia
(e) Hyperglycemia, hypoglycemia, or diabetic ketoacidosis

89. The onset of _______ is slow, requiring 12 to 24 hours before significant signs and symptoms are apparent.

(a) hyperglycemia or diabetic ketoacidosis
(b) hypoglycemia or diabetic ketoacidosis
(c) hyperglycemia
(d) hypoglycemia
(e) hyperglycemia, hypoglycemia, or diabetic ketoacidosis

90. Excessive fluid intake and urine output is associated with

(a) hyperglycemia or diabetic ketoacidosis.
(b) hypoglycemia or diabetic ketoacidosis.
(c) hyperglycemia.
(d) hypoglycemia.
(e) hyperglycemia, hypoglycemia, or diabetic ketoacidosis.

91. An altered mentation or decreased level of consciousness in any patient may indicate a life-threat secondary to

(a) hyperglycemia or diabetic ketoacidosis.
(b) hypoglycemia or diabetic ketoacidosis.
(c) hyperglycemia.
(d) hypoglycemia.
(e) hyperglycemia, hypoglycemia, or diabetic ketoacidosis.

92. A rapid onset of signs and symptoms is associated with

(a) hyperglycemia or diabetic ketoacidosis.
(b) hypoglycemia or diabetic ketoacidosis.

(c) hyperglycemia.
(d) hypoglycemia.
(e) hyperglycemia, hypoglycemia, or diabetic ketoacidosis.

93. An abnormally large food intake or complaints of excessive hunger is associated with

(a) hyperglycemia or diabetic ketoacidosis.
(b) hypoglycemia or diabetic ketoacidosis.
(c) hyperglycemia.
(d) hypoglycemia.
(e) hyperglycemia, hypoglycemia, or diabetic ketoacidosis.

94. Complaints of abdominal pain are associated with

(a) hyperglycemia or diabetic ketoacidosis.
(b) hypoglycemia or diabetic ketoacidosis.
(c) hyperglycemia.
(d) hypoglycemia.
(e) hyperglycemia, hypoglycemia, or diabetic ketoacidosis.

95. Complaints of headache are associated with

(a) hyperglycemia or diabetic ketoacidosis.
(b) hypoglycemia or diabetic ketoacidosis.
(c) hyperglycemia.
(d) hypoglycemia.
(e) hyperglycemia, hypoglycemia, or diabetic ketoacidosis.

96. Tachycardia is associated with

(a) hyperglycemia or diabetic ketoacidosis.
(b) hypoglycemia or diabetic ketoacidosis.
(c) hyperglycemia.
(d) hypoglycemia.
(e) hyperglycemia, hypoglycemia, or diabetic ketoacidosis.

97. Cool, diaphoretic skin is associated with

(a) hyperglycemia or diabetic ketoacidosis.
(b) hypoglycemia or diabetic ketoacidosis.
(c) hyperglycemia.
(d) hypoglycemia.
(e) hyperglycemia, hypoglycemia, or diabetic ketoacidosis.

98. Warm, dry skin is associated with
- (a) hyperglycemia or diabetic ketoacidosis.
- (b) hypoglycemia or diabetic ketoacidosis.
- (c) hyperglycemia.
- (d) hypoglycemia.
- (e) hyperglycemia, hypoglycemia, or diabetic ketoacidosis.

99. Kussmaul's respirations are associated with
- (a) hyperglycemia or diabetic ketoacidosis.
- (b) hypoglycemia or diabetic ketoacidosis.
- (c) hyperglycemia.
- (d) hypoglycemia.
- (e) hyperglycemia, hypoglycemia, or diabetic ketoacidosis.

100. A fruity, acetonelike breath odor is associated with
- (a) hyperglycemia or diabetic ketoacidosis.
- (b) hypoglycemia or diabetic ketoacidosis.
- (c) hyperglycemia.
- (d) hypoglycemia.
- (e) hyperglycemia, hypoglycemia, or diabetic ketoacidosis.

101. Seizures may be precipitated by
- (a) hyperglycemia or diabetic ketoacidosis.
- (b) hypoglycemia or diabetic ketoacidosis.
- (c) hyperglycemia.
- (d) hypoglycemia.
- (e) hyperglycemia, hypoglycemia, or diabetic ketoacidosis.

102. Hypokalemia is frequently associated with
- (a) hyperglycemia or diabetic ketoacidosis.
- (b) hypoglycemia or diabetic ketoacidosis.
- (c) hyperglycemia.
- (d) hypoglycemia.
- (e) hyperglycemia, hypoglycemia, or diabetic ketoacidosis.

103. Alcohol abuse is frequently associated with
- (a) hyperglycemia or diabetic ketoacidosis.
- (b) hypoglycemia or diabetic ketoacidosis.
- (c) hyperglycemia.
- (d) hypoglycemia.
- (e) hyperglycemia, hypoglycemia, or diabetic ketoacidosis.

104. Nonketotic hyperosmolar coma is associated with

(a) hyperglycemia or diabetic ketoacidosis.
(b) hypoglycemia or diabetic ketoacidosis.
(c) hyperglycemia.
(d) hypoglycemia.
(e) hyperglycemia, hypoglycemia, or diabetic ketoacidosis.

105. The medical term for an abnormally large food intake (complaint of excessive hunger) is

(a) polytrophia.
(b) polyphagia.
(c) polyopsia.
(d) polyfrasia.
(e) polygastria.

106. The medical term for an excessive urine output is

(a) polyhydruria.
(b) polysaccharose.
(c) polygastria.
(d) polydipsia.
(e) polyuria.

107. The medical term for an abnormally large fluid intake (complaint of excessive thirst) is

(a) polyhydrosis.
(b) polydipsia.
(c) polyhydruria.
(d) polyuria.
(e) polygastria.

108. The structural and functional unit of the nervous system is the nerve cell, also called

(a) a nucleus.
(b) a dendrite.
(c) an axon.
(d) a ganglia.
(e) a neuron.

109. Each nerve cell has branches that receive impulses and carry them to the cell body. Each of theses branches is called

(a) a nucleus.
(b) a dendrite.
(c) an axon.
(d) a ganglia.
(e) a neuron.

110. Each nerve cell also has at least one branch that carries impulses away from the cell body. This branch is called

(a) a nucleus.
(b) a dendrite.
(c) an axon.
(d) a ganglia.
(e) a neuron.

111. Impulses are communicated from nerve cell to nerve cell by means of

(a) interconnected myelinated junctions that intertwine and provide direct transfer of impulses from cell branch to cell branch.
(b) interconnected synapses that provide direct transfer of impulses from the "sending" cell body to the "receiving" cell branch.
(c) endoplasmic reticulum, providing direct contact of nerves at connecting synapses.
(d) a chemical neurotransmitter release that bridges the gap between a "sending" branch and transfers the impulse across the synapse to the "receiving" branches.
(e) None of the above.

112. The spinal cord communicates with the brain at the brain stem. The brain stem consists of three segments, called the

(a) cerebrum, pons, and medulla oblongata.
(b) cerebellum, diencephalon, and medulla oblongata.
(c) midbrain, pons, and medulla oblongata.
(d) mesencephalon, diencephalon, and medulla oblongata.
(e) cerebrum, cerebellum, and the diencephalon.

113. The seat of consciousness and center of higher mental faculties (memory, learning, judgment) is located in the largest part of the brain. This part of the brain is called the

(a) cerebellum.
(b) medulla oblongata.
(c) cerebrum.
(d) pons.
(e) diencephalon.

114. The primary control of balance and coordination is located in the brain portion that is called the

(a) cerebellum.
(b) medulla oblongata.
(c) cerebrum.
(d) pons.
(e) diencephalon.

115. Life-sustaining, involuntary functions of the respiratory and cardiovascular systems are primarily controlled by centers within the

(a) cerebellum.
(b) medulla oblongata.
(c) cerebrum.
(d) pons.
(e) diencephalon.

116. Within the brain, the speech center is located in the

(a) temporal lobes of the cerebrum.
(b) cerebellum.
(c) frontal lobe of the cerebrum.
(d) parietal lobes of the cerebrum.
(e) None of the above.

117. The portion of the brain that specializes in aspects of the individual's personality is the

(a) temporal lobes of the cerebrum.
(b) cerebellum.
(c) frontal lobe of the cerebrum.
(d) parietal lobes of the cerebrum.
(e) None of the above.

118. The spinal cord is 17 to 18 inches long and ends approximately at the level of the

(a) first lumbar vertebra.
(b) third lumbar vertebra.
(c) fifth lumbar vertebra.
(d) seventh lumbar vertebra.
(e) ninth lumbar vertebra.

119. The dorsal roots of the 31 pairs of spinal nerves contain afferent fibers. Afferent impulses travel from the

(a) body to the spinal cord only.
(b) body to the brain.
(c) brain to the body.
(d) spinal cord to the body only.
(e) None of the above.

120. The ventral roots of the 31 pairs of spinal nerves contain efferent fibers. Efferent impulses travel from the

(a) body to the spinal cord only.
(b) body to the brain.
(c) brain to the body.
(d) spinal cord to the body only.
(e) None of the above.

121. There are ___ pairs of cranial nerves that originate in the brain and innervate structures outside the brain.

(a) 9
(b) 12
(c) 5
(d) 7
(e) 14

122. Peripheral nerves are divided into categories consisting of

(a) somatic motor nerves.
(b) somatic sensory nerves.
(c) visceral motor and sensory nerves.
(d) Answers (a) and (b) only.
(e) Answers (a), (b), and (c).

123. Afferent impulses communicating pain, temperature, and position or muscle sense are transmitted by

(a) somatic motor nerves.
(b) somatic sensory nerves.

(c) visceral motor and sensory nerves.
(d) Answers (a) and (b) only.
(e) Answers (a), (b), and (c).

124. There is a collection of nerves located at the posterior base of the neck and extending bilaterally to the axilla. Injury to this nerve collection will result in permanent disability. This nerve collection is called the

(a) cervical plexus.
(b) solar plexus.
(c) brachial plexus.
(d) phrenic plexus.
(e) vertebral plexus.

125. Alcohol intoxication (overdose) represents one of the _______ causes of coma or altered level of consciousness.

(a) neurologic
(b) metabolic
(c) drug use
(d) cardiovascular
(e) respiratory

126. Hypoglycemia and diabetic ketoacidosis are examples of the _______ causes of coma or altered level of consciousness.

(a) structural
(b) metabolic
(c) drug use
(d) cardiovascular
(e) respiratory

127. Toxic inhalation (for example, CO poisoning) represents one of the _______ causes of coma or altered level of consciousness.

(a) neurologic
(b) metabolic
(c) drug use
(d) cardiovascular
(e) respiratory

128. Intracranial bleeding or brain tumors are examples of the _______ causes of coma or altered level of consciousness.

(a) structural
(b) metabolic
(c) drug use
(d) cardiovascular
(e) respiratory

129. Kidney and liver failure are examples of the _______ causes of coma or altered level of consciousness.

(a) structural
(b) metabolic
(c) drug use
(d) cardiovascular
(e) respiratory

130. Hypovolemic shock is an example of the _______ causes of coma or altered level of consciousness.

(a) neurologic
(b) metabolic
(c) drug use
(d) cardiovascular
(e) respiratory

131. Which of the following statements regarding administration of 50% dextrose in water is false?

(a) All adult patients who are unresponsive should receive 50 cc of $D_{50}W$, IV.
(b) All pediatric patients who are unresponsive should receive $D_{25}W$, IV.
(c) Extravasation of $D_{50}W$ will cause tissue necrosis.
(d) All adult patients who are unresponsive should receive 25 grams of $D_{50}W$, IV.
(e) $D_{50}W$ should be withheld if a reagent strip indicates hypoglycemia in an unconscious patient.

132. Which of the following statements regarding administration of naloxone is false?

(a) All patients who are unresponsive for unknown reasons should receive IV naloxone.
(b) Only unresponsive patients with known narcotic exposure should receive IV naloxone.

(c) Alcohol-induced coma may be an indication for naloxone administration.
(d) Naloxone administration may precipitate violent withdrawal signs and symptoms.
(e) Naloxone may be administered via IV, IM, SQ, or ET routes.

133. Which of the following statements regarding administration of thiamine is false?

(a) Thiamine is a B vitamin utilized in carbohydrate metabolism.
(b) Alcoholics frequently have a thiamine deficiency that causes Wernicke's and Korsakoff's syndromes.
(c) Wernicke's syndrome is an acute and reversible encephalopathy characterized by ataxia, eye muscle weakness (diplopia and nystagmus), and mental derangements.
(d) Korsakoff's psychosis is an acute and easily reversible memory disorder.
(e) Administration of $D_{50}W$ may precipitate signs and symptoms of Wernicke's or Korsakoff's syndromes in the alcoholic patient.

134. Seizures most commonly are attributed to

(a) hypoxia, hypoglycemia, infections, and other metabolic disorders.
(b) brain tumors, strokes, and other vascular disorders.
(c) head trauma.
(d) toxin exposure (including alcohol or other drug ingestion or withdrawal).
(e) idiopathic epilepsy.

135. Seizures that include a loss of consciousness and incontinence of urine or feces would be called

(a) Jacksonian seizures.
(b) focal motor seizures.
(c) grand mal seizures.
(d) petit mal seizures.
(e) hysterical seizures.

136. Seizures productive of sharp and bizarre movements that can be interrupted by firm command would be called

(a) Jacksonian seizures.
(b) psychomotor seizures.
(c) grand mal seizures.
(d) petit mal seizures.
(e) hysterical seizures.

137. Seizures that frequently go unnoticed because of their brief duration and lack of overt motor movement are

(a) Jacksonian seizures.
(b) focal motor seizures.
(c) grand mal seizures.
(d) petit mal seizures.
(e) hysterical seizures.

138. Unexplained attacks of rage or bizarre behavior are indicative of

(a) Jacksonian seizures.
(b) psychomotor seizures.
(c) grand mal seizures.
(d) petit mal seizures.
(e) hysterical seizures.

139. A child who abruptly stares off into space for a few seconds, then returns immediately to consciousness without demonstrating any motor symptoms may be experiencing

(a) Jacksonian seizures.
(b) psychomotor seizures.
(c) grand mal seizures.
(d) petit mal seizures.
(e) hysterical seizures.

140. Generalized, full-body tonic-clonic activity, occasionally resulting in tongue biting, is indicative of

(a) Jacksonian seizures.
(b) focal motor seizures.
(c) grand mal seizures.
(d) petit mal seizures.
(e) hysterical seizures.

141. Twitching of one body part that frequently progresses to generalized body seizures characterizes

(a) Jacksonian seizures.
(b) focal motor seizures.

(c) grand mal seizures.
(d) petit mal seizures.
(e) hysterical seizures.

142. Seizure activity involving one side of the body only is indicative of

(a) Jacksonian seizures.
(b) psychomotor seizures.
(c) grand mal seizures.
(d) petit mal seizures.
(e) hysterical seizures.

143. Temporal lobe seizures produce altered personality states and are also called

(a) Jacksonian seizures.
(b) psychomotor seizures.
(c) grand mal seizures.
(d) petit mal seizures.
(e) hysterical seizures.

144. The peculiar metallic taste that frequently precedes temporal lobe seizures can be called

(a) an auditory aura.
(b) a visual aura.
(c) an olfactory aura.
(d) a gustatory aura.
(e) a tactile aura.

145. An "odd" feeling in part of the body preceding a seizure can be called

(a) an auditory aura.
(b) a visual aura.
(c) an olfactory aura.
(d) a gustatory aura.
(e) a tactile aura.

146. Smelling a specific odor prior to a seizure can be called

(a) an auditory aura.
(b) a visual aura.
(c) an olfactory aura.
(d) a gustatory aura.
(e) a tactile aura.

147. Muscle rigidity alternating with relaxation characterizes the

(a) hyperclonic phase of a grand mal seizure.
(b) clonic phase of a grand mal seizure.
(c) tonic phase of a grand mal seizure.
(d) hypertonic phase of a grand mal seizure.
(e) None of the above.

148. Extreme muscular rigidity and hyperextension of the back characterizes the

(a) hyperclonic phase of a grand mal seizure.
(b) clonic phase of a grand mal seizure.
(c) tonic phase of a grand mal seizure.
(d) hypertonic phase of a grand mal seizure.
(e) None of the above.

149. Massive autonomic discharge with hyperventilation, frothy salivation, and tachycardia accompanies the

(a) hyperclonic phase of a grand mal seizure.
(b) clonic phase of a grand mal seizure.
(c) tonic phase of a grand mal seizure.
(d) hypertonic phase of a grand mal seizure.
(e) None of the above.

150. Continuous motor tension (contraction) of the muscles characterizes the

(a) hyperclonic phase of a grand mal seizure.
(b) clonic phase of a grand mal seizure.
(c) tonic phase of a grand mal seizure.
(d) hypertonic phase of a grand mal seizure.
(e) None of the above.

151. Although the time of duration for any particular phase of a grand mal seizure varies from patient to patient, the order of phase progression remains basically the same. Using the following phase descriptions, indicate the correct phase progression for a typical grand mal seizure (not all phase descriptions must be used).

(1) hypertonic phase
(2) hyperclonic phase
(3) aura
(4) clonic phase
(5) tonic phase
(6) loss of consciousness
(7) tongue swallowing
(8) confusion, fatigue, and/or headache.

(a) 6, 3, 4, 2, 5, 8
(b) 3, 6, 4, 2, 7, 5, 8
(c) 3, 5, 1, 4, 6, 7, 8
(d) 6, 3, 5, 4, 2, 7, 8
(e) 3, 6, 5, 1, 4, 8

152. Vasovagal syncope differs from seizures, in that

(a) it may occur in any position (that is, standing, sitting, or lying).
(b) it may occur without warning.
(c) the patient regains consciousness almost immediately upon becoming supine.
(d) Answers (a) and (b) only.
(e) Answers (a), (b), and (c).

153. Which of the following is included in the correct management of an isolated seizure?

(a) Airway protection: using the cross-finger technique, part the teeth to insert a padded tongue blade or bite block to prevent tongue trauma or broken teeth during the seizure.
(b) High-flow oxygen (15 LPM) via a nonrebreather mask.
(c) Soft restraints as needed, to prevent self-injury from flailing extremities.
(d) All of the above.
(e) None of the above.

154. *Status epilepticus* is defined as

(a) any seizure lasting 30 seconds or longer.
(b) two or more seizures within one hour.
(c) two or more seizures without an intervening period of consciousness.
(d) Any of the above.
(e) None of the above.

155. Status epilepticus is a life-threatening emergency because it may result in

(a) respiratory inadequacy, hypoxia, or respiratory arrest.
(b) hypoxic brain damage, necrosis of heart muscle, or severe dehydration.
(c) aspiration of secretions or vomitus.
(d) All of the above.
(e) None of the above.

156. The most common cause of status epilepticus in adults is
- (a) head trauma.
- (b) failure to take prescribed anticonvulsant medications.
- (c) infection.
- (d) hypoglycemia.
- (e) cardiovascular disease.

157. Management of status epilepticus includes airway management, supplemental oxygen, assisted ventilations as needed, and
- (a) 25 grams of 50% dextrose in water, IV.
- (b) diazepam administration.
- (c) morphine sulfate administration (with naloxone prepared and ready for administration should respiratory arrest ensue).
- (d) Answers (a) and (b) only.
- (e) Answers (a), (b), and (c).

158. The adult dosage for diazepam when one is treating status epilepticus is
- (a) 5 to 10 mg IV.
- (b) 5 to 15 mg IV.
- (c) 2 to 5 mg IV.
- (d) 0.5 to 1 mg IV.
- (e) 0.5 to 1 mg/kg IV.

159. Diazepam administration can cause
- (a) respiratory depression or arrest.
- (b) hypotension.
- (c) acute hypersensitivity reactions.
- (d) Answers (a) and (b) only.
- (e) Answers (a), (b), and (c).

160. A stroke, or cerebrovascular accident (CVA), may be caused by all of the following, except
- (a) thrombus occlusion of cerebral vasculature.
- (b) air or fat embolus occlusion of cerebral vasculature.
- (c) atherosclerotic plaque or tumor tissue occlusion of cerebral vasculature.
- (d) subarachnoid hemorrhage.
- (e) aortic aneurysm.

161. Strokes are the

(a) second most common cause of death in adults.
(b) third most common cause of death in adults.
(c) least common cause of death in adults.
(d) least common cause of death, but most common cause of disability in adults.
(e) None of the above.

162. Predisposing factors that increase the incidence of stroke include all of the following, except

(a) hypertension and some cardiac dysrhythmias.
(b) diabetes and sickle cell disease.
(c) long-bone fractures (especially prior to immobilization).
(d) use of oral contraceptives.
(e) abnormal blood lipid levels.

163. Transient ischemic attacks (TIAs) are defined as episodes of cerebral dysfunction with signs and symptoms similar to that of CVAs, lasting

(a) only a few minutes, never more than one hour.
(b) a week to ten days, with complete recovery achieved at the end of that time period.
(c) anywhere from two or three minutes to several hours, but always less than 24 hours.
(d) no longer than three to four days.
(e) two to three hours only.

164. Signs and symptoms of a patient experiencing a TIA or CVA include all of the following, except

(a) paraparesis or paraplegia.
(b) speech disturbances (dysarthria, aphasia).
(c) altered level of consciousness, confusion, or agitation.
(d) vision disturbances.
(e) unresponsiveness.

165. The preferred position of transport for the patient suffering from CVA or TIA is

(a) left laterally recumbent.
(b) right laterally recumbent.
(c) supine, with feet elevated.
(d) the Trendelenburg position.
(e) supine, with head elevated.

166. Treatment of the CVA/TIA victim includes airway management, oxygen administration, ventilatory assistance, and
- (a) NTG administration if acute hypertension is present.
- (b) administration of 50% dextrose in water if hypoglycemia is suspected.
- (c) diazepam administration if the patient is acutely combative.
- (d) Answers (a) and (b) only.
- (e) Answers (a), (b), and (c).

167. Digestive enzymes are secreted by
- (a) salivary glands.
- (b) the pancreas.
- (c) the liver.
- (d) Answers (b) and (c) only.
- (e) Answers (a), (b), and (c).

168. Fat digestion is accomplished with bile, which is secreted by
- (a) the kidneys.
- (b) the pancreas.
- (c) the liver.
- (d) Answers (a) and (b) only.
- (e) Answers (a), (b), and (c).

169. Glucose storage (as glycogen) occurs in the
- (a) the kidneys.
- (b) the pancreas.
- (c) the liver.
- (d) Answers (b) and (c) only.
- (e) Answers (a), (b), and (c).

170. Causes of acute abdominal pain are frequently divided into "hemorrhagic" and "nonhemorrhagic" groups. However, many causes of hemorrhagic abdominal pain first produce pain while in a nonhemorrhagic state. All of the following causes of abdominal pain may be either hemorrhagic or nonhemorrhagic, except
- (a) peptic ulcer.
- (b) diverticulitis.
- (c) ectopic pregnancy.
- (d) appendicitis.
- (e) duodenal ulcer.

171. All of the following are nonhemorrhagic causes of abdominal pain, except

(a) pyelonephritis.
(b) esophageal varices.
(c) renal calculus.
(d) pelvic inflammatory disease.
(e) ovarian cyst.

172. All of the following are hemorrhagic causes of abdominal pain, except

(a) pyelonephritis.
(b) esophageal varices.
(c) carcinoma of the colon.
(d) aortic aneurysm.
(e) perforated abdominal viscus.

173. The acute onset of periumbilical pain with radiation (or migration) to the right lower quadrant of the abdomen is frequently associated with

(a) diverticulitis.
(b) pyelonephritis.
(c) appendicitis.
(d) an aortic aneurysm.
(e) a renal calculus.

174. The acute onset of unilateral flank pain with radiation (or migration) into the lower abdominal quadrant of that side and into the genitals is frequently associated with

(a) diverticulitis.
(b) pyelonephritis.
(c) appendicitis.
(d) an aortic aneurysm.
(e) a renal calculus.

175. Fever accompanied by complaint of lower back pain with urinary burning and frequency is often associated with

(a) diverticulitis.
(b) pyelonephritis.
(c) appendicitis.
(d) an aortic aneurysm.
(e) a renal calculus.

176. The patient who complains of lower back pain, describing it as "burning" or "tearing" in nature, should be considered to have

(a) diverticulitis.
(b) pyelonephritis.
(c) appendicitis.
(d) an aortic aneurysm.
(e) a renal calculus.

177. Bleeding of the upper GI tract should be suspected when the patient reports or exhibits

(a) bright red emesis.
(b) coffee ground emesis.
(c) wine-colored stool or melena.
(d) Answers (a) and (b) only.
(e) Answers (a), (b), and (c).

178. Bleeding of the lower GI tract should be suspected when the patient reports or exhibits

(a) bright red blood in the stool.
(b) wine-colored stool.
(c) melena.
(d) Answers (a) and (b) only.
(e) Answers (a), (b), and (c).

179. The presence of abdominal pain can be suspected when your patient is observed to prefer being transported in

(a) a supine position with the legs extended.
(b) a prone position with the legs extended.
(c) a position with the knees flexed and drawn toward the chest.
(d) a Trendelenburg position.
(e) Any of the above.

180. The presence of "rebound tenderness" indicates

(a) peritoneal irritation.
(b) a guarded abdomen.
(c) a bowel obstruction.
(d) an aortic aneurysm.
(e) None of the above.

181. Rebound tenderness is present when the patient complains of

(a) generalized abdominal pain and guards the abdomen.
(b) pain on deep palpation of the abdomen.
(c) pain on rapid release of deep abdominal palpation.
(d) Any of the above.
(e) None of the above.

182. Orthostatic vital sign changes (also called the "tilt test") are measured first when the patient is supine, and then sitting and/or standing. Hypovolemia is indicated if the patient's pulse ______ as the patient moves to a sitting or standing position.

(a) increases by 5 or more beats per minute
(b) increases by 15 or more beats per minute
(c) decreases by 5 or more beats per minute
(d) decreases by 15 or more beats per minute
(e) None of the above.

183. Hypovolemia is indicated if the patient's blood pressure ______ as the patient moves to a sitting or standing position.

(a) increases by 5 or more mmHg
(b) increases by 15 or more mmHg
(c) decreases by 5 or more mmHg
(d) decreases by 15 or more mmHg
(e) None of the above.

184. Functions of the kidneys include

(a) maintenance of the body's fluid volume.
(b) elimination of metabolic waste products.
(c) maintenance of blood pH and body fluid composition.
(d) Answers (a) and (b) only.
(e) Answers (a), (b), and (c).

185. Renal failure can be caused by any of the following, except

(a) shock states or direct trauma.
(b) testicular torsion.
(c) dehydration or infection.
(d) prostate enlargement.
(e) bladder or ureter obstruction.

186. Renal failure results in an increased blood level of urea, a chemical produced by the metabolism of protein. This increased level of urea in the blood is called

(a) uremia.
(b) hematuria.
(c) hematourea.
(d) polyurea.
(e) ureamegolly.

187. Renal failure can cause

(a) fluid retention, hypertension, and hyperkalemia.
(b) fluid retention, hypotension, and hypokalemia.
(c) fluid retention, hypertension, and hypokalemia.
(d) dehydration, hypertension, and hyperkalemia.
(e) dehydration, hypotension, and hypokalemia.

188. The accumulation of serous fluid (edema) in the peritoneal cavity is called

(a) peritonitis.
(b) peristalsis.
(c) angiogastrotic edema.
(d) ascites.
(e) chylosus.

189. Signs and symptoms of renal failure include all of the following, except

(a) pulmonary edema and jugular vein distention.
(b) jaundice.
(c) peripheral and peritoneal edema.
(d) polyuria.
(e) cardiac dysrhythmias.

190. A patient with renal failure

(a) requires higher initial doses of medication than a patient with normally functioning kidneys.
(b) requires more frequent maintenance doses to sustain a therapeutic blood level of medication than does the patient with normally functioning kidneys.
(c) is more susceptible to toxic accumulation of drugs than the patient with normally functioning kidneys, even at normal doses.
(d) Answers (a) and (b) only.
(e) Answers (a), (b), and (c).

191. Management of the dialysis patient includes
- (a) measuring blood pressure only on the arm without the dialysis shunt.
- (b) IV access, obtained only in the arm without the dialysis shunt.
- (c) medical treatment protocols the same as for any other patient.
- (d) Answers (a) and (b) only.
- (e) Answers (a), (b), and (c).

192. Which of the following statements regarding kidney stones is false?
- (a) A kidney stone may also be called a urinary stone or a renal calculus.
- (b) Kidney stones more commonly occur in men than in women.
- (c) The causes of kidney stone formation are unknown.
- (d) Urinary inflammation, infection, or obstruction may result from kidney stone formation.
- (e) A kidney stone may appear anywhere within the urinary tract but will produce pain only during its passage through the ureter.

193. Signs and symptoms of kidney stone formation include all of the following, except
- (a) severe dyspnea.
- (b) nausea and vomiting.
- (c) excruciating flank pain which may or may not radiate to the groin.
- (d) hematuria.
- (e) dysuria.

194. Which of the following statements regarding urinary tract infection (UTI) is false?
- (a) UTI occurs more frequently in men due to the greater length of the male urethra.
- (b) UTI may produce lower abdominal pain, especially on urination.
- (c) UTI may produce urine discoloration.
- (d) Complaints of pain, burning, or difficulty on attempts to urinate commonly accompany UTI.
- (e) No prehospital care is indicated for UTI.

195. Methods of renal dialysis include hemodialysis (via an external arteriovenous shunt or an internal fistula) and

(a) gastrointestinal dialysis.
(b) peritoneal dialysis.
(c) cerebrodialysis.
(d) Answers (a) and (b) only.
(e) Answers (a), (b), and (c).

196. Complications related to renal dialysis include

(a) hypotension and/or chest pain.
(b) air embolism (dyspnea, cyanosis, hypotension).
(c) lethargy and/or seizures.
(d) Answers (a) and (b) only.
(e) Answers (a), (b), and (c).

The answer key to Section Five begins on page 387.

6

Test Section Six

Test Section Six covers the following subjects:

* Cardiovascular Disorders
* Electrophysiology of the Heart
* Basic Dysrhythmia Identification (using Lead II)

EMT - Paramedic
National Standards Review Self Test
Third Edition

1. The outermost layer of heart wall muscle is called the
(a) endocardium.
(b) epicardium.
(c) pericardium.
(d) myocardium.
(e) intercardium.

2. The thick middle layer of heart wall muscle is called the
(a) endocardium.
(b) epicardium.
(c) pericardium.
(d) myocardium.
(e) intercardium.

3. The smooth inner layer of heart wall muscle is called the
(a) endocardium.
(b) epicardium.
(c) pericardium.
(d) myocardium.
(e) intercardium.

4. Which of the following statements regarding myocardial muscle is false?
(a) Myocardial muscle is composed of specialized muscle cells found only in the heart.
(b) Myocardial muscle is striated like skeletal muscle.
(c) Myocardial muscle has electrical properties similar to smooth muscle.
(d) All of the above are false.
(e) None of the above is false.

5. Which of the following statements regarding the chambers of the heart is false?
(a) The atria are the right and left superior chambers of the heart.
(b) The interatrial septum separates the atria.
(c) The ventricles are located at the base of the heart.
(d) The ventricles are more muscular than the atria.
(e) The left ventricle is thicker than the right ventricle.

6. Beginning with the return of blood to the heart from the peripheral circulation, place the following in order, describing the normal flow of blood.

(1) right atrium
(2) left atrium
(3) right ventricle
(4) left ventricle
(5) pulmonary arteries
(6) aorta
(7) inferior and superior vena cava
(8) bicuspid (mitral) valve
(9) tricuspid valve
(10) pulmonic valve
(11) aortic valve
(12) pulmonary veins
(13) pulmonary capillaries

(a) 7, 1, 8, 3, 10, 5, 13, 12, 2, 9, 4, 11, 6
(b) 7, 1, 8, 3, 10, 12, 13, 5, 2, 9, 4, 11, 6
(c) 12, 2, 9, 4, 11, 6, 7, 1, 8, 3, 10, 5, 13
(d) 7, 1, 9, 3, 10, 5, 13, 12, 2, 8, 4, 11, 6
(e) 7, 1, 9, 3, 10, 12, 13, 5, 2, 8, 4, 11, 6

7. The coronary arteries are the exclusive arterial blood supply to the heart muscle and its electrical conduction system. The coronary arteries originate in the

(a) coronary sinus of the right atrium.
(b) pulmonary artery just above the leaflets of the pulmonic valve.
(c) pulmonary vein just above the leaflets of the aortic valve.
(d) aorta just above the leaflets of the aortic valve.
(e) aorta just above the leaflets of the pulmonic valve.

8. The cavity within a blood vessel, the diameter of which varies greatly, is called

(a) the container.
(b) the vascular hollow.
(c) the lumen.
(d) an os.
(e) a sinus.

9. The smooth, single-cell layer that is the innermost lining of a blood vessel wall is called the

(a) tunica adventitia.
(b) tunica intima.
(c) tunica terres.
(d) tunica arteriosum.
(e) tunica media.

10. The middle layer of a blood vessel wall is composed of elastic fibers and muscle. This layer gives strength and recoil to vessels, is thicker in arteries than in veins, and is called the

(a) tunica adventitia.
(b) tunica intima.
(c) tunica terres.
(d) tunica arteriosum.
(e) tunica media.

11. A protective fibrous tissue covers the exterior of a blood vessel, providing strength to withstand high pressures within. This outer layer is called the

(a) tunica adventitia.
(b) tunica intima.
(c) tunica terres.
(d) tunica arteriosum.
(e) tunica media.

12. Blood is carried away from the heart by

(a) capillaries.
(b) veins.
(c) arteries.
(d) All of the above.
(e) None of the above.

13. All fluid, gas, and nutrient exchange occurs in

(a) capillaries.
(b) veins.
(c) arteries.
(d) All of the above.
(e) None of the above.

14. One-way valves aid the direction of blood flow within

(a) capillaries.
(b) veins.
(c) arteries.
(d) All of the above.
(e) None of the above.

15. The right and left atria contract together during the cardiac cycle called

(a) systole.
(b) diastole.

(c) asystole.
(d) Either answer (a) or (b).
(e) Answer (a), (b), or (c).

16. The ventricular contraction phase of a cardiac cycle is called
(a) systole.
(b) diastole.
(c) asystole.
((d) Either answer (a) or (b).
(e) Answer (a), (b), or (c).

17. The ventricular relaxation phase of a cardiac cycle is called
(a) systole.
(b) diastole.
(c) asystole.
(d) Either answer (a) or (b).
(e) Answer (a), (b), or (c).

18. The bicuspid (mitral) and tricuspid valves are open during
(a) systole.
(b) diastole.
(c) asystole.
(d) Either answer (a) or (b).
(e) Answer (a), (b), or (c).

19. The pulmonic and aortic valves are open during
(a) systole.
(b) diastole.
(c) asystole.
(d) Either answer (a) or (b).
(e) Answer (a), (b), or (c).

20. The phase during which most coronary artery filling occurs (about 70 percent) is
(a) systole.
(b) diastole.
(c) asystole.
(d) Either answer (a) or (b).
(e) Answer (a), (b), or (c).

21. Stroke volume is the amount of blood ejected from a ventricle with one contraction. The average adult stroke volume is
- (a) 120 milliliters.
- (b) 15 milliliters.
- (c) 7.45 milliliters.
- (d) 40 milliliters.
- (e) 70 milliliters.

22. Preload influences stroke volume and is defined as
- (a) the volume and pressure available to the atrium for cardiac pumping.
- (b) the pressure under which the ventricle fills, influenced by venous return.
- (c) the resistance against which the ventricle contracts, determined by arterial resistance.
- (d) Both answers (a) and (b).
- (e) Both answers (a) and (c).

23. Afterload influences stroke volume and is defined as
- (a) the volume and pressure available to the atrium for cardiac pumping.
- (b) the pressure under which the ventricle fills, influenced by venous return.
- (c) the resistance against which the ventricle contracts, determined by arterial resistance.
- (d) Both answers (a) and (b).
- (e) Both answers (a) and (c).

24. Starling's law of the heart states that up to a limit,
- (a) the stronger the force of cardiac contraction, the greater will be the rebound chamber filling.
- (b) the more a myocardial muscle is stretched, the greater will be its force of contraction.
- (c) the greater the volume of chamber filling, the greater will be its stroke volume.
- (d) Both answers (a) and (c).
- (e) Both answers (b) and (c).

25. An increase in peripheral vascular resistance will
- (a) increase preload and increase stroke volume.
- (b) increase afterload and increase stroke volume.

(c) decrease preload and increase stroke volume.
(d) increase afterload and decrease stroke volume.
(e) have no effect on stroke volume.

26. Cardiac output is the volume of blood pumped through the circulatory system during
(a) one cardiac cycle.
(b) ten cardiac cycles.
(c) one minute.
(d) Any of the above.
(e) None of the above.

27. Which of the following statements regarding the nervous system's control of the heart is false?
(a) The autonomic nervous system influences rate, conductivity, and contractility of the heart.
(b) Parasympathetic stimulation of the heart occurs via the vagus nerve and its chemical neurotransmitter, acetylcholine.
(c) Acetylcholine release increases heart rate and AV conduction.
(d) Carotid sinus massage produces vagus nerve stimulation.
(e) Valsalva maneuvers decrease heart rate and AV conduction.

28. Sympathetic stimulation of the heart is produced by the chemical neurotransmitter norepinephrine, which has both alpha and beta effects. Alpha effects on the heart produce
(a) an increased heart rate.
(b) an increased myocardial contractility.
(c) an increased conductivity.
(d) All of the above.
(e) None of the above.

29. Beta effects on the heart produce
(a) an increased heart rate.
(b) an increased myocardial contractility.
(c) an increased conductivity.
(d) All of the above.
(e) None of the above.

30. The most important role of the electrolyte sodium is its action relating to

(a) the depolarization phase of myocardial cells.
(b) the force of myocardial contraction.
(c) the repolarization phase of myocardial cells.
(d) Both answers (a) and (b).
(e) Both answers (b) and (c).

31. The most important role of the electrolyte potassium is its action relating to

(a) the depolarization phase of myocardial cells.
(b) the force of myocardial contraction.
(c) the repolarization phase of myocardial cells.
(d) Both answers (a) and (b).
(e) Both answers (b) and (c).

32. Calcium plays a major role in

(a) the depolarization phase of myocardial cells.
(b) the force of myocardial contraction.
(c) the repolarization phase of myocardial cells.
(d) Both answers (a) and (b).
(e) Both answers (b) and (c).

33. A property of the heart's pacemaker cells is the ability to generate an electrical impulse without stimulation from another source. This ability is called

(a) excitability.
(b) reciprocity.
(c) automaticity.
(d) conjunctivity.
(e) conductivity.

34. A property of all myocardial cells is the ability to respond to an electrical stimulus. This ability is called

(a) excitability.
(b) reciprocity.
(c) automaticity.
(d) conjunctivity.
(e) conductivity.

35. The ability to propagate an impulse from cell to cell is called

(a) excitability.
(b) reciprocity.

(c) automaticity.
(d) conjunctivity.
(e) conductivity.

36. Place the following in the correct sequence, showing the normal electrical conduction of the heart.

(1) AV node
(2) SA node
(3) Purkinje fibers
(4) internodal and interatrial tracts
(5) bundle branches
(6) bundle of His

(a) 2, 1, 6, 5, 3, 4
(b) 2, 4, 1, 6, 5, 3
(c) 1, 5, 2, 6, 3, 4
(d) 1, 4, 2, 5, 6, 3
(e) 2, 4, 1, 5, 6, 3

37. The normal intrinsic rate of spontaneous discharge of the sinoatrial node is

(a) 60–80 beats per minute.
(b) 20–40 beats per minute.
(c) 40–60 beats per minute.
(d) 60–100 beats per minute.
(e) 80–120 beats per minute.

38. The normal intrinsic rate of spontaneous discharge of the atrioventricular node is

(a) 60–80 beats per minute.
(b) 20–40 beats per minute.
(c) 40–60 beats per minute.
(d) 60–100 beats per minute.
(e) 80–120 beats per minute.

39. The normal intrinsic rate of spontaneous discharge of the ventricles (bundle branches and Purkinje fibers) is

(a) 60–80 beats per minute.
(b) 20–40 beats per minute.
(c) 40–60 beats per minute.
(d) 60–100 beats per minute.
(e) 80–120 beats per minute.

40. Which of the following statements regarding ECG monitoring is false?

(a) The ECG is a record of the electrical activity of the heart as sensed by electrodes on the body surface.
(b) In the absence of radial pulses, the ECG provides information as to whether or not the heart is still pumping.
(c) The isoelectric line on the ECG indicates an absence of net electrical activity.
(d) All of the above are false.
(e) None of the above is false.

41. On standard ECG paper, one small box horizontally represents

(a) 0.02 seconds of time.
(b) 0.04 seconds of time.
(c) 0.08 seconds of time.
(d) 0.12 seconds of time.
(e) 0.20 seconds of time.

42. On standard ECG paper, one large box horizontally represents

(a) 0.02 seconds of time.
(b) 0.04 seconds of time.
(c) 0.08 seconds of time.
(d) 0.12 seconds of time.
(e) 0.20 seconds of time.

43. Depolarization of the atria is ______ of a normal ECG tracing.

(a) indicated by the P wave
(b) indicated by the T wave
(c) indicated by the QRS complex
(d) indicated by the P–R interval
(e) hidden within the QRS

44. Repolarization of the atria is ______ of a normal ECG tracing.

(a) indicated by the P wave
(b) indicated by the T wave
(c) indicated by the QRS complex
(d) indicated by the P–R interval
(e) hidden within the QRS

45. Depolarization of the ventricles is _______ of a normal ECG tracing.

(a) indicated by the P wave
(b) indicated by the T wave
(c) indicated by the QRS complex
(d) indicated by the P–R interval
(e) hidden within the QRS

46. Repolarization of the ventricles is _______ of a normal ECG tracing.

(a) indicated by the P wave
(b) indicated by the T wave
(c) indicated by the QRS complex
(d) indicated by the P–R interval
(e) hidden within the QRS

47. The amount of time required for an atrial impulse to reach the ventricles is _______ of a normal ECG tracing.

(a) indicated by the P wave
(b) indicated by the T wave
(c) indicated by the QRS complex
(d) indicated by the P–R interval
(e) indicated by the S–T segment

48. The period of time when cells have been depolarized and have not yet returned to a polarized state is called the

(a) resting state.
(b) unpolarized state.
(c) polarizing period.
(d) depolarized state.
(e) refractory period.

49. During repolarization of depolarized cells, the period of time when no amount of stimulation can produce early depolarization is called the

(a) absolute repolarizing state.
(b) absolute unpolarized state.
(c) primary repolarizing period.
(d) secondary repolarizing period.
(e) absolute refractory period.

50. The period of time mentioned in question 49 is observed on the ECG during the

(a) QRS complex.
(b) first half of the T wave.
(c) second half of the T wave.
(d) Both answers (a) and (b).
(e) Both answers (a) and (c).

51. During repolarization of depolarized cells, the period when cells are close enough to repolarization that a sufficiently strong stimulus may produce premature depolarization is called the

(a) relative repolarizing state.
(b) partially unpolarized state.
(c) partially polarized period.
(d) susceptibility period.
(e) relative refractory period.

52. The period of time mentioned in question 51 is observed on the ECG during the

(a) QRS complex.
(b) first half of the T wave.
(c) second half of the T wave.
(d) Both answers (a) and (b).
(e) Both answers (a) and (c).

53. The normal duration of a P-R interval is

(a) 0.04–0.08 seconds.
(b) 0.04–0.12 seconds.
(c) 0.08–0.20 seconds.
(d) 0.12–0.20 seconds.
(e) 0.14–0.22 seconds.

54. The normal duration of a QRS complex is

(a) 0.04–0.08 seconds.
(b) 0.04–0.12 seconds.
(c) 0.08–0.20 seconds.
(d) 0.12–0.20 seconds.
(e) 0.14–0.22 seconds.

55. Your patient has an irregular pulse. On the ECG monitor you note regularly irregular groups of QRS complexes without any ectopy present. There are more P waves present than QRSs and the R–R intervals progressively decrease until a P wave without a QRS occurs. Your patient has a

(a) first-degree AV block.
(b) second-degree AV block, type I.
(c) second-degree AV block, type II.
(d) third-degree AV block.
(e) None of the above.

56. Your patient has a bradycardic pulse. On the ECG monitor you note regular R–R intervals, PRIs of 0.14 seconds, and QRSs of 0.14 seconds. There are twice as many P waves as QRSs and there is no ectopy present. Your patient has a

(a) first-degree AV block.
(b) second-degree AV block, type I.
(c) second-degree AV block, type II.
(d) third-degree AV block.
(e) None of the above.

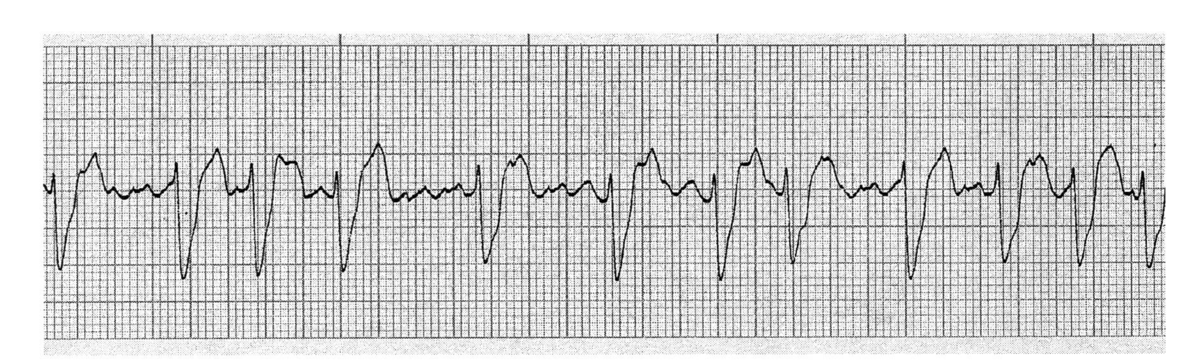

FIGURE 6-1

57. Which of the following descriptions best identifies the ECG strip in Figure 6-1?

(a) Sinus rhythm with aberrantly conducted PACs.
(b) Sinus rhythm with frequent unifocal PVCs.
(c) Atrial flutter with variable ventricular response.
(d) Atrial fibrillation with a bundle branch block.
(e) Atrial fibrillation with aberrantly conducted PACs.

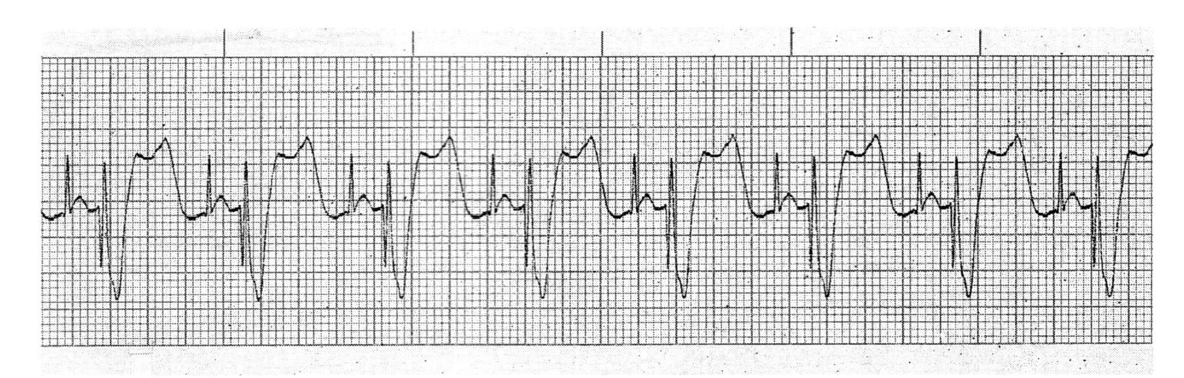

FIGURE 6-2

58. Which of the following descriptions best identifies the ECG strip in Figure 6-2?

(a) Sinus rhythm with bigeminal, unifocal PVCs.
(b) Sinus rhythm with bigeminal, multifocal PVCs.
(c) AV sequential pacemaker rhythm.
(d) Pacemaker rhythm with underlying sinus tachycardia.
(e) Junctional rhythm with bigeminal, unifocal PVCs.

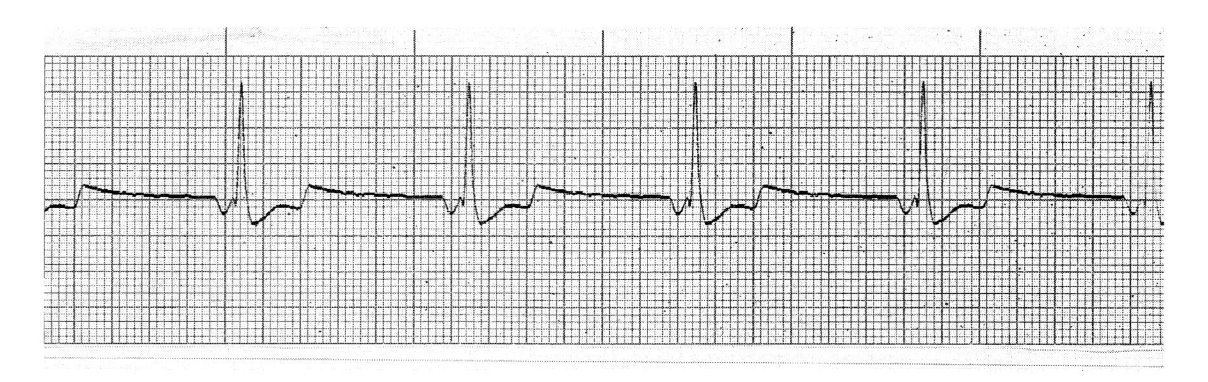

FIGURE 6-3

59. Which of the following descriptions best identifies the ECG strip in Figure 6-3?

(a) Junctional rhythm.
(b) Sinus bradycardia.
(c) Accelerated idioventricular rhythm.
(d) Wandering atrial pacemaker.
(e) Atrial bradycardia.

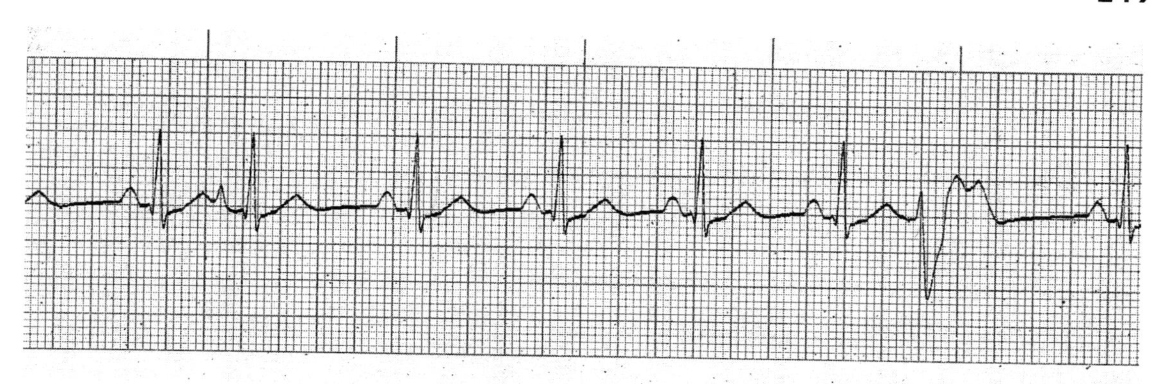

FIGURE 6-4

60. Which of the following descriptions best identifies the ECG strip in Figure 6-4?

(a) Sinus rhythm with one PVC.
(b) Sinus rhythm with multifocal PVCs.
(c) Sinus rhythm with one PAC.
(d) Sinus rhythm with multifocal PACs.
(e) Sinus rhythm with one PAC and one PVC.

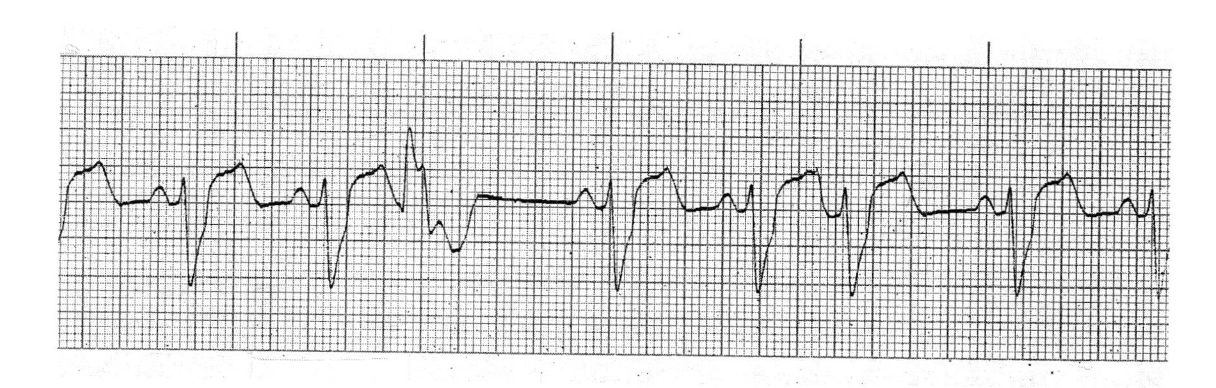

FIGURE 6-5

61. Which of the following descriptions best identifies the ECG strip in Figure 6-5?

(a) Sinus dysrhythmia with one PVC.
(b) Sinus dysrhythmia with multifocal PVCs.
(c) Sinus dysrhythmia with one PAC.
(d) Wenkebach, second-degree AV block with one PVC.
(e) Sinus rhythm with one PAC and one PVC.

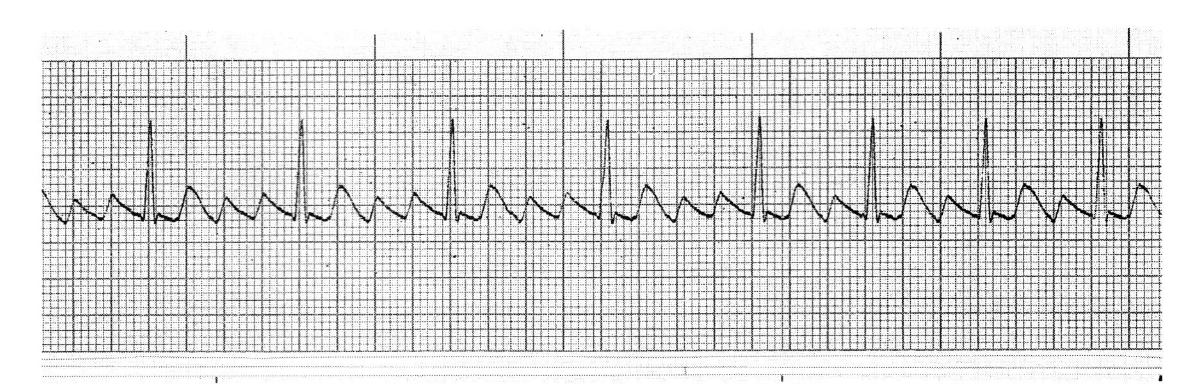

FIGURE 6-6

62. Which of the following descriptions best identifies the ECG strip in Figure 6-6?

(a) Atrial flutter with variable ventricular response.
(b) Atrial fibrillation.
(c) Atrial flutter with 3:1 and 2:1 ventricular response.
(d) Classic second-degree AV block with variable ventricular response.
(e) Atrial fibrillation with uncontrolled ventricular response.

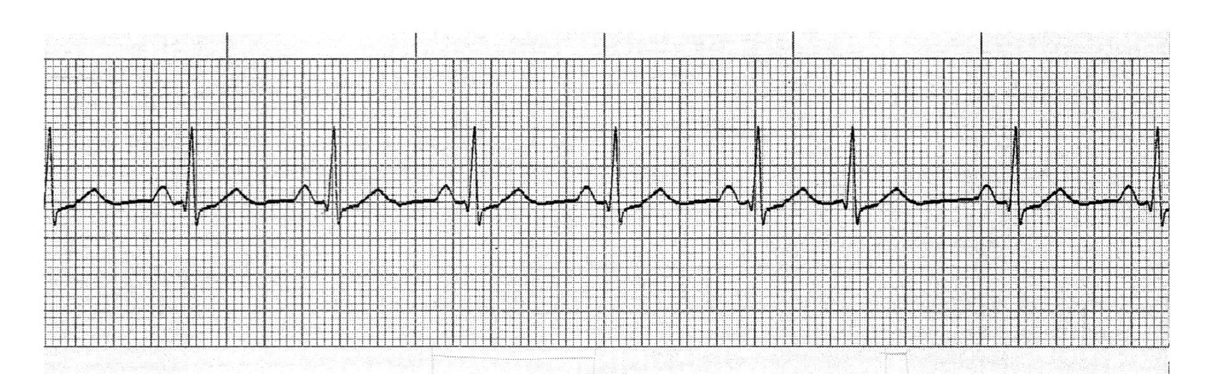

FIGURE 6-7

63. Which of the following descriptions best identifies the ECG strip in Figure 6-7?

(a) Normal sinus rhythm.
(b) Sinus rhythm with one PAC.
(c) Sinus rhythm with one PJC.
(d) Sinus rhythm with one PVC.
(e) Sinus dysrhythmia.

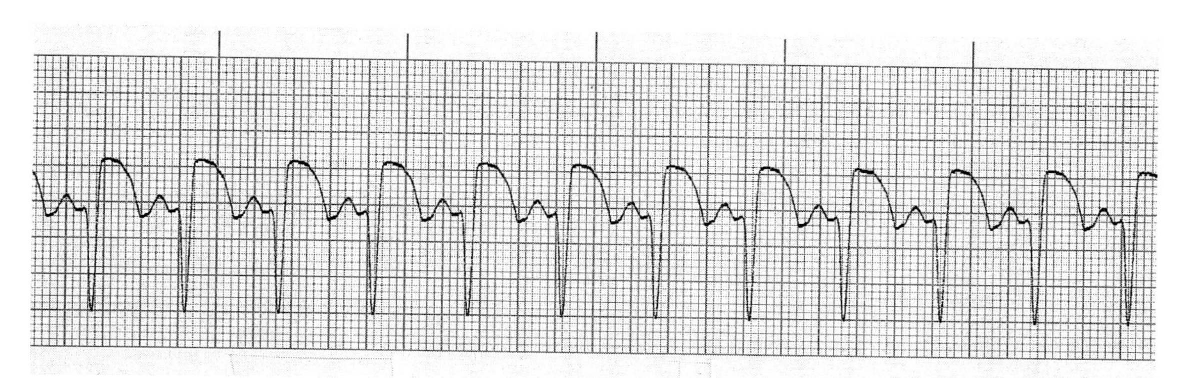

FIGURE 6-8

64. Which of the following descriptions best identifies the ECG strip in Figure 6-8?

(a) Sinus tachycardia with an elevated S–T segment.
(b) Atrial tachycardia with an elevated S–T segment.
(c) Junctional tachycardia with an elevated S–T segment.
(d) Ventricular tachycardia.
(e) Runaway pacemaker rhythm.

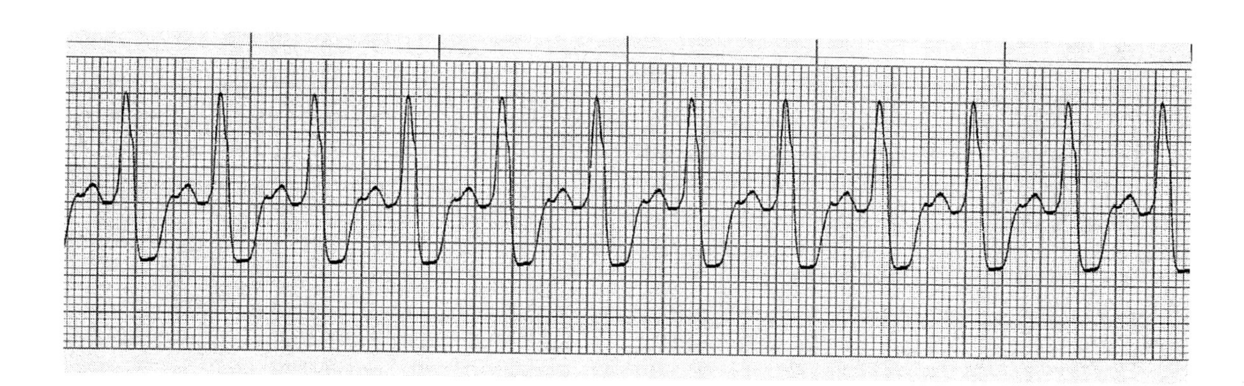

FIGURE 6-9

65. Which of the following descriptions best identifies the ECG strip in Figure 6-9?

(a) Sinus tachycardia with a BBB and an inverted T wave.
(b) Atrial tachycardia with a BBB and an inverted T wave.
(c) Junctional tachycardia with a BBB and an inverted T wave.
(d) Ventricular tachycardia.
(e) Runaway pacemaker rhythm.

In each of the following four scenarios you are out of radio contact, without access to a phone, and have standing orders to follow ACLS protocols when unable to communicate with your base physician.

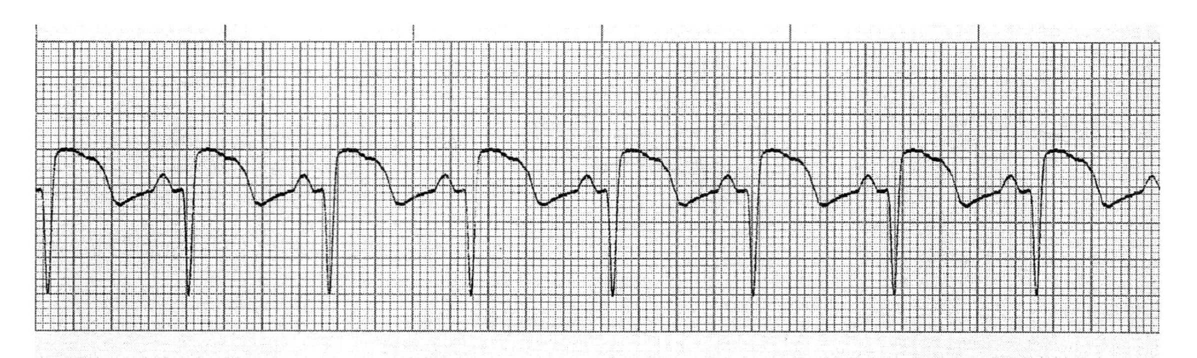

FIGURE 6-10

66. The patient whose ECG strip is shown in Figure 6-10 is a 45-year-old female; AAOx3 (alert and oriented to person, place, and time); with cool, pale, and diaphoretic skin. Her chief complaint is substernal chest pain radiating to her left shoulder for the past two hours. P: 88; B/P: 110/70; R: 24. You administer oxygen and start an IV lifeline. Which of the following would you elect to perform?

(a) Administration of sublingual nitroglycerine.
(b) Administration of IV morphine sulfate (if NTG unsuccessful).
(c) Administration of a bolus of IV lidocaine, followed by a maintenance infusion.
(d) Answers (a) and (b) only.
(e) Answers (a), (b), and (c).

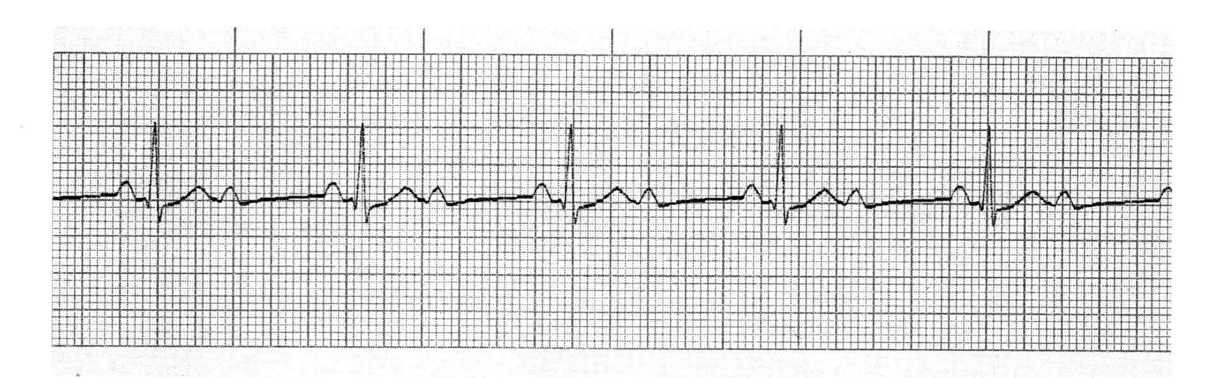

FIGURE 6-11

67. The patient whose ECG strip is shown in Figure 6-11 is a 65-year-old male; AAOx3; with warm, pale, and dry skin. His only complaint is of feeling "weaker than usual today." P: 50; B/P: 110/70; R: 20 You administer oxygen and start an IV lifeline. Which of the following would you elect to perform?

(a) Administration of atropine in 0.5-1.0 mg increments, every 3 to 5 minutes, to a maximum administration of 0.04 mg/kg.
(b) Administration of an isoproterenol IV infusion, titrated to effect.
(c) Administration of an IV lidocaine bolus of 1.5 mg/kg, followed by a maintenance infusion.
(d) Answers (a) and (b) only.
(e) None of the above; transportation only.

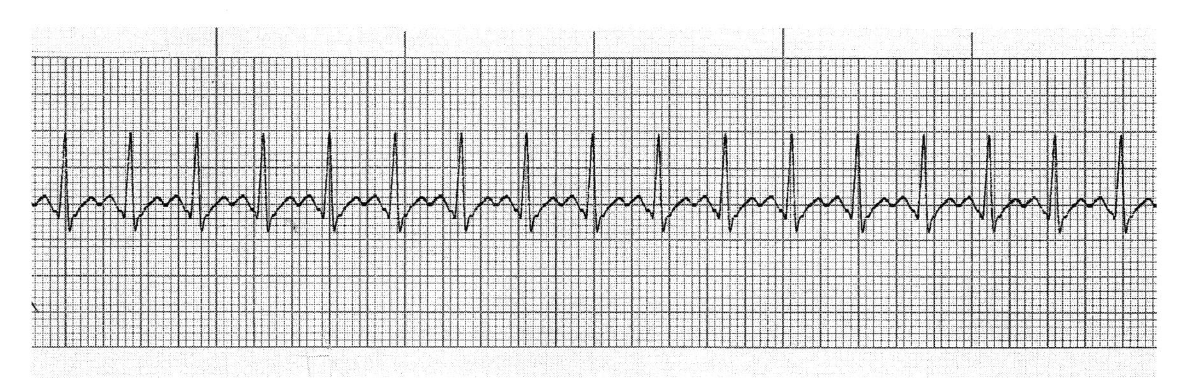

FIGURE 6-12

68. The patient whose ECG strip is shown in Figure 6-12 is a 70-year-old female; anxious, fearful, and somewhat disoriented; with cool, pale, and diaphoretic skin. She is complaining of chest pain, nausea, and shortness of breath. Her husband denies any history of trauma. Her breath sounds are clear and equal bilaterally. P: weak and thready at 170; B/P: 80/P; R: 30 and shallow. You administer oxygen and start an IV lifeline. Which of the following would you elect to perform?

(a) Administration of 5 mg of IV diazepam, followed by synchronized cardioversion at 100 ws.
(b) Administration of a fluid challenge of 1000 cc NS.
(c) Administration of an IV dopamine infusion, titrated to a blood pressure of 100 systolic.
(d) Administration of sublingual nitroglycerine, followed by IV morphine sulfate if 3 NTG are unsuccessful.
(e) None of the above; transportation only.

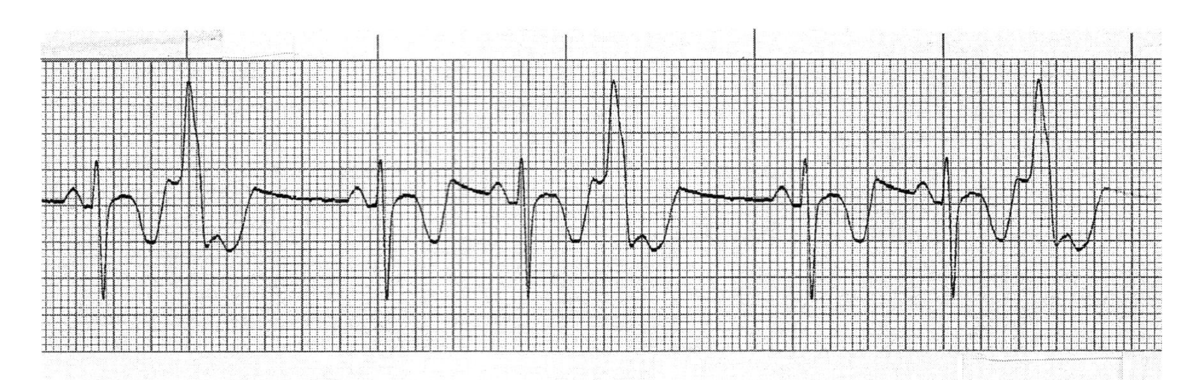

FIGURE 6-13

69. The patient whose ECG strip is shown in Figure 6-13 is a 47-year-old male; AAOx3; with warm, pale, and dry skin. He is complaining of nausea without vomiting for approximately two hours. He denies shortness of breath and chest pain. P: 50; B/P: 100/P; R: 24. You administer oxygen and start an IV lifeline. Which of the following would you elect to perform?

(a) Administration of 1.5 mg/kg IV lidocaine, followed every 5 to 10 minutes by a 0.5-0.75 mg/kg bolus, to a maximum dose of 3 mg/kg.
(b) Administration of 0.5-1.0 mg IV atropine, repeated every 3 to 5 minutes as needed, to a maximum of 0.04 mg.
(c) Administration of an isoproterenol infusion (if the atropine is unsuccessful) titrated to a heart rate of 80.
(d) Answers (b) and (c) only.
(e) None of the above; transport only.

70. The complaint of substernal chest pressure that does not radiate to the shoulders, arms, neck, or jaw

(a) is usually "stable angina" only.
(b) may be "stable" or "unstable" angina.
(c) is always to be considered "unstable" or preinfarction angina.
(d) is rarely related to cardiac ischemia.
(e) is never related to cardiac ischemia (indicates esophageal disease only).

71. Onset of chest pain during physical stress, rapidly relieved by rest,

(a) is usually "stable angina" only.
(b) may be "stable" or "unstable" angina.

(c) is always to be considered "unstable" or preinfarction angina.
(d) is rarely related to cardiac ischemia.
(e) is never related to cardiac ischemia (indicates esophageal disease only).

72. Onset of chest pain during emotional stress
(a) is usually "stable angina" only.
(b) may be "stable" or "unstable" angina.
(c) is always to be considered "unstable" or preinfarction angina.
(d) is rarely related to cardiac ischemia.
(e) is never related to cardiac ischemia (indicates esophageal disease only).

73. Onset of chest pain during rest
(a) is usually "stable angina" only.
(b) may be "stable" or "unstable" angina.
(c) is always to be considered "unstable" or preinfarction angina.
(d) is rarely related to cardiac ischemia.
(e) is never related to cardiac ischemia (indicates esophageal disease only).

74. A myocardial infarction (MI) has occurred when
(a) the patient begins to complain of radiating angina.
(b) myocardial tissue begins to become ischemic.
(c) myocardial tissue begins to become necrotic.
(d) Any of the above.
(e) None of the above.

75. The most common cause of acute myocardial infarction (AMI) in the adult is
(a) cardiac dysrhythmias.
(b) coronary artery spasm.
(c) acute volume overload.
(d) acute respiratory failure.
(e) coronary thrombosis.

76. The most common cause of death from AMI is
(a) cardiac dysrhythmias.
(b) coronary artery spasm.
(c) acute volume overload.
(d) acute respiratory failure.
(e) coronary thrombosis.

77. The location and size of an infarct is dependent on the site of coronary vessel obstruction. The majority of infarcts, however, involve the

(a) right ventricle.
(b) right atrium.
(c) left atrium.
(d) left ventricle.
(e) aorta.

78. An infarction of only a partial thickness of the heart wall is called

(a) a subendocardial infarction.
(b) a submural infarction.
(c) a second-degree infarction.
(d) a transendocardial infarction.
(e) a transmural infarction.

79. An infarction of the full thickness of the heart wall is called

(a) a subendocardial infarction.
(b) a submural infarction.
(c) a third-degree infarction.
(d) a transendocardial infarction.
(e) a transmural infarction.

80. The most common complication in the first few hours of AMI is

(a) nausea and vomiting.
(b) congestive heart failure.
(c) hypertension.
(d) hypotension.
(e) dysrhythmias.

81. Prehospital management of the uncomplicated MI includes all of the following, except

(a) high-flow oxygen and IV initiation.
(b) assisting the patient to self-administer any of her or his prescription cardiac medications.
(c) ECG and vital sign monitoring.
(d) administration of NTG and/or MS.
(e) administration of diazepam for extreme agitation or apprehension.

82. Left ventricular failure is caused by various types of heart disease, including

(a) AMI or valvular disease.
(b) chronic hypertension.
(c) dysrhythmias.
(d) Answers (a) and (b) only.
(e) Answers (a), (b), and (c).

83. Which of the following statements regarding left ventricular failure is false?

(a) As the left ventricle fails, left atrial pressure rises and is transmitted to pulmonary veins and capillaries.
(b) When pulmonary capillary pressure becomes too high, the capillaries burst, resulting in the hemorrhage that produces pulmonary edema.
(c) Progressive pulmonary congestion will lead to death from hypoxia unless intervention occurs.
(d) Since AMI is a common cause of left ventricular failure, all patients in pulmonary edema must be presumed to also be having AMI.
(e) None of the above is false.

84. Lung sounds associated with pulmonary edema caused by left ventricular failure include

(a) rales.
(b) rhonchi.
(c) wheezes.
(d) All of the above.
(e) Answers (a) and (b) only.

85. Fluid in the larger airways indicates a more severe degree of pulmonary edema and produces

(a) rales.
(b) rhonchi.
(c) wheezes.
(d) stridor.
(e) snoring.

86. The bronchoconstriction that occurs with pulmonary edema produces

(a) rales.
(b) rhonchi.
(c) wheezes.
(d) stridor.
(e) snoring.

87. Fluid in the alveoli produces

(a) rales.
(b) rhonchi.
(c) wheezes.
(d) stridor.
(e) snoring.

88. Jugular vein distention

(a) is a direct result of left ventricular failure.
(b) is associated only with right ventricular failure.
(c) is associated only with pericardial tamponade and tension pneumothorax.
(d) may be present if back-pressure from left ventricular failure reflects all the way through the right heart to the venous system.
(e) Both answers (a) and (d).

89. During left ventricular failure there is usually an intense sympathetic discharge in attempt to help the body compensate. This produces the characteristic vital signs associated with left ventricular failure:

(a) hypotension, tachycardia, and labored tachypnea.
(b) hypotension, bradycardia, and labored tachypnea.
(c) elevated blood pressure, tachycardia, and labored tachypnea.
(d) elevated blood pressure, bradycardia, and labored tachypnea.
(e) elevated blood pressure, bradycardia, and Kusmaul's respirations.

90. Which of the following statements regarding prehospital management of acute left ventricular failure with pulmonary edema is false?

(a) Administer high-flow oxygen, using positive pressure assistance if the patient can cooperate or has altered mentation.
(b) Rotating tourniquets may be applied to enhance venous pooling, especially if you are unable to establish an IV.
(c) Establish an IV of D_5W on a microdrip infusion set.
(d) If cerebral hypoxia is present (as evidenced by agitation or combativeness), the patient must be placed supine to facilitate cerebral perfusion.
(e) Medication administration includes NTG, MS, furosemide, and aminophylline.

91. The most common cause of right ventricular failure is

(a) left ventricular failure.
(b) chronic hypertension.
(c) COPD.
(d) pulmonary embolus.
(e) infarct of right atrium or ventricle.

92. Venous congestion from right ventricular failure produces

(a) organ engorgement (tender right upper abdominal quadrant from liver engorgement).
(b) peripheral edema.
(c) fluid accumulation in serous cavities (ascites, pleural effusions, pericardial effusion).
(d) All of the above.
(e) Answers (b) and (c) only.

93. Which of the following statements regarding prehospital management of right ventricular failure is true?

(a) Administer nasal cannula or venturi mask oxygen at low flow rates (right ventricular failure is related to COPD).
(b) Establish IV access of NS or LR with a large bore catheter (relative dehydration is present as fluid has shifted to serous cavities).
(c) Right ventricular failure is usually not a medical emergency unless accompanied by left ventricular failure and pulmonary edema.
(d) Both answers (a) and (b) are true.
(e) None of the above is true.

94. Cardiogenic shock is defined as

(a) shock that persists after correction of existing dysrhythmias.
(b) shock that persists after correction of hypovolemia.
(c) shock that occurs as compensatory mechanisms are exhausted.
(d) the most extreme form of pump failure.
(e) All of the above.

95. Which of the following statements regarding cardiogenic shock is false?

(a) Cardiogenic shock occurs when left ventricular function is so compromised that the heart cannot meet the metabolic needs of the body.
(b) Cardiogenic shock is usually due to extensive MI or diffuse ischemia involving 40 percent or more of the left ventricle.
(c) The mortality rate for cardiogenic shock is 80 to 90 percent despite any kind of treatment.
(d) Any of the signs and symptoms of AMI may accompany cardiogenic shock.
(e) None of the above is false.

96. Management of cardiogenic shock includes all of the following, except

(a) careful and prolonged stabilization on scene prior to risking transport.
(b) securing an open airway and administering high-flow oxygen.
(c) supine positioning and IV access.
(d) ECG monitoring with appropriate pharmacologic dysrhythmia intervention.
(e) IV infusion of dopamine.

97. The definition of sudden death requires that death occur

(a) without any warning signs and symptoms.
(b) within two minutes of the onset of signs and symptoms.
(c) within one hour of the onset of signs and symptoms.
(d) within two hours of the onset of signs and symptoms.
(e) within the first 24 hours following the onset of signs and symptoms.

98. Which of the following statements regarding the management of cardiac arrest is false?

(a) Ensuring correct performance of BLS remains an essential responsibility of the ACLS provider.
(b) Rapid defibrillation of the patient in V-fib provides the best chance for successful resuscitation.

(c) The most sophisticated airway management (endotracheal intubation) is not always needed immediately.
(d) It is more difficult to abolish V-fib when it is the primary cause of the arrest (as opposed to V-fib that occurs secondary to the cause of the arrest).
(e) External pacing may be used for bradycardias, asystole, or immediately postdefibrillation of V-fib.

99. Which of the following statements regarding pulseless electrical activity (PEA) [once called electromechanical dissociation (EMD)] is false?

(a) PEA/EMD is defined as an organized rhythm without a pulse, and as such, has a better prognosis for resuscitation than that of asystole.
(b) PEA/EMD may be caused by massive myocardial damage or cardiac rupture.
(c) PEA/EMD may be caused by hypovolemia, cardiac tamponade, or acute pulmonary embolism.
(d) Consider use of MAST/PASG and IV fluid challenge for treatment of PEA/EMD.
(e) PEA/EMD requires earlier consideration of transport than other medical cardiac arrest situations.

100. The most common site for an abdominal aneurysm is

(a) the ascending aorta.
(b) the aortic arch.
(c) the descending aorta, where it passes through the diaphragm.
(d) the abdominal aorta, below the renal arteries and above the common iliac bifurcation.
(e) the abdominal aorta, below the iliac arteries and above the renal artery bifurcation.

101. The most common site for a thoracic aneurysm is

(a) the ascending aorta.
(b) the aortic arch.
(c) the descending aorta, where it passes through the diaphragm.
(d) the abdominal aorta, below the renal arteries and above the common iliac bifurcation.
(e) the abdominal aorta, below the iliac arteries and above the renal artery bifurcation.

102. In addition to complaints of abdominal, back, and/or flank pain, signs and symptoms of abdominal aortic aneurysms include all of the following, except

(a) hypotension.
(b) bilateral cramping of the lower extremities.
(c) the urge to defecate.
(d) a pulsating abdominal mass.
(e) decreased femoral pulses.

103. Which of the following statements regarding aortic aneurysm is false?

(a) Abdominal aneurysm is ten times more common in men than in women.
(b) Abdominal aneurysm is most prevalent in ages 60 to 70.
(c) Once begun, an aneurysm may extend to involve all of the thoracic and abdominal aorta, as well as aortic tributaries, coronary arteries, the aortic valve, carotid and subclavian arteries.
(d) Aneurysms do not rupture without a precipitating strain or exertion.
(e) The most common cause of dissecting aortic aneurysm is hypertension.

104. Signs and symptoms of dissecting aortic aneurysm include all of the following, except

(a) intermittent, nonradiating chest pain, characteristically described as "sharp, stabbing, twinges."
(b) elevated blood pressure.
(c) syncope or CVA.
(d) absent or reduced pulses.
(e) pericardial tamponade.

105. Malignant hypertension (also called hypertensive emergency or hypertensive crisis) occurs in less than one percent of patients with hypertension. Signs and symptoms of malignant hypertension include

(a) restlessness, confusion, or somnolence (prolonged drowsiness or fatigue).
(b) blurred vision and/or headache.
(c) nausea and vomiting.
(d) Answers (a) and (b) only.
(e) Answers (a), (b), and (c).

106. Malignant hypertension is marked by a rapid

(a) decrease in diastolic blood pressure (with widening pulse pressure).
(b) increase in diastolic blood pressure (usually equal to or greater than 130 mm Hg).
(c) decrease in systolic blood pressure with increased diastolic blood pressure (a narrowing pulse pressure).
(d) Both answers (b) and (c).
(e) None of the above.

107. Hypertension-related emergencies include all of the following, except

(a) pulmonary edema from left ventricular failure.
(b) deep vein thrombophlebitis.
(c) dissecting aortic aneurysm.
(d) toxemia of pregnancy.
(e) cerebral vascular accident.

108. Your patient is regularly taking the prescription medication Lanoxin. This medication indicates a medical history that may include all of the following, except

(a) congestive heart failure.
(b) atrial fibrillation.
(c) atrial flutter.
(d) chronic ventricular ectopy.
(e) paroxysmal supraventricular tachycardias.

109. Your patient is regularly taking the prescription medication Inderal. This medication indicates a medical history that may include all of the following, except

(a) congestive heart failure or heart blocks.
(b) atrial dysrhythmias.
(c) ventricular dysrhythmias.
(d) angina pectoris.
(e) hypertension.

110. Which of the following statements regarding the precordial thump is false?

(a) The precordial thump is sometimes effective in causing ventricular depolarization and resumption of organized rhythm.
(b) The precordial thump is recommended for the witnessed onset of ventricular tachycardia.
(c) Defibrillation of the pediatric myocardium may result in tissue injury. Therefore, a precordial thump is attempted prior to defibrillation in all pediatric cardiac arrests.
(d) The precordial thump is recommended in complete AV block with ventricular asystole when a pulse can be produced with rhythmic thumps.
(e) The precordial thump is recommended for the witnessed onset of ventricular fibrillation.

111. The precordial thump is delivered to the mid-sternum with a fist (thumb up) from a height of

(a) 2 to 4 inches.
(b) 4 to 8 inches.
(c) 8 to 10 inches.
(d) 10 to 12 inches.
(e) 14 to 20 inches.

112. Which of the following statements regarding defibrillation is false?

(a) Successful conversion after defibrillation is less likely in the presence of hypoxia, acidosis, hypothermia, electrolyte imbalance, or drug toxicity.
(b) A larger (obese) adult will always require a higher energy setting for defibrillation than an emaciated or normal-weight adult.
(c) Transthoracic resistance decreases with repeated countershocks, allowing more energy to be delivered to the heart at the same energy setting.
(d) Larger paddles are thought to be more effective and cause less myocardial damage in the adult patient.
(e) The creams or pastes used must be those made specifically for defibrillation, not for ECG monitoring.

113. Which of the following statements regarding transthoracic paddle placement is false?

(a) One paddle is positioned just below the right clavicle, lateral to the upper sternum (not ON the sternum).
(b) One paddle is positioned lateral to the left nipple in the anterior axillary line (over the apex of the heart).
(c) The paddle marked "apex" is the positive electrode.
(d) The paddle marked "sternum" is the negative electrode.
(e) Reversing the polarity (by reversing the placement of the "apex" and "sternum" paddles) will produce ineffective defibrillation and increased myocardial damage.

114. The energy recommendation for the initial defibrillation of an adult in ventricular fibrillation is

(a) 10 joules per kilogram.
(b) 4 joules per kilogram.
(c) 2 joules per kilogram.
(d) 360 joules.
(e) 200 to 300 joules.

115. The energy recommendation for the second defibrillation of an adult in ventricular fibrillation is

(a) 10 joules per kilogram.
(b) 4 joules per kilogram.
(c) 2 joules per kilogram.
(d) 360 joules.
(e) 200 to 300 joules.

116. The energy recommendation for the third defibrillation of an adult in ventricular fibrillation is

(a) 10 joules per kilogram.
(b) 4 joules per kilogram.
(c) 2 joules per kilogram.
(d) 360 joules.
(e) 200 to 300 joules.

117. The energy recommendation for the initial defibrillation of the pediatric patient in ventricular fibrillation is

(a) 10 joules per kilogram.
(b) 4 joules per kilogram.
(c) 2 joules per kilogram.
(d) 360 joules.
(e) 200 to 300 joules.

118. The energy recommendation for the second defibrillation of the pediatric patient in ventricular fibrillation is

(a) 10 joules per kilogram.
(b) 4 joules per kilogram.
(c) 2 joules per kilogram.
(d) 360 joules.
(e) 200 to 300 joules.

119. Emergency synchronized cardioversion is indicated for treatment of a patient decompensating secondary to any of the following, except

(a) perfusing ventricular tachycardia.
(b) nonperfusing ventricular tachycardia.
(c) paroxysmal supraventricular tachycardias.
(d) rapid atrial fibrillation.
(e) 2:1 atrial flutter.

120. The synchronizing circuit in a defibrillator allows delivery of a countershock programmed to coincide with the occurrence of the

(a) P wave.
(b) Q wave of the QRS.
(c) R wave of the QRS.
(d) S wave of the QRS.
(e) T wave.

121. The energy setting for the initial synchronized countershock when one is treating unstable PSVT is

(a) 25 joules.
(b) 200 joules.
(c) 50 joules.
(d) 100 joules.
(e) 360 joules.

122. The energy setting for the initial synchronized countershock when one is treating unstable ventricular tachycardia is

(a) 25 joules.
(b) 200 joules.
(c) 50 joules.
(d) 100 joules.
(e) 360 joules.

123. Carotid sinus massage is used to convert a paroxysmal supraventricular tachycardia into a sinus rhythm by stimulation of

(a) the baroreceptors in the carotid artery, resulting in increased vagal tone.
(b) the baroreceptors in the carotid artery, resulting in decreased vagal tone.
(c) the blood brain barrier, resulting in increased vagal tone.
(d) the blood brain barrier, resulting in decreased vagal tone.
(e) the sympathetic nervous system, resulting in a slowing of the tachycardia.

124. Which of the following statements regarding carotid artery massage is false?

(a) Never massage both carotids simultaneously.
(b) When the carotid pulses are unequal, massage the side with the stronger pulse.
(c) Massage is contraindicated if carotid bruits are present.
(d) Massage is contraindicated if the patient has a history of CVA.
(e) Massage should continue for no longer than 15 to 20 seconds.

125. Complications of correctly performed carotid sinus massage include

(a) production of dysrhythmias (PVCs, V-tach or V-fib, asystole).
(b) interference with cerebral circulation resulting in syncope, seizure, or CVA.
(c) increased parasympathetic tone resulting in bradycardias, nausea, or vomiting.
(d) Correctly performed carotid sinus massage does not cause complications.
(e) Answers (a), (b), and (c).

126. Transcutaneous cardiac pacing (TCP) is recommended for symptomatic

(a) bradycardia.
(b) high-degree AV blocks.
(c) atrial fibrillation with slow ventricular response.
(d) Answers (a) and (b) only.
(e) Answers (a), (b), and (c).

The following test section consists of ECG Figures 6-14 through 6-65. Because the multiple-choice-answer format does not lend itself well to ECG strip identification, the authors have developed an exercise called "Name That Strip."

Each ECG strip is of six seconds in duration. You do not have a patient present, so you need not provide a pulse rate, but will need to indicate the rhythm's rate at times.

Carefully observe each strip for
- Rate
- Rhythm/regularity
- P waves
- P–R intervals
- QRS complexes
- S–T segments
- Ectopic beats

Describe each strip as completely and concisely as possible.

For example; Figure 6-14 can most completely be described as a complete (or third-degree) AV block with an accelerated ventricular escape rhythm at a rate of 60/min.

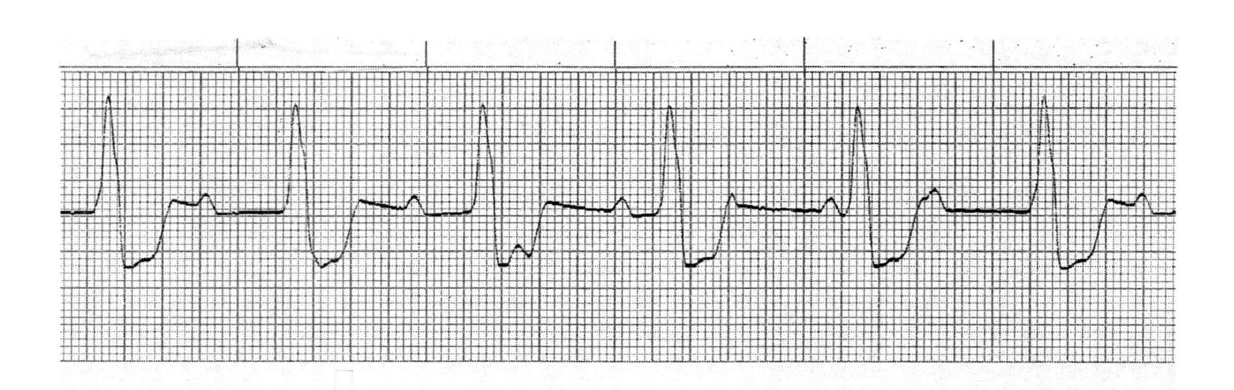

FIGURE 6-14

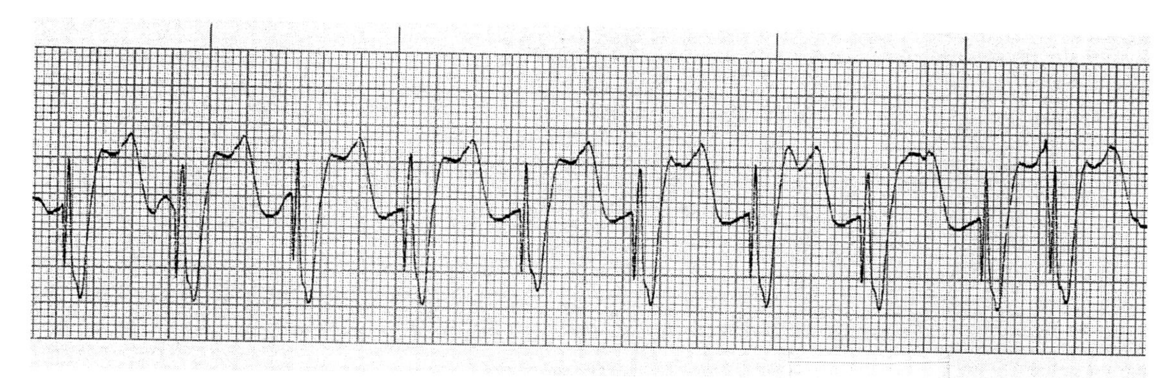

FIGURE 6-15

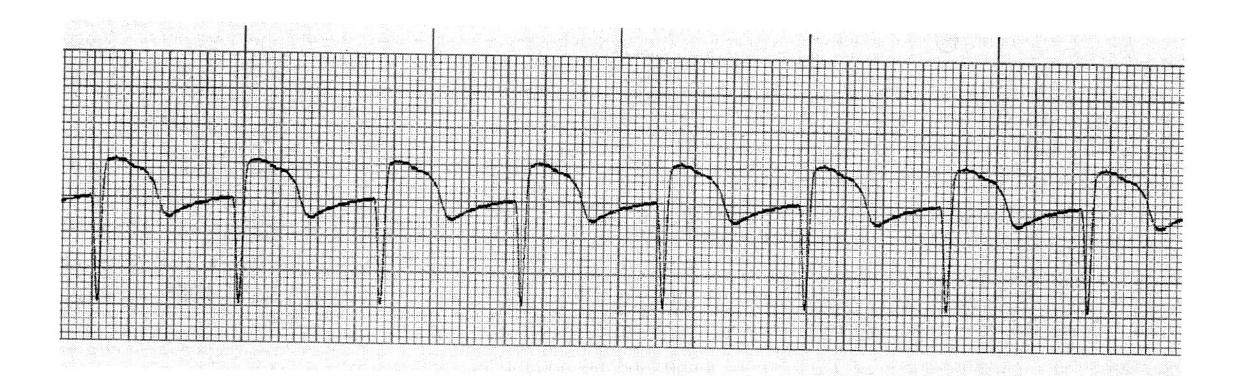

FIGURE 6-16

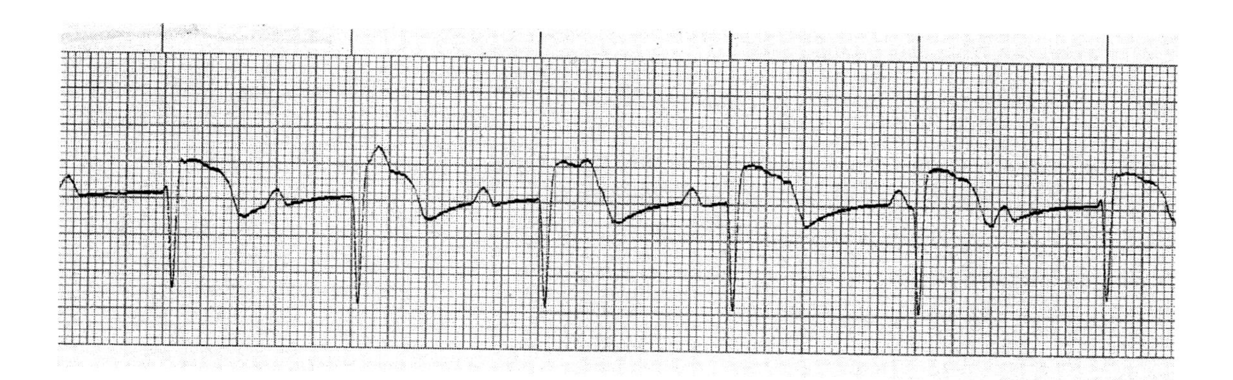

FIGURE 6-17

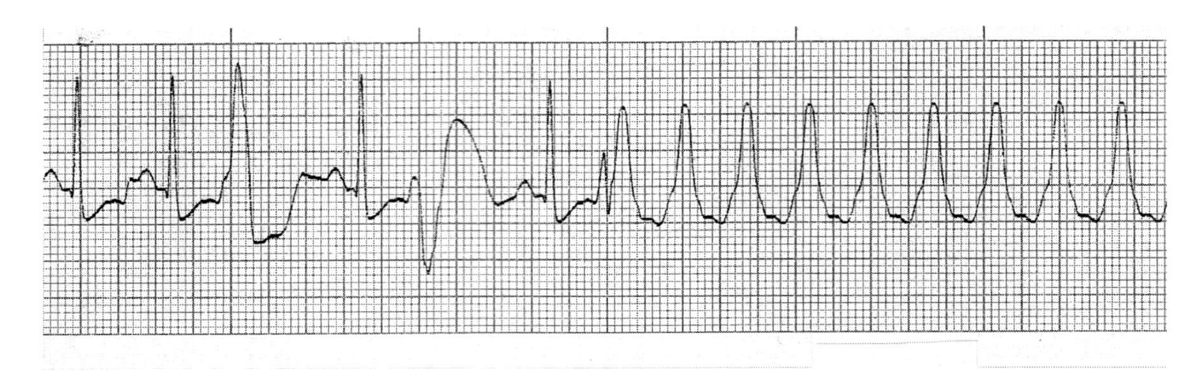

FIGURE 6-18

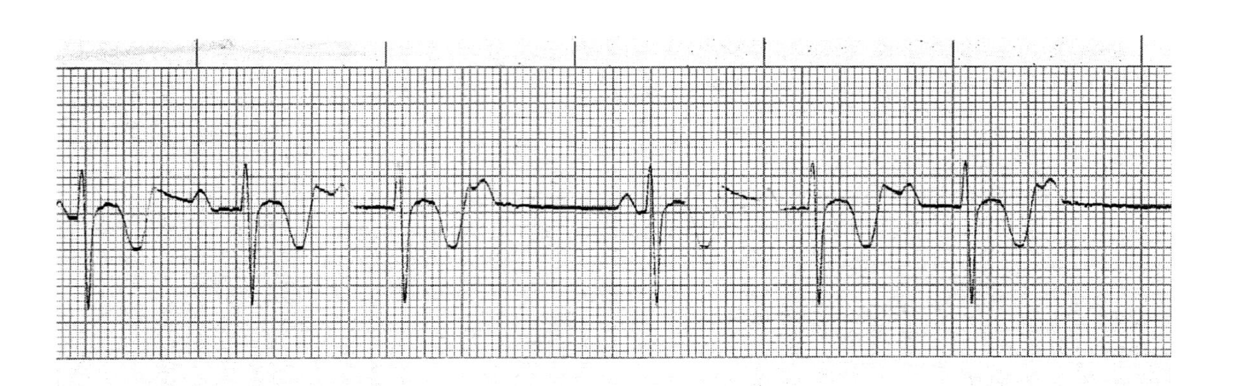

FIGURE 6-19

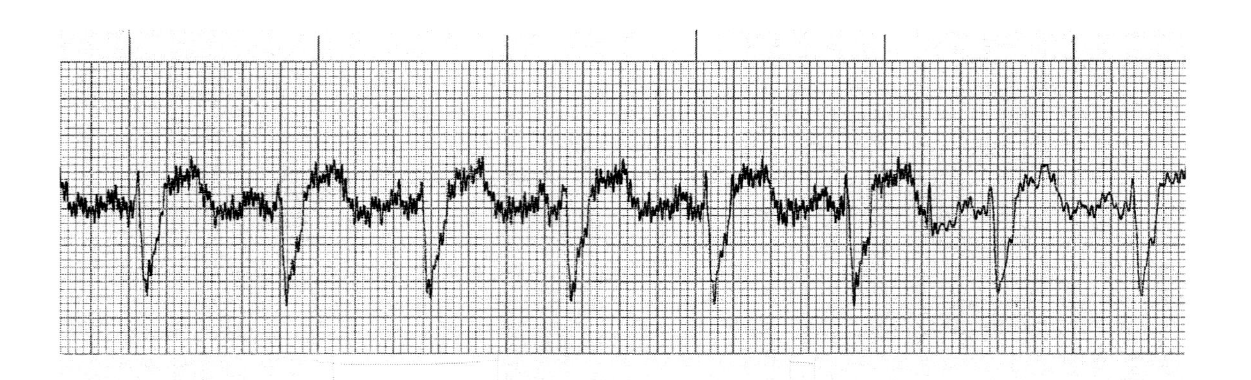

FIGURE 6-20

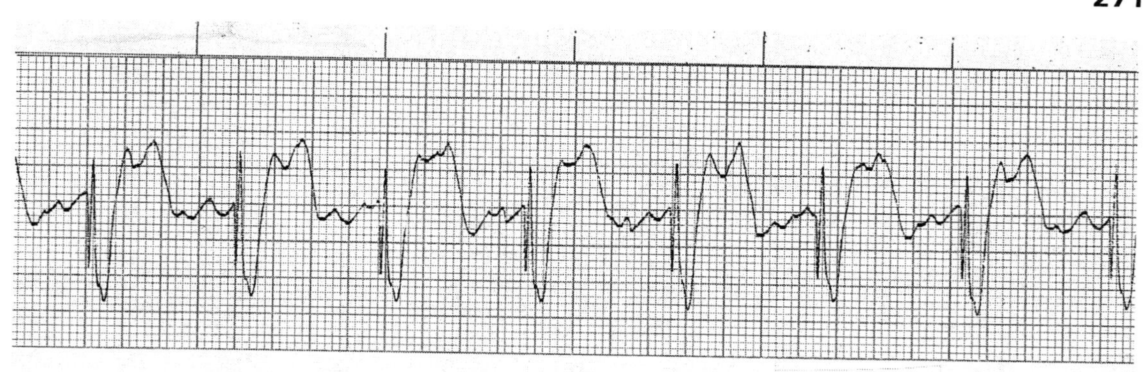

FIGURE 6-21

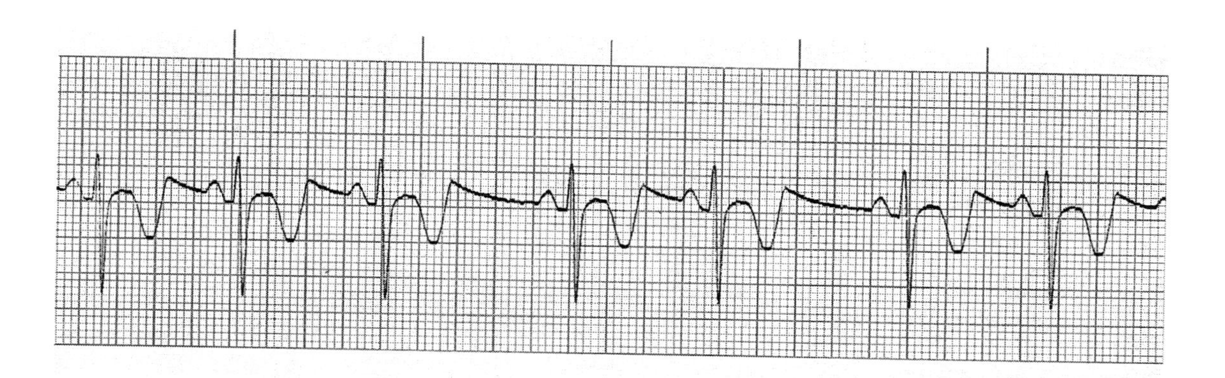

FIGURE 6-22

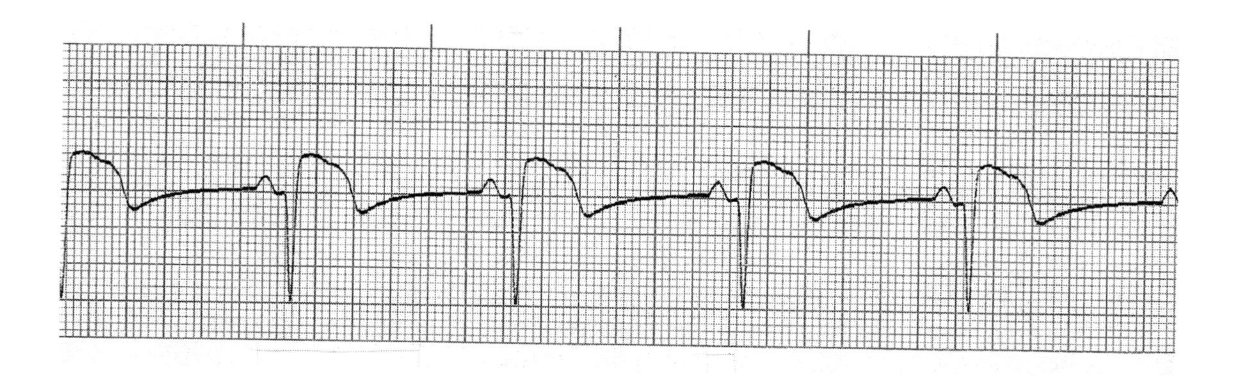

FIGURE 6-23

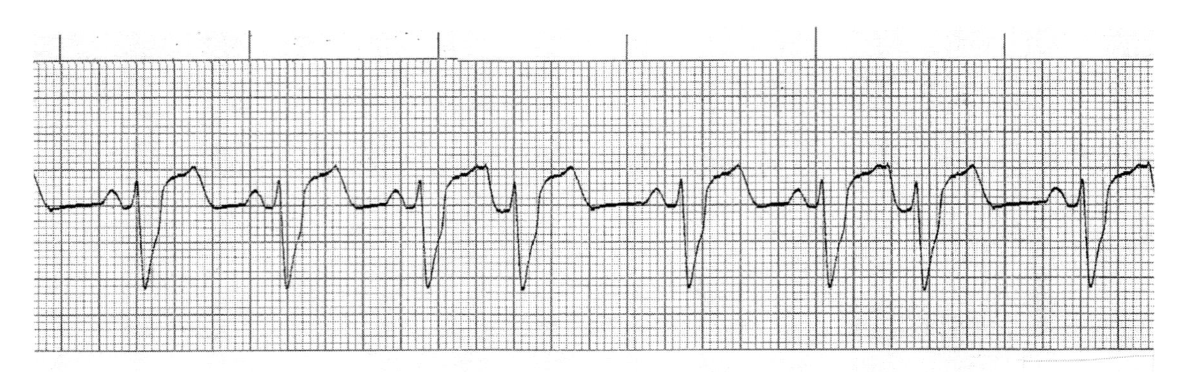

FIGURE 6-24

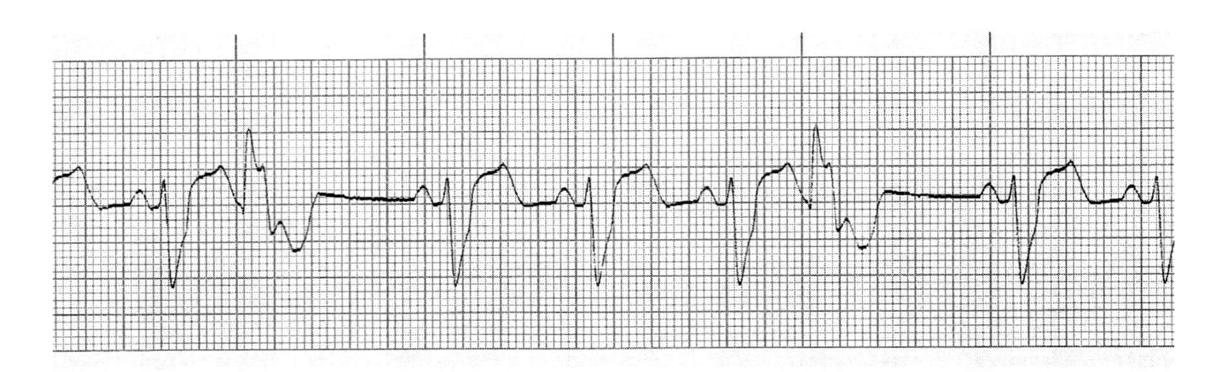

FIGURE 6-25

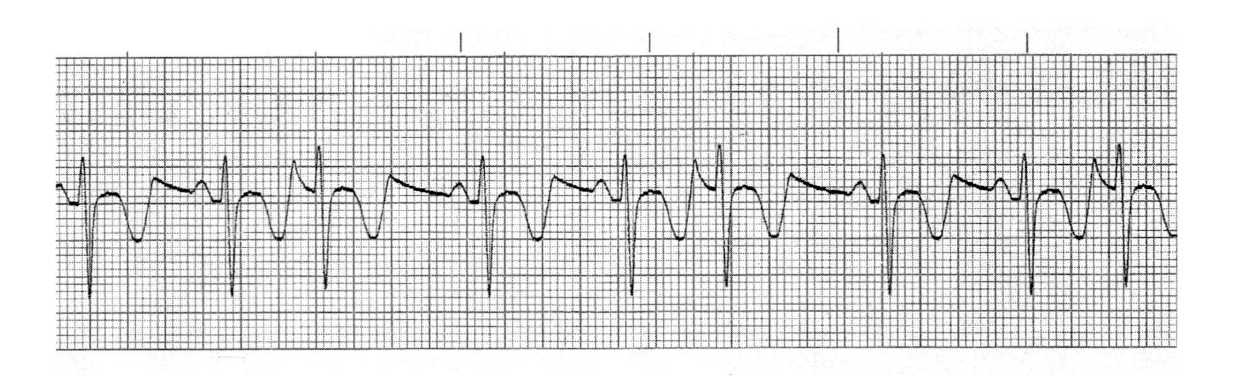

FIGURE 6-26

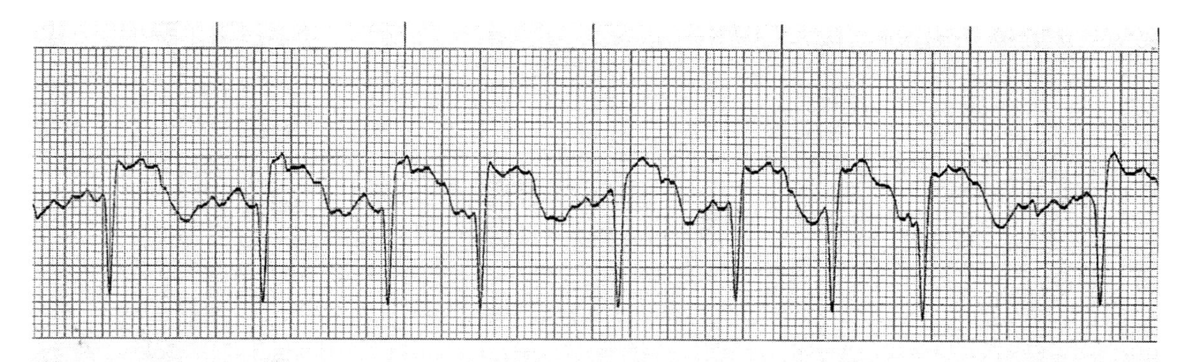

FIGURE 6-27

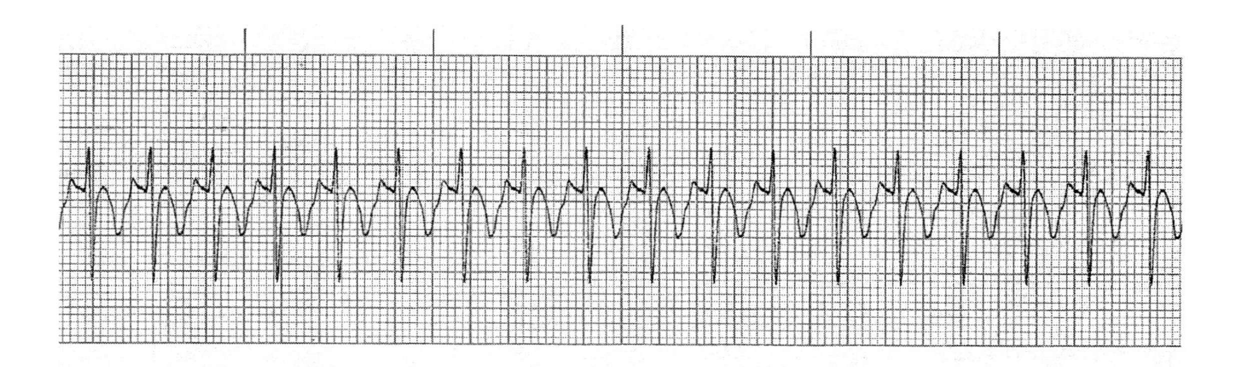

FIGURE 6-28

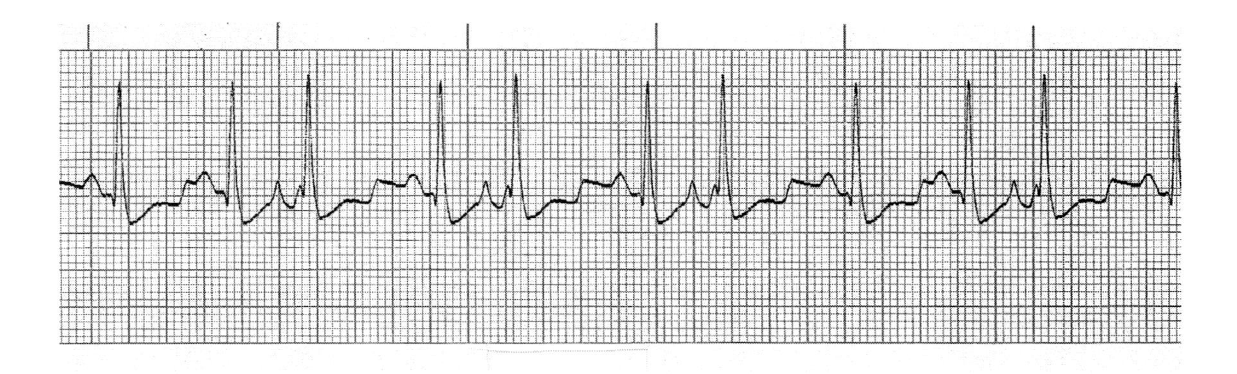

FIGURE 6-29

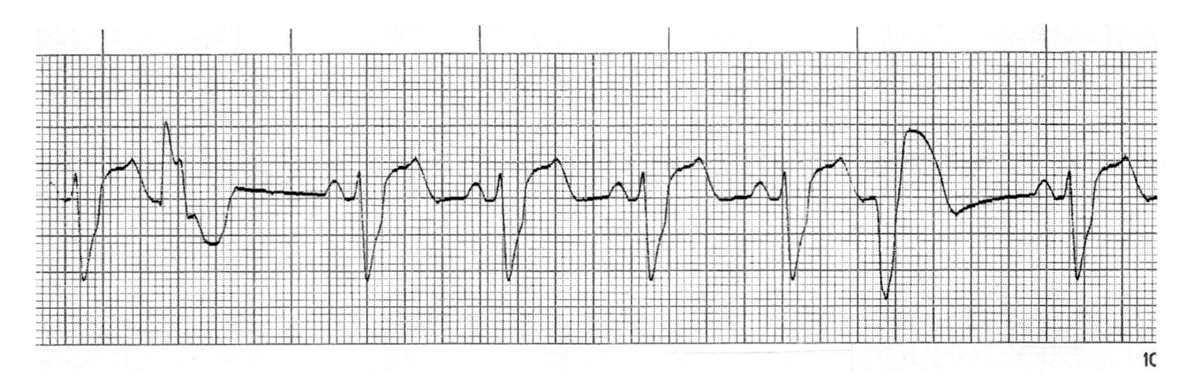

FIGURE 6-30

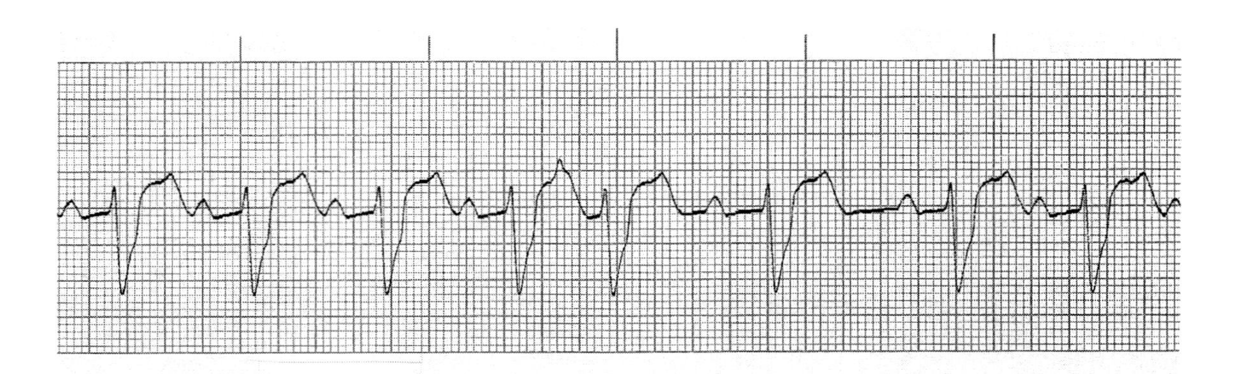

FIGURE 6-31

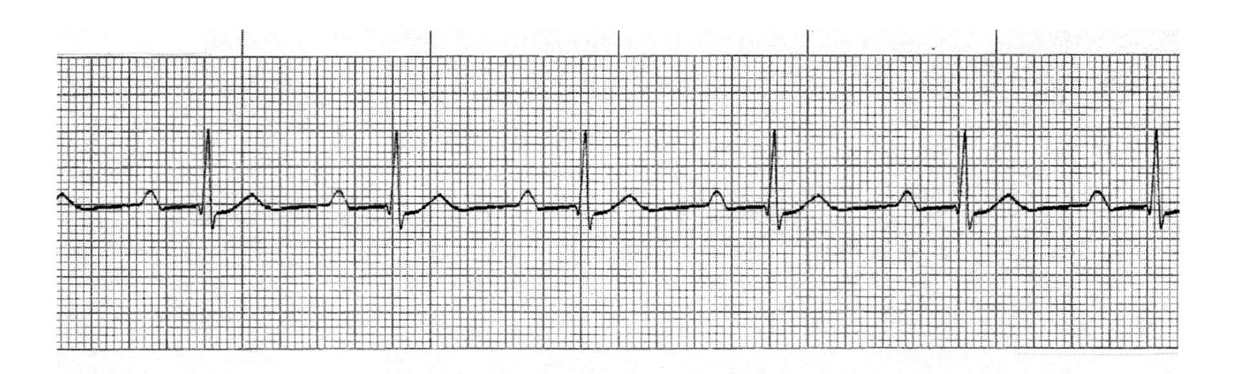

FIGURE 6-32

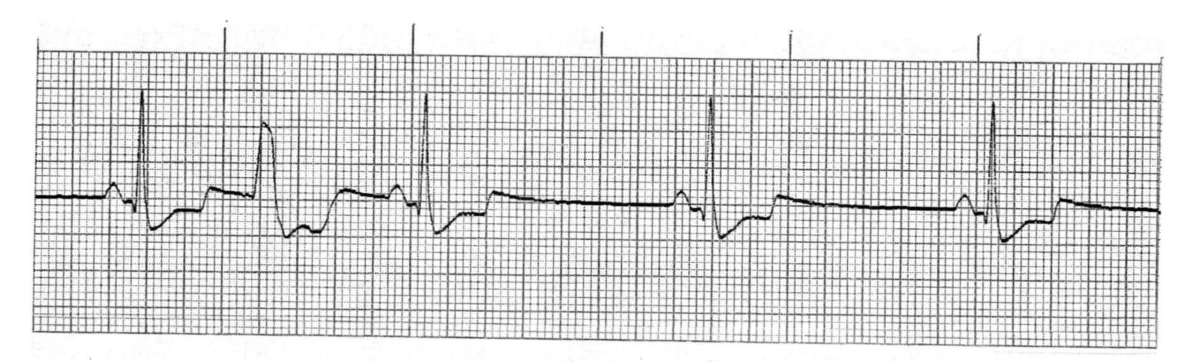

FIGURE 6-33

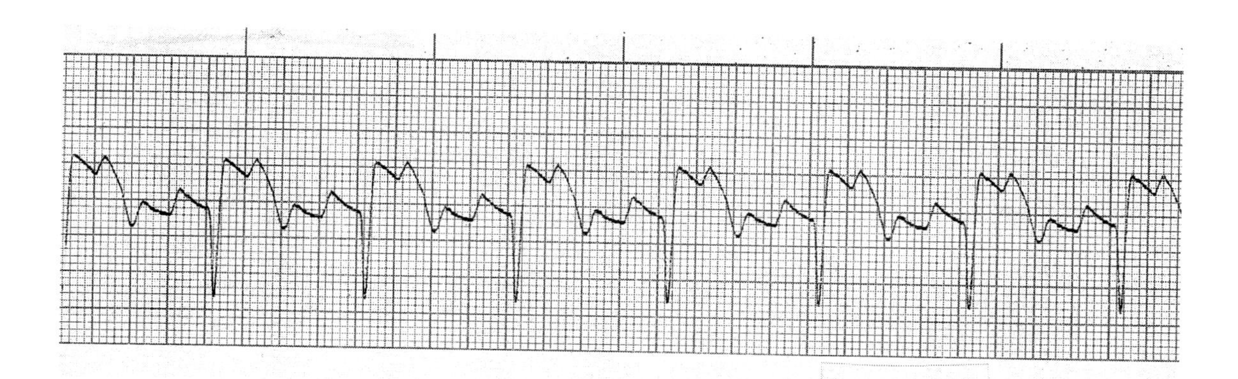

FIGURE 6-34

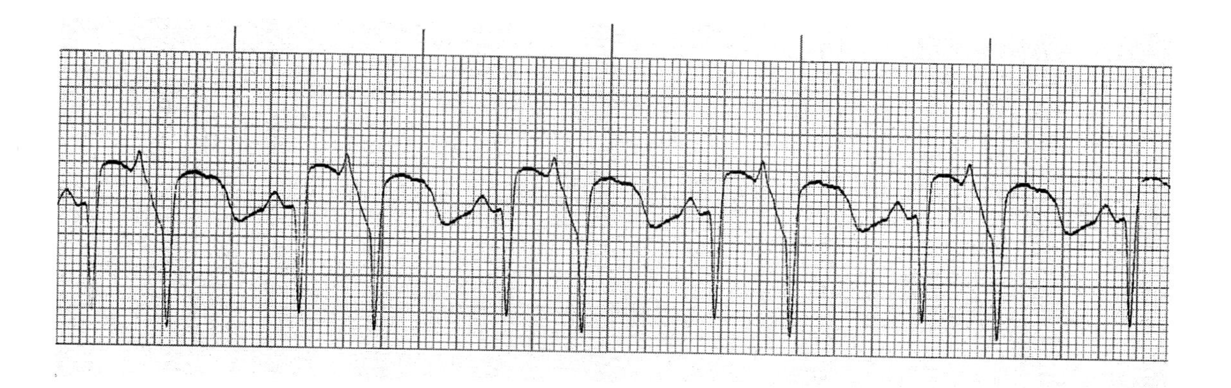

FIGURE 6-35

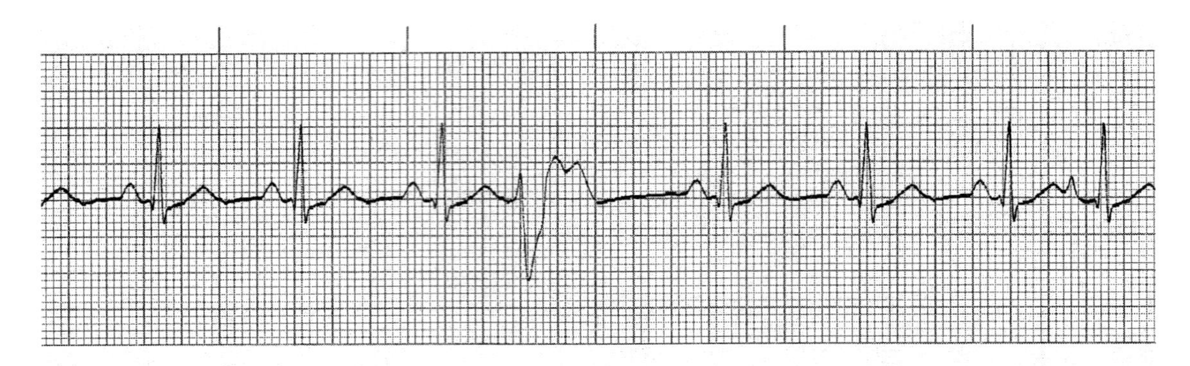

FIGURE 6-36

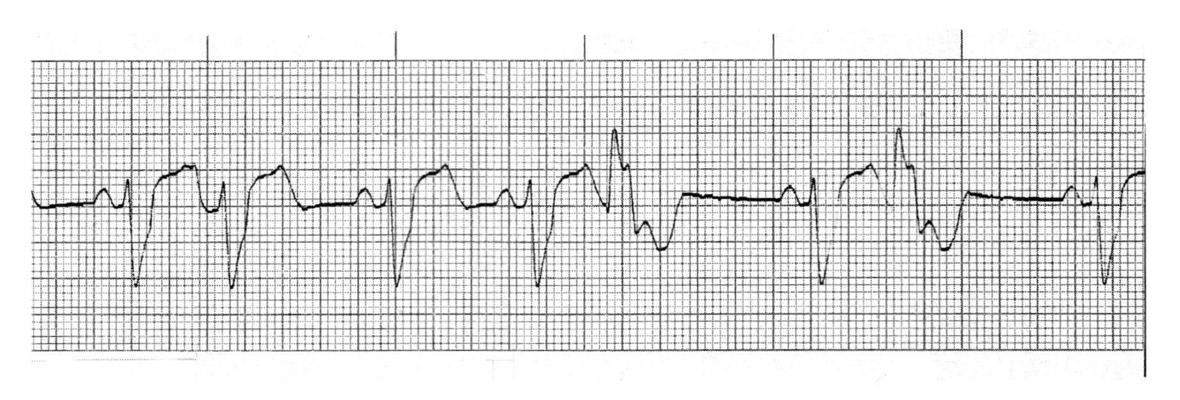

FIGURE 6-37

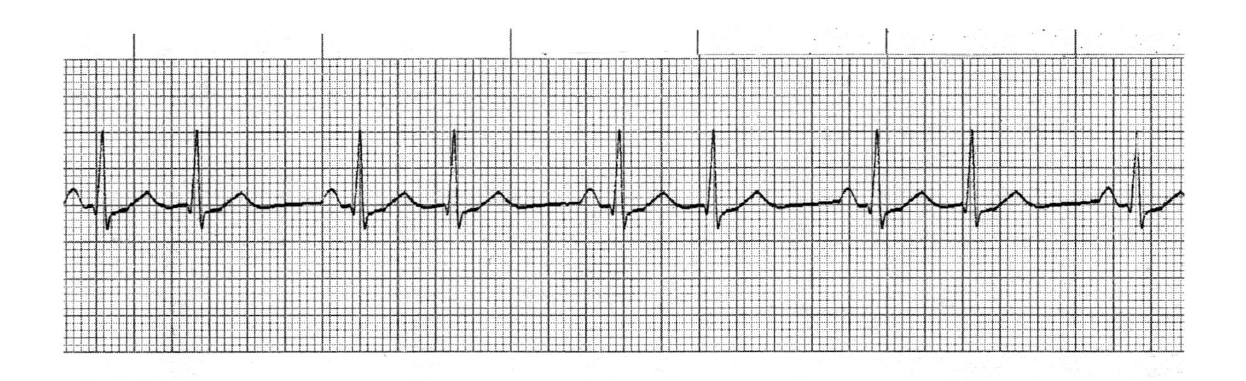

FIGURE 6-38

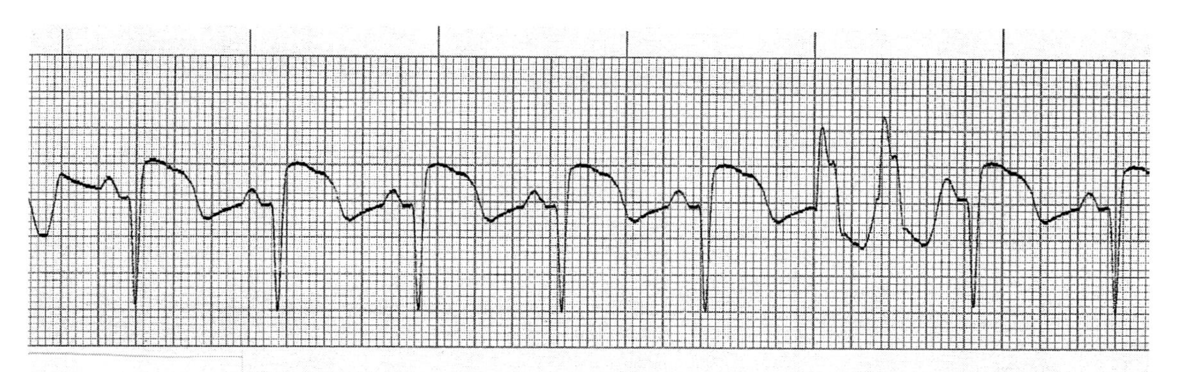

FIGURE 6-39

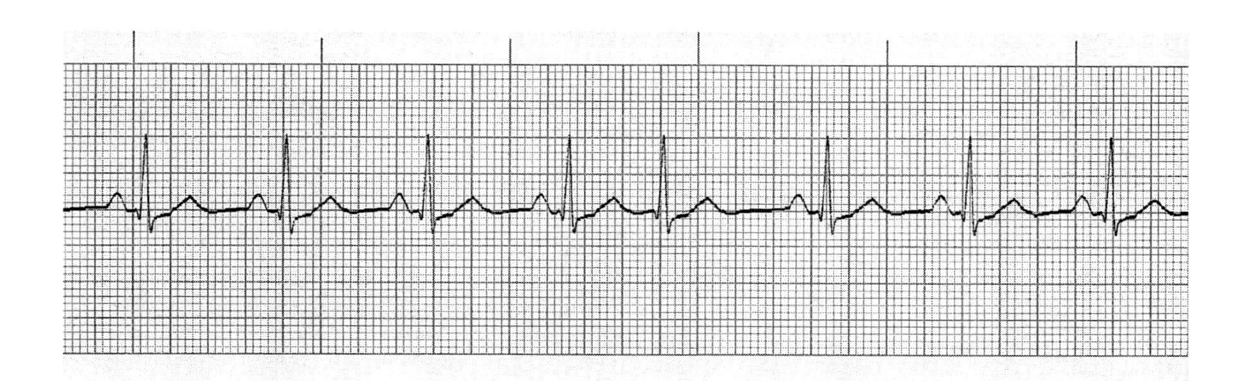

FIGURE 6-40

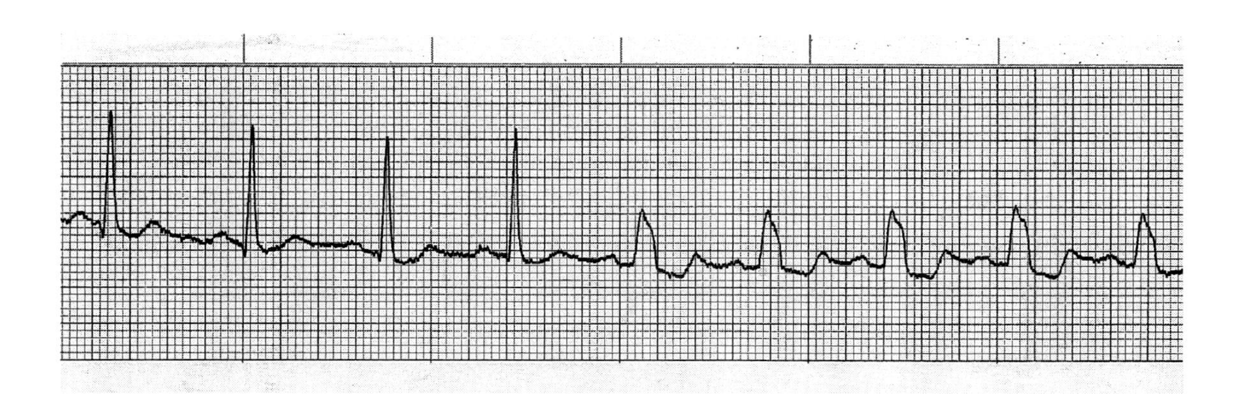

FIGURE 6-41

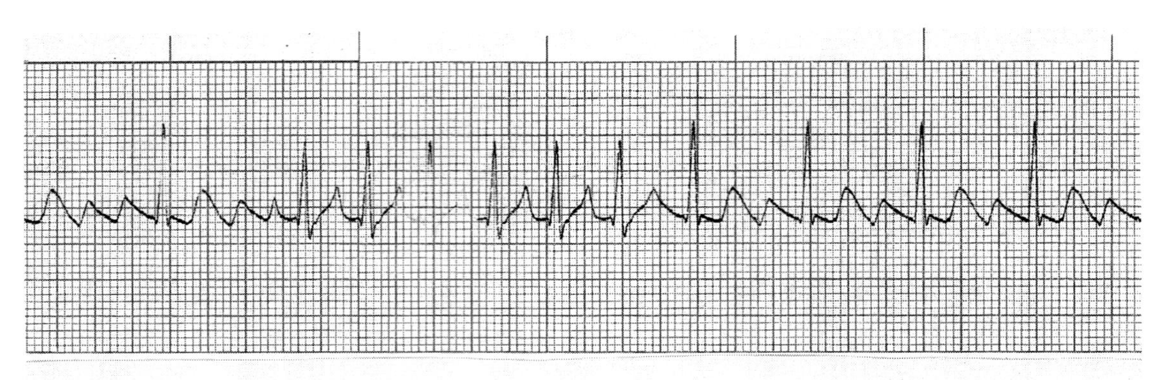

FIGURE 6-42

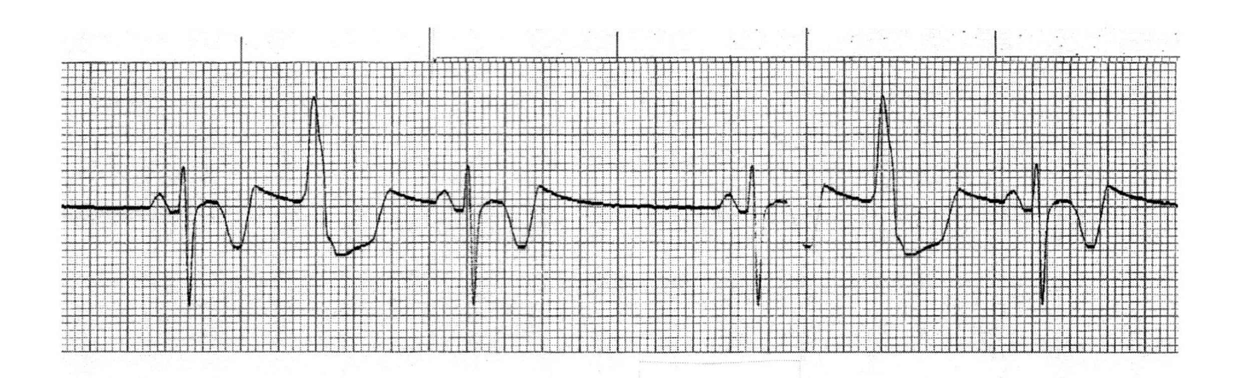

FIGURE 6-43

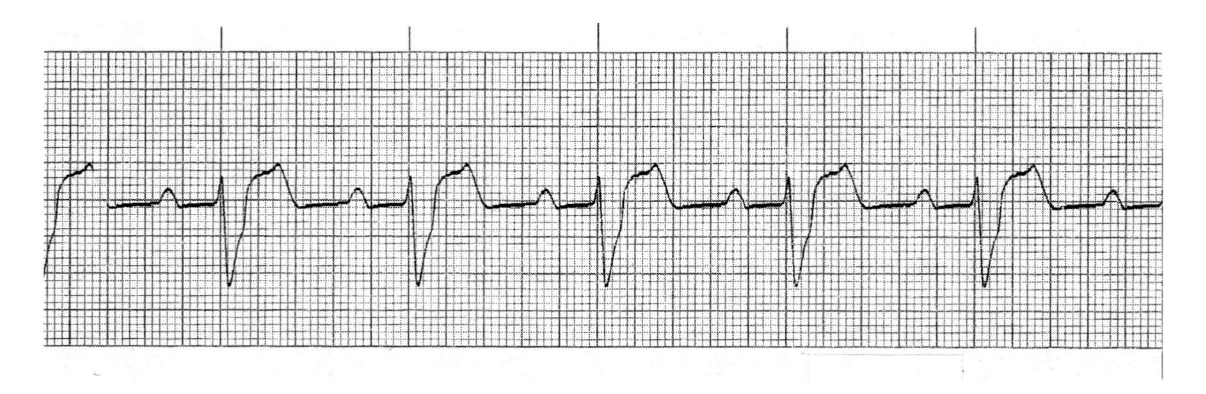

FIGURE 6-44

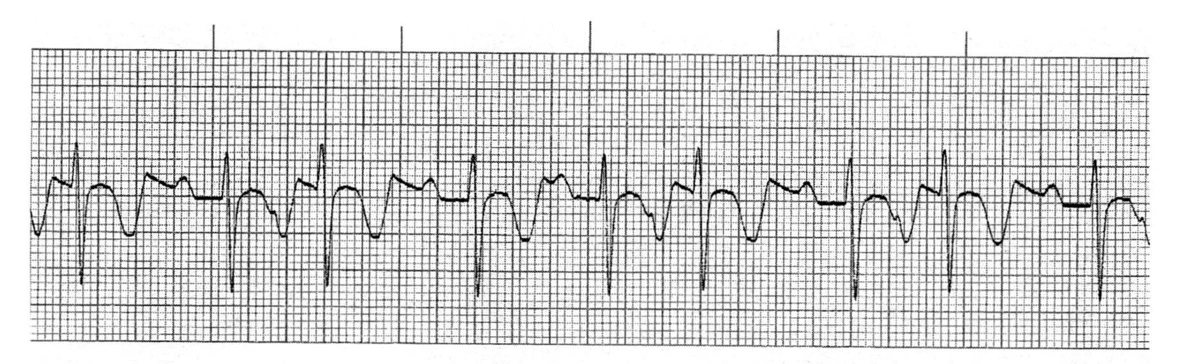

FIGURE 6-45

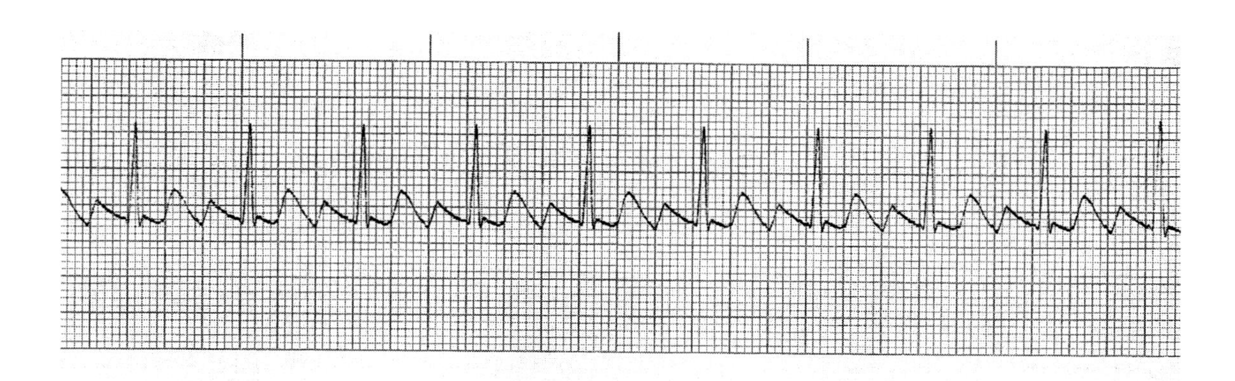

FIGURE 6-46

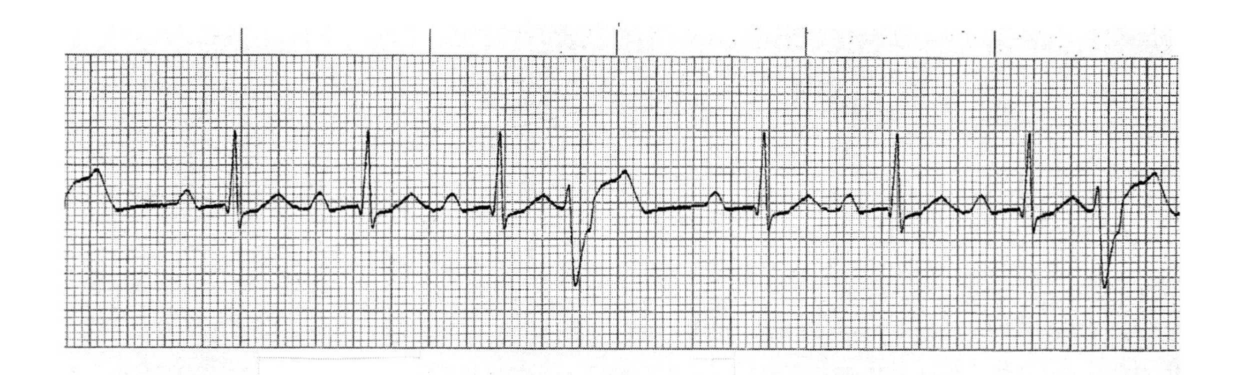

FIGURE 6-47

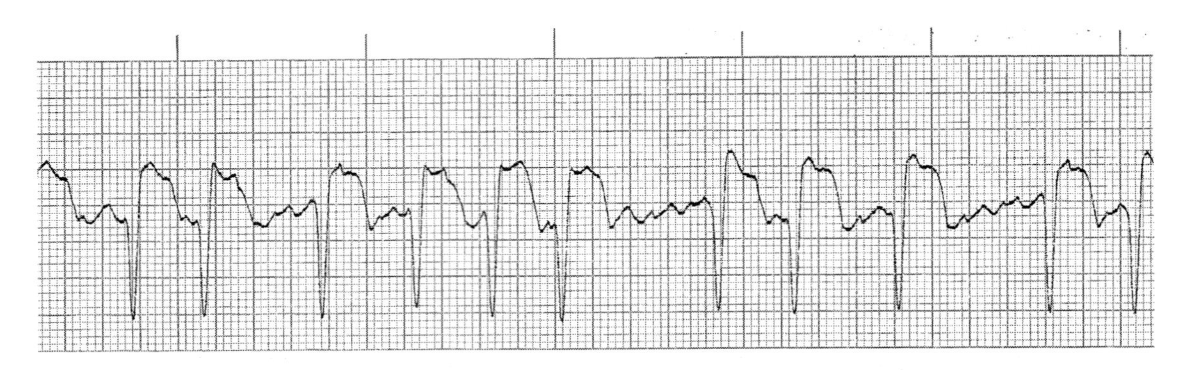

FIGURE 6-48

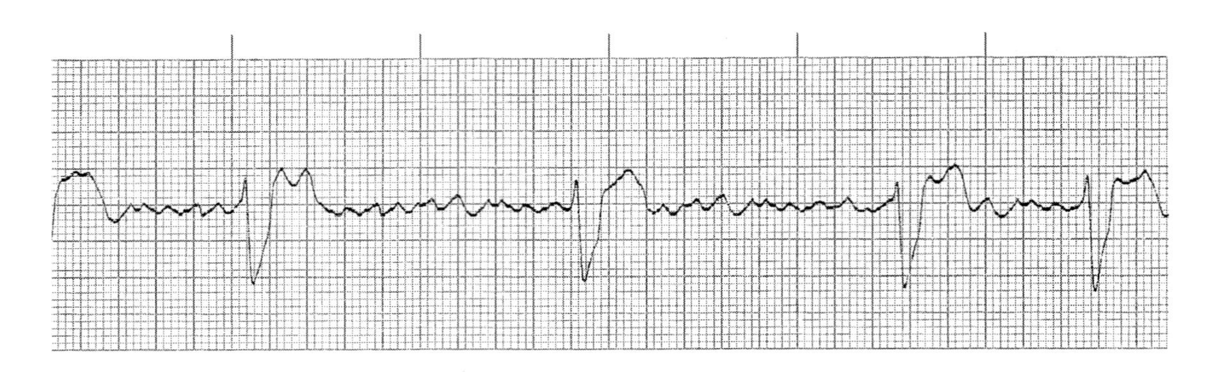

FIGURE 6-49

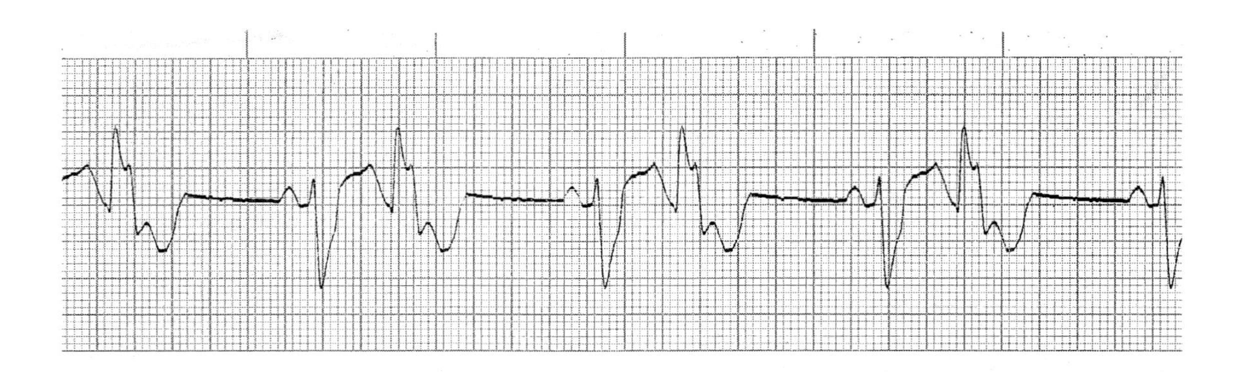

FIGURE 6-50

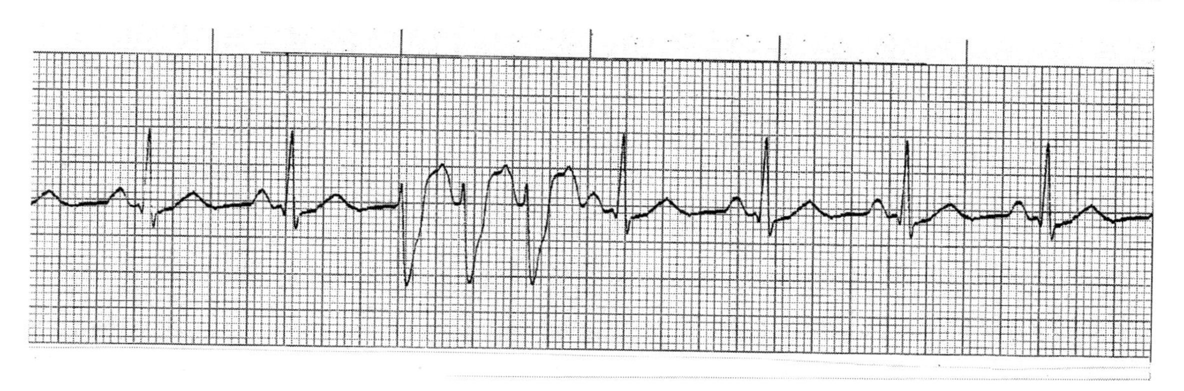

FIGURE 6-51

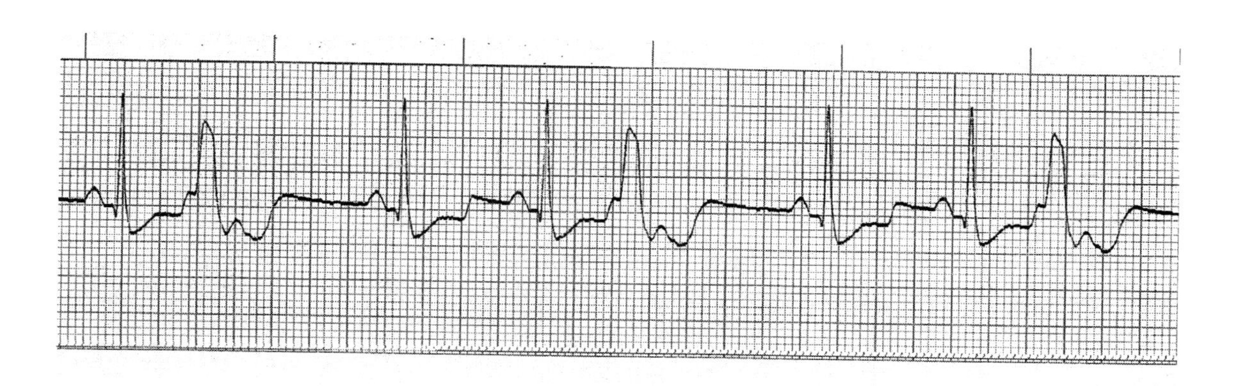

FIGURE 6-52

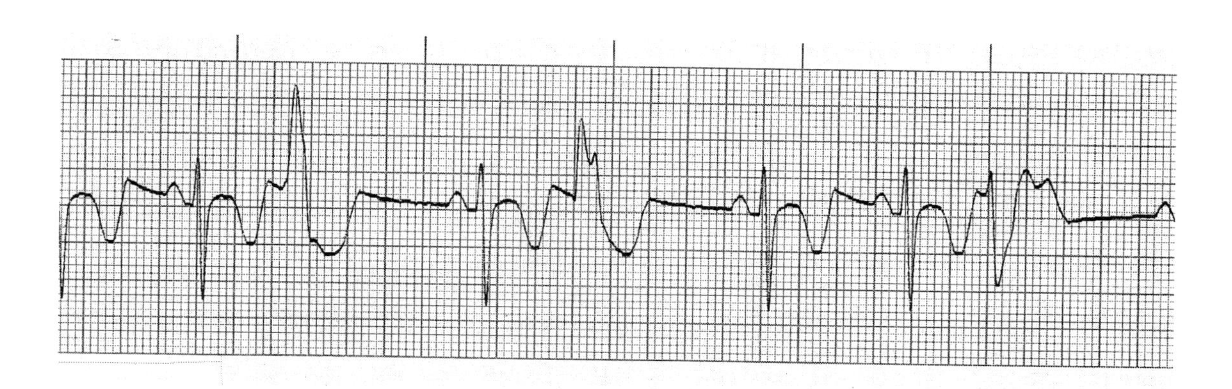

FIGURE 6-53

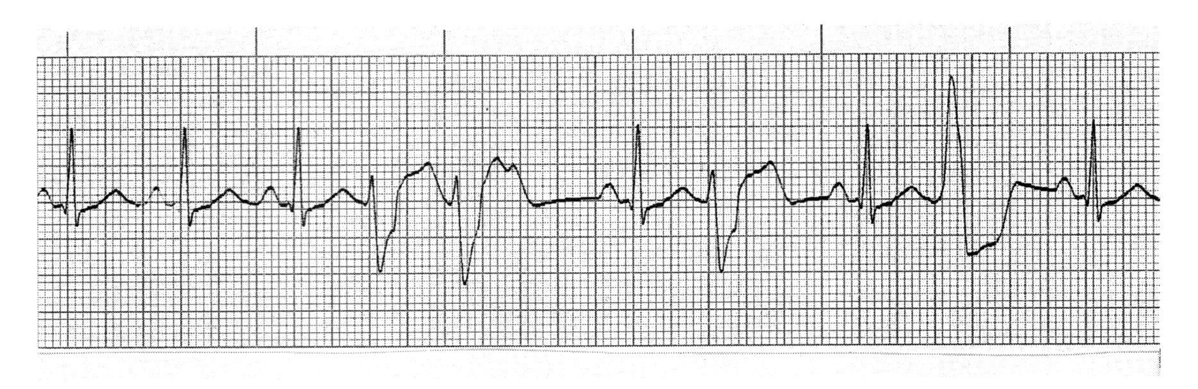

FIGURE 6-54

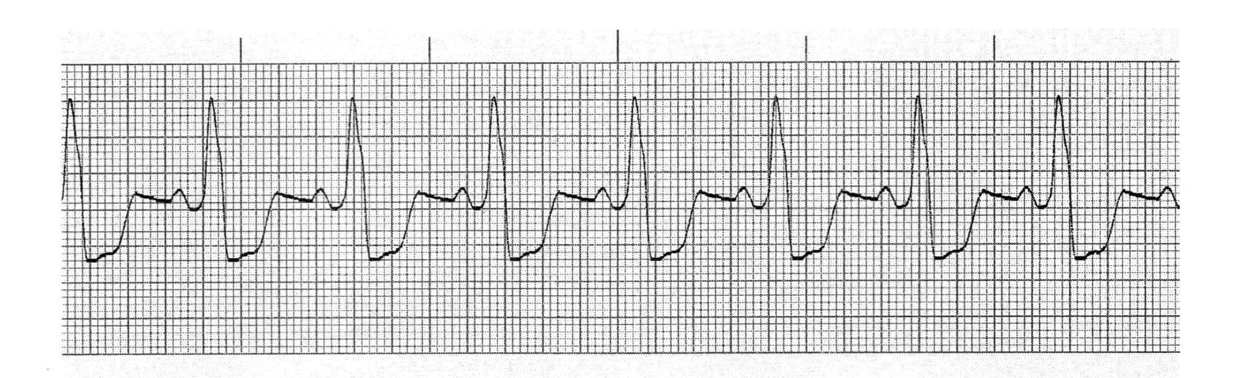

FIGURE 6-55

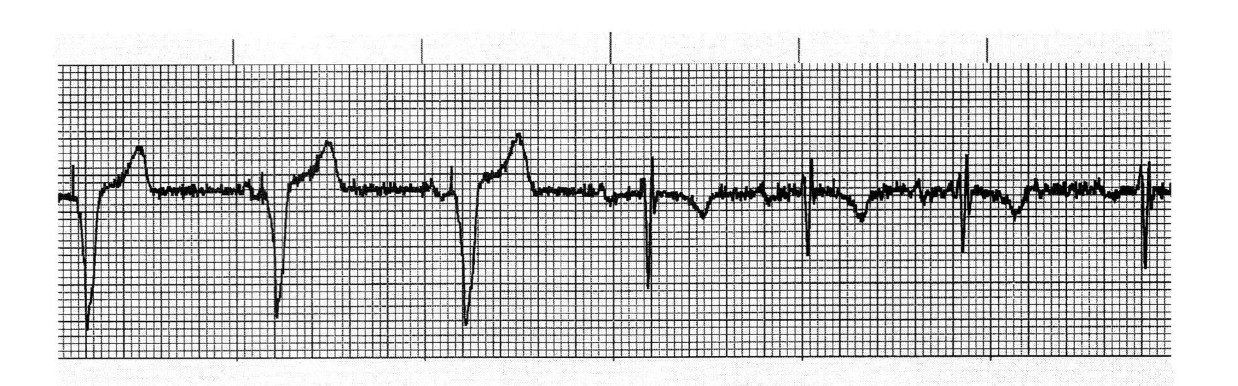

FIGURE 6-56

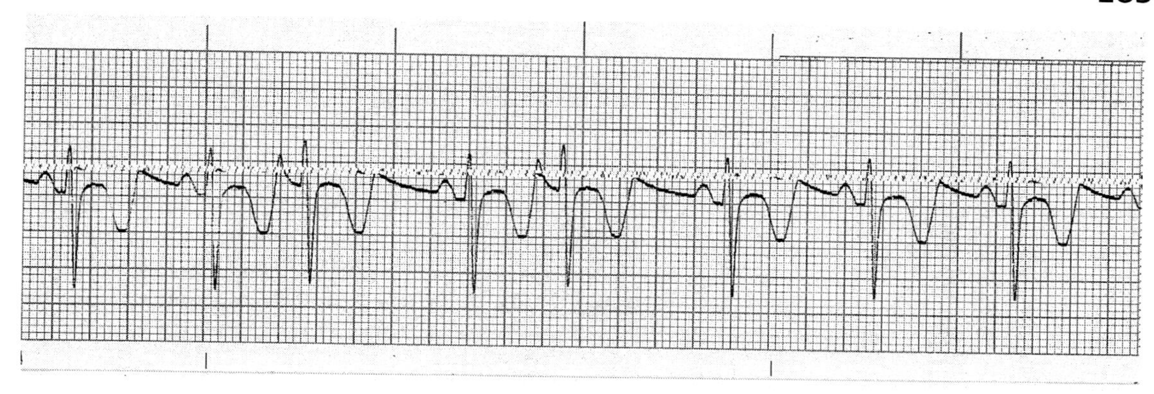

FIGURE 6-57

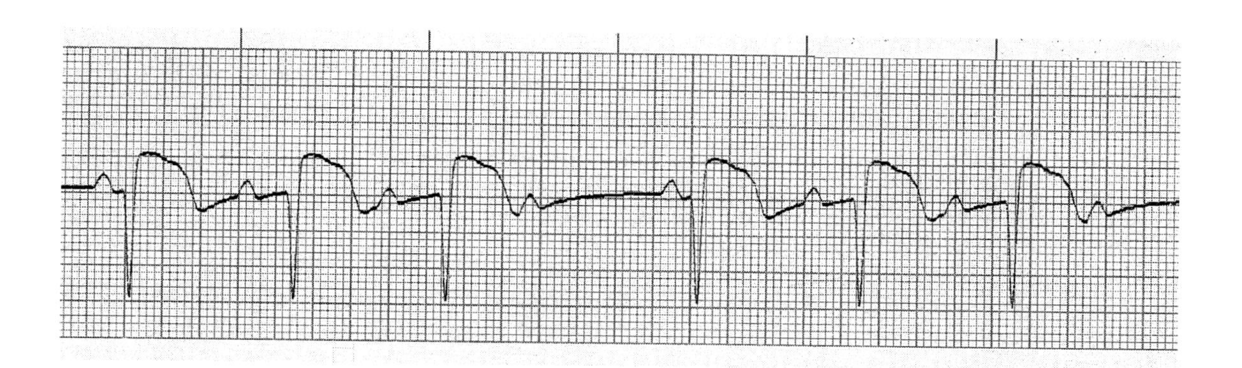

FIGURE 6-58

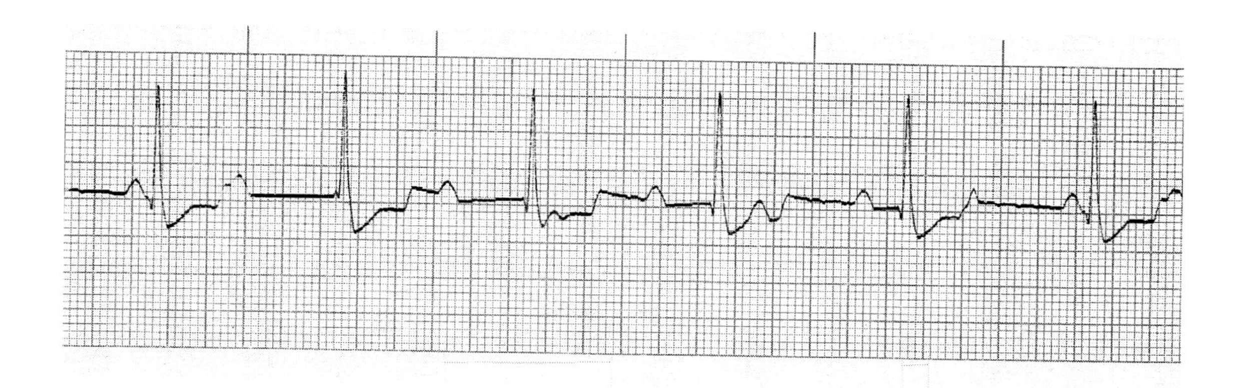

FIGURE 6-59

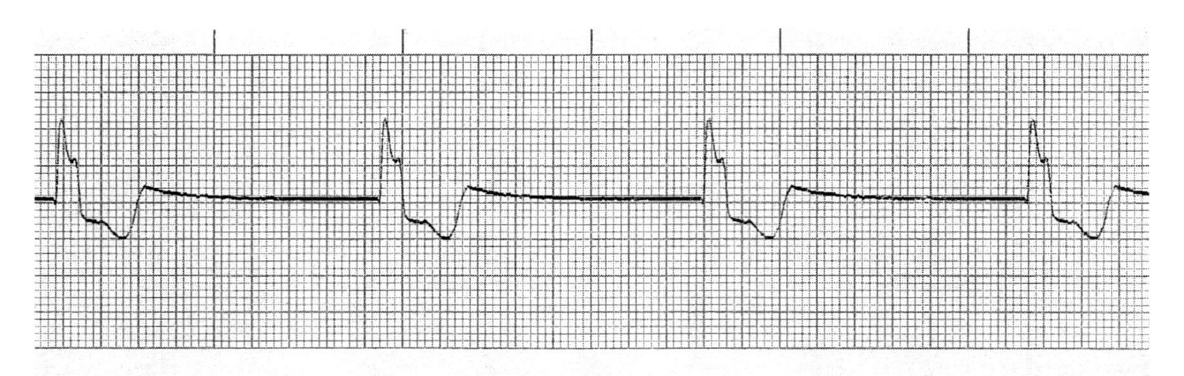

FIGURE 6-60

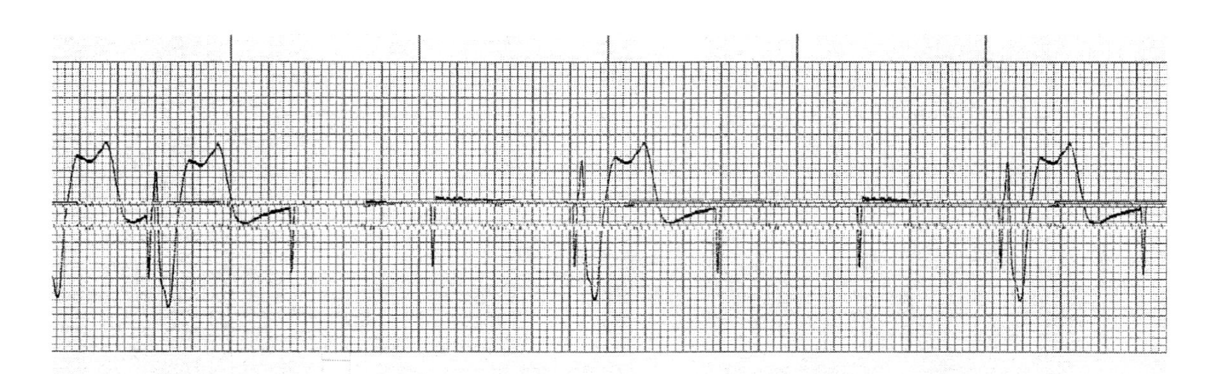

FIGURE 6-61

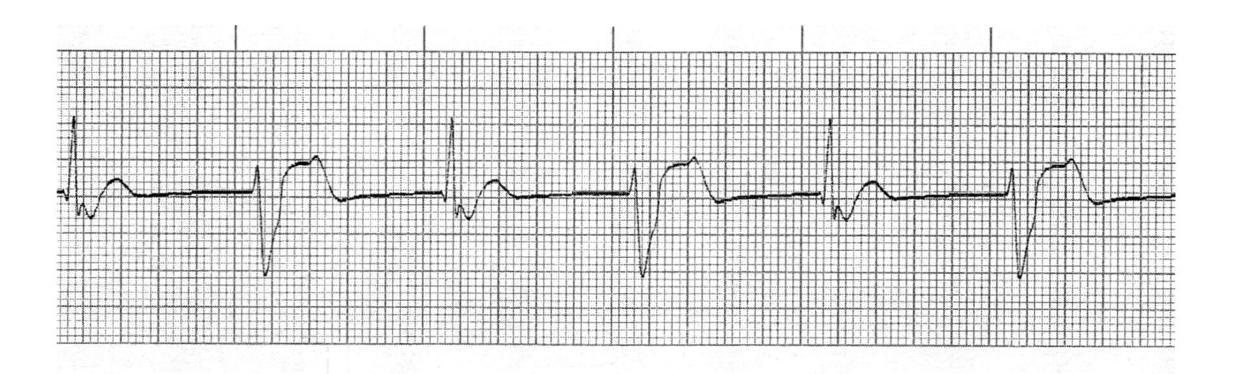

FIGURE 6-62

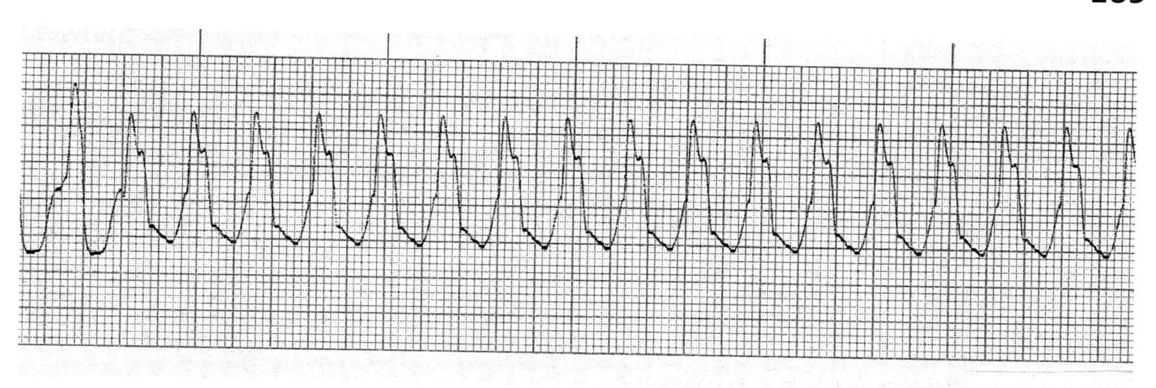

FIGURE 6-63

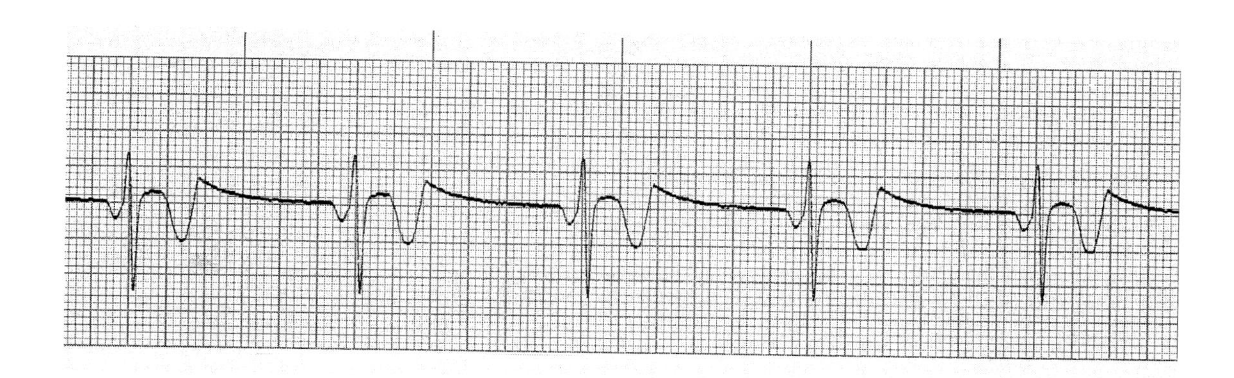

FIGURE 6-64

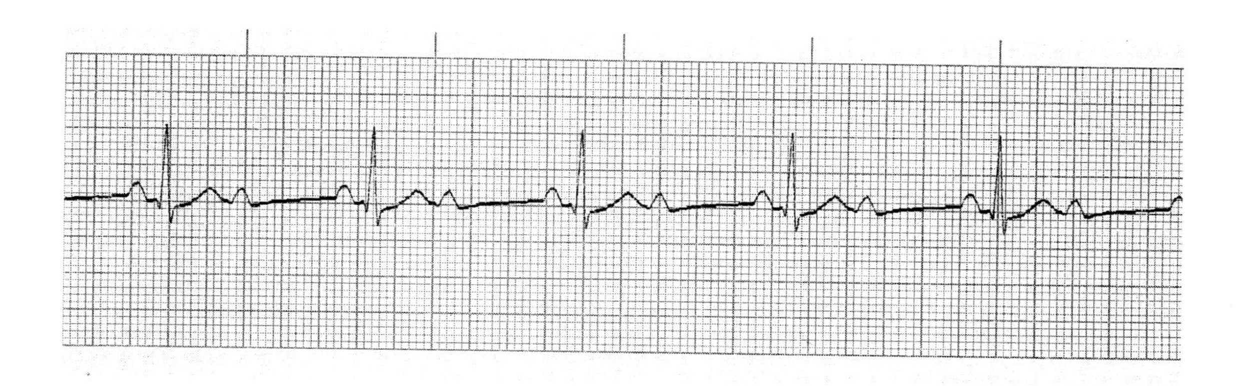

FIGURE 6-65

The answer key to Section Six begins on page 392.

7

Test Section Seven

Test Section Seven covers the following subjects:

* Anaphylaxis
* Toxicology
* Chemical Injuries
* Radiation Injuries
* Alcoholism and Drug Abuse
* Infectious Diseases
* Environmental Injuries
* Geriatrics/Gerontology
* Pediatrics
* OB/GYN/Neonatal Care
* Behavioral Emergencies

EMT - Paramedic National Standards Review Self Test Third Edition

1. Which of the following statements regarding the pathophysiology of an antigen–antibody reaction is false?
- (a) Upon introduction to the body, an antigen causes the production of antibodies.
- (b) Antibodies are produced to eliminate antigens from the body.
- (c) During sensitization, antibodies specific to the sensitizing antigen attach to mast cells.
- (d) The first time the antigen is introduced to the body it becomes attached to the corresponding antibody on the mast cells and can cause an anaphylactic reaction.
- (e) Antigen attachment to an antibody on a mast cell causes the mast cell to release histamine.

2. Antigens may be introduced to the body by way of
- (a) injection or ingestion.
- (b) inhalation.
- (c) absorption.
- (d) Answers (a) and (b) only.
- (e) Answers (a), (b), and (c).

3. Histamine causes constriction of
- (a) arterioles.
- (b) capillaries.
- (c) bronchial muscles.
- (d) All of the above.
- (e) None of the above.

4. Which of the following statements regarding myocardial muscle is false?
- (a) arterioles.
- (b) capillaries.
- (c) bronchial muscles.
- (d) All of the above.
- (e) None of the above.

5. Interstitial edema occurs because histamine increases the permeability of
- (a) arterioles.
- (b) capillaries.

(c) bronchial muscles.
(d) All of the above.
(e) None of the above.

6. Anaphylaxis is a massive antigen-antibody reaction that may produce signs and symptoms that include all of the following, except

(a) dyspnea, sneezing, coughing, or stridor.
(b) wheezing, rales, or total respiratory obstruction.
(c) peripheral vasoconstriction and hypertension.
(d) tachycardia.
(e) abdominal cramping, nausea, vomiting, or diarrhea.

7. In addition to the signs and symptoms mentioned in question 6, anaphylaxis may produce any of the following, except

(a) headache.
(b) seizures.
(c) cyanosis.
(d) facial edema.
(e) pitting pedal edema.

8. The medical term for the hives or wheals (raised areas of edema about the skin) associated with anaphylaxis is

(a) angioedema.
(b) urticaria.
(c) hyphema.
(d) polyps.
(e) lesions.

9. Which of the following statements regarding airway management for an anaphylaxis patient is true?

(a) If the patient has a history of COPD, oxygenation must be limited to a nasal cannula or venturi mask to prevent further respiratory compromise.
(b) IV medications to reverse the anaphylaxis should always be administered prior to any attempts at endotracheal intubation.
(c) Endotracheal intubation is rarely required in anaphylactic shock, as it is excessively damaging to respiratory mucosa.
(d) The presence or onset of stridor indicates impending total airway occlusion and is a valid clue that endotracheal intubation should be performed.
(e) None of the above is true.

10. Which of the following solutions should be used when initiating IV access for the treatment of anaphylaxis?

(a) Lactated Ringers solution
(b) Normal Saline (0.9% sodium chloride)
(c) 5% Dextrose in water
(d) Answers (a) or (b) only.
(e) Either answer (a), (b), or (c).

11. Place the following medications in order of their importance (rapidity of onset of action) when treating a *life-threatening* anaphylactic reaction (not all forms of epinephrine must be used).

(1) epinephrine 1:1000
(2) diphenhydramine
(3) dexamethasone, methylprednisolone, or hydrocortisone
(4) epinephrine 1:10,000

(a) 4, 2, 3
(b) 1, 2, 3, 5
(c) 1, 5, 2, 3
(d) 4, 2, 3, 5
(e) 4, 3, 2

12. The dosage of epinephrine 1:1000 when one is treating allergic reactions is

(a) 0.3 to 0.5 mg SQ.
(b) 0.3 to 0.5 mg IV.
(c) 3 to 5 ml IV.
(d) Either answer (a) or (b).
(e) Either answer (b) or c).

13. The dosage of epinephrine 1:10,000 when one is treating allergic reactions is

(a) 0.3 to 0.5 mg SQ.
(b) 0.3 to 0.5 mg IV.
(c) 3 to 5 ml IV.
(d) Either answer (a) or (b).
(e) Either answer (b) or (c).

14. The most common type of poisoning occurs via

(a) inhalation.
(b) injection.

(c) ingestion.
(d) absorption.
(e) None of the above.

15. Both immediate and delayed effects may be encountered when a poisoning occurs via
(a) inhalation.
(b) injection.
(c) ingestion.
(d) absorption.
(e) All of the above.

16. Situations when vomiting should not be induced include all of the following, except when
(a) non-organophosphate pesticides were ingested.
(b) petroleum products were ingested.
(c) the patient is pregnant.
(d) the patient has had a seizure.
(e) the patient has signs and symptoms of AMI.

17. Your 4-year-old patient is alert (has an intact gag reflex) and has ingested approximately 6 oz. of unleaded gasoline. You should consider administration of
(a) ipecac to induce vomiting.
(b) activated charcoal (without ipecac) or transport with supportive treatment only.
(c) an acidic substance to neutralize the ingested alkaline substance.
(d) an alkaline substance to neutralize the ingested acidic substance.
(e) copious amounts of milk to dilute the ingested substance.

18. The adult dose of activated charcoal is______mixed with water.
(a) 50-100 grams
(b) 100-200 grams
(c) 20-50 grams
(d) 100 mg/kg
(e) 200 mg/kg

19. The pediatric dose of activated charcoal is ______mixed with water.

(a) 50-100 grams
(b) 100-200 grams
(c) 20-50 grams
(d) 100 mg/kg
(e) 200 mg/kg

20. Your 4-year-old patient is alert (has an intact gag reflex) and has ingested six ounces of bleach. You should consider administration of

(a) ipecac to induce vomiting.
(b) activated charcoal (without ipecac) or transport with supportive treatment only.
(c) an acidic substance to neutralize the ingested alkaline substance.
(d) an alkaline substance to neutralize the ingested acidic substance.
(e) copious amounts of milk to dilute the ingested substance.

21. The dosage of syrup of ipecac for a child over the age of one is

(a) 10 cc, followed by 1 cup of water.
(b) 15 cc, followed by 1 to 2 glasses of water.
(c) 20 cc, followed by 2 to 3 glasses of water.
(d) 30 cc, followed by 2 to 3 glasses of water.
(e) 40 cc, followed by 3 to 4 glasses of water.

22. The dosage of syrup of ipecac for an adult is

(a) 10 cc, followed by 1 cup of water.
(b) 15 cc, followed by 1 to 2 glasses of water.
(c) 20 cc, followed by 2 to 3 glasses of water.
(d) 30 cc, followed by 2 to 3 glasses of water.
(e) 40 cc, followed by 3 to 4 glasses of water.

23. When managing the victim of a toxic inhalation, your highest priority should be to

(a) intubate and hyperventilate the unconscious patient.
(b) remove contaminated clothing.
(c) perform the usual primary and secondary exam.
(d) remove the patient from the toxic environment.
(e) ensure personal safety prior to access of the unconscious patient.

24. Which of the following statements regarding treatment of insect bites and stings is true?

(a) Remove the stinger by using forceps or tweezers (not your fingers) and squeezing the area below the venom sac.
(b) Apply ice directly to the injection/bite site.
(c) Be alert for allergic reactions and anaphylactic shock.
(d) Both answers (a) and (c) are true.
(e) All of the above answers are true.

25. Observation of a spider with a violin-shaped marking is associated with the identification of

(a) a black widow spider.
(b) a brown recluse spider.
(c) a scorpion.
(d) Answers (a) and (b) only.
(e) Answers (a), (b), and (c).

26. Observation of a spider with a yellow-orange- or red-colored hourglass marking is associated with the identification of

(a) a black widow spider.
(b) a brown recluse spider.
(c) a scorpion.
(d) Answers (a) and (b) only.
(e) Answers (a), (b), and (c).

27. There is no specific antivenin or antiserum for

(a) a black widow spider bite.
(b) a brown recluse spider bite.
(c) a scorpion sting.
(d) Answers (a) and (b) only.
(e) Answers (a), (b), and (c).

28. Administration of analgesics will increase the venom toxicity of

(a) a black widow spider bite.
(b) a brown recluse spider bite.
(c) a scorpion sting.
(d) Answers (a) and (b) only.
(e) Answers (a), (b), and (c).

29. Consider administration of diazepam or calcium gluconate for treatment of severe muscle spasms secondary to

(a) a black widow spider bite.
(b) a brown recluse spider bite.
(c) a scorpion sting.
(d) Answers (a) and (b) only.
(e) Answers (a), (b), and (c).

30. The venom of the ______ contains a neurotoxin that can produce slurred speech, excessive salivation, tongue or larynx paralysis, loss of consciousness, seizures, or respiratory arrest.

(a) rattlesnake
(b) copperhead snake
(c) cottonmouth snake
(d) coral snake
(e) water moccasin

31. Which of the following statements regarding treatment of snake bite is true?

(a) Apply an arterial tourniquet proximal to the bite and a venous tourniquet distal to the bite.
(b) Apply ice, a cold pack, or freon spray to the wound.
(c) Immobilize the bitten limb with a splint.
(d) Elevate the bitten limb to diminish edema.
(e) As soon as possible (regardless of proximity to the receiving hospital), make an "X" incision over each fang mark and apply suction, using a commercial snake bite suction cup.

32. Surface exposure to ______ requires removal by gentle brushing prior to copious water lavage.

(a) dry lime
(b) phenol
(c) sodium metal
(d) All of the above.
(e) None of the above.

33. Surface exposure to ______ requires immediate application of a neutralizing agent.

(a) dry lime
(b) phenol
(c) sodium metal
(d) All of the above.
(e) None of the above.

34. Surface exposure to _______ requires first a thorough alcohol lavage, followed by a copious water lavage.

(a) dry lime
(b) phenol
(c) sodium metal
(d) All of the above.
(e) None of the above.

35. Surface exposure to _______ requires removal by gentle brushing prior to application of an oil coverage.

(a) dry lime
(b) phenol
(c) sodium metal
(d) All of the above.
(e) None of the above.

36. Surface exposure to organophosphate chemicals (used in insecticides and some chemical warfare agents) can be extremely toxic to both victim and responder. Organophosphate poisoning stimulates the parasympathetic nervous system, causing signs and symptoms that include all of the following, except

(a) excessive salivation.
(b) nausea/vomiting/diarrhea and diaphoresis.
(c) blurred vision and constricted pupils.
(d) tachycardia.
(e) hypotension.

37. In addition to removal of clothing and decontamination with copious amounts of water, treatment of symptomatic organophosphate poisoning includes oxygen administration, IV access,cardiac monitoring, and

(a) epinephrine 1:1000, 0.3–0.5 mg IV.
(b) atropine IV in 0.5-1.0 mg increments every 3 to 5 minutes, to a maximum dose of 0.04 mg/kg IV.
(c) epinephrine 1:10,000 IV in 1.0 mg increments every 5 minutes, to a maximum dose of 10 mg IV.
(d) atropine IV, in 2–5 mg increments every 10 to 15 minutes as needed.
(e) transport only.

38. Opium-based drugs are called *opiates* or *narcotics*. All of the following drugs are narcotics, except

(a) heroin.
(b) Demerol.
(c) codeine.
(d) Darvon.
(e) cocaine.

39. Larger than average doses of naloxone will be required to manage an overdose of

(a) heroin.
(b) Demerol.
(c) codeine.
(d) Darvon.
(e) cocaine.

40. Which of the following statements regarding alcohol withdrawal syndrome is false?

(a) Signs and symptoms of withdrawal can occur within several hours after sudden abstinence and can last several days.
(b) Signs and symptoms include an increase in sympathetic tone, producing bradycardia, sweating, and hypotension.
(c) Delirium tremens (DTs) usually develops on the second or third day of withdrawal.
(d) Seizures ("rum fits") may occur, usually within the first 24 to 36 hours of abstinence.
(e) Seizures or DTs are signs of a significant emergency and may require administration of IV diazepam.

41. The immune system is the body's defense against disease. All of the following are major components of the immune system, except

(a) leukocytes.
(b) lymphocytes.
(c) antigens.
(d) antibodies.
(e) macrophages.

42. Interstitial fluid is defined as

(a) a clear, watery fluid found in lymphatic vessels.
(b) the fluid that fills each cell.

(c) the fluid that fills the space between cells.
(d) Both answers (a) and (b).
(e) Both answers (a) and (c).

43. Lymph is defined as

(a) a clear, watery fluid found in lymphatic vessels.
(b) the fluid that fills each cell.
(c) the fluid that fills the space between cells.
(d) Both answers (a) and (b).
(e) Both answers (a) and (c).

44. The primary organ of the lymphatic system is

(a) the kidney.
(b) the liver.
(c) the spleen.
(d) the thyroid.
(e) the pancreas.

45. Pneumonia is an infectious respiratory disease that is caused

(a) by bacteria.
(b) by a virus.
(c) by fungi.
(d) by either bacteria or a virus.
(e) by bacteria, viruses, or fungi.

46. Meningitis is an infectious nervous system disease that is caused

(a) by bacteria.
(b) by a virus.
(c) by fungi.
(d) by either bacteria or a virus.
(e) by bacteria, viruses, or fungi.

47. Tuberculosis is an infectious respiratory disease that is caused

(a) by bacteria.
(b) by a virus.
(c) by fungi.
(d) by either bacteria or a virus.
(e) by bacteria, viruses, or fungi.

48. The most common form of viral hepatitis is

(a) hepatitis C.
(b) hepatitis B.
(c) hepatitis A.
(d) All of the above.
(e) None of the above.

49. An infectious form of viral hepatitis is

(a) hepatitis C.
(b) hepatitis B.
(c) hepatitis A.
(d) All of the above.
(e) None of the above.

50. Serum hepatitis is transmitted via infected blood or urine and is also called

(a) hepatitis C.
(b) hepatitis B.
(c) hepatitis A.
(d) All of the above.
(e) None of the above.

51. A patient who complains of fever, nausea/vomiting, headache, and a stiff neck should be considered to be suffering from

(a) tuberculosis.
(b) AIDS.
(c) meningitis.
(d) Either answer (a) or (b).
(e) Either answer (b) or (c).

52. A patient who complains of fever, night sweats, and recent weight loss should be considered to be suffering from

(a) tuberculosis.
(b) AIDS.
(c) meningitis.
(d) Either answer (a) or (b).
(e) Either answer (b) or (c).

53. AIDS is transmitted via

(a) blood contact.
(b) semen or vaginal secretions contact.

(c) skin surface contact.
(d) Answers (a) and (b) only.
(e) Answers (a), (b), and (c).

54. Many AIDS patients develop Kaposi's sarcoma, which is
(a) evidenced by jaundiced skin lesions.
(b) evidenced by red- or purple-colored skin lesions.
(c) a severe form of leukemia.
(d) a severe form of pneumonia.
(e) None of the above.

55. Many AIDS patients develop Pneumocystis carinii, which is
(a) evidenced by jaundiced skin lesions.
(b) evidenced by red- or purple-colored skin lesions.
(c) a severe form of leukemia.
(d) a severe form of pneumonia.
(e) None of the above.

56. Common forms of sexually transmitted diseases (STDs) include all of the following, except
(a) syphilis.
(b) gonorrhea.
(c) chlamydia.
(d) herpes.
(e) varicella.

57. The universal precautions recommended for all health care providers to maximize protection from infectious diseases include all of the following, except
(a) the wearing of disposable gloves and face masks.
(b) thorough soap and water hand washing, both before and after patient contact.
(c) the wearing of protective eye gear or face shields.
(d) recapping of all used needles or scalpels.
(e) the wearing of gowns or aprons.

58. The body's thermoregulatory control mechanism is located in the
(a) cerebrum.
(b) cerebellum.
(c) pons.
(d) medulla oblongata.
(e) hypothalamus.

59. Compensatory mechanisms employed by the body to eliminate heat include all of the following, except

(a) peripheral vasodilation.
(b) sweating.
(c) increased cardiac output.
(d) increased respiratory rate.
(e) increased metabolic rate.

60. Compensatory mechanisms employed by the body to preserve and generate heat include all of the following, except

(a) peripheral vasoconstriction.
(b) thermogenesis (sweating).
(c) parasympathetic stimulation.
(d) piloerection ("goose flesh").
(e) release of epinephrine and norepinephrine.

61. Heat loss from the body that occurs because of sweating is attributed to the mechanism of

(a) radiation.
(b) conduction.
(c) convection.
(d) evaporation.
(e) respiration.

62. Cold packs placed directly upon the body surface effect body heat loss via the mechanism of

(a) radiation.
(b) conduction.
(c) convection.
(d) evaporation.
(e) respiration.

63. Heat loss from the body that is enhanced by air currents is attributed to the mechanism of

(a) radiation.
(b) conduction.
(c) convection.
(d) evaporation.
(e) respiration.

64. Diaphoresis (without exertion) accompanies

(a) heat stroke.
(b) heat cramps.

(c) heat exhaustion.
(d) Both answers (a) and (c).
(e) Both answers (b) and (c).

65. Confusion or altered level of consciousness accompanies
(a) heat stroke.
(b) heat cramps.
(c) heat exhaustion.
(d) Both answers (a) and (c).
(e) Both answers (b) and (c).

66. Hyperkalemia may accompany
(a) heat stroke.
(b) heat cramps.
(c) heat exhaustion.
(d) Both answers (a) and (c).
(e) Both answers (b) and (c).

67. Oral administration of a "sports" drink or other beverage containing sodium is appropriate to the treatment of
(a) heat stroke.
(b) heat cramps.
(c) heat exhaustion.
(d) Both answers (a) and (c).
(e) Both answers (b) and (c).

68. Treatment of the most severe form of heat-related illness includes all of the following, except
(a) rapid cooling accomplished with ice packs, ice-water-soaked sheets, and/or fans while en route to the ER.
(b) oxygen administration.
(c) two sites of IV access with NS or LR with the flow wide open.
(d) initiation of a piggybacked dopamine infusion, titrated to a systolic blood pressure of 100.
(e) cardiac monitoring.

69. When measured orally, the normal body temperature is
(a) 37 degrees centigrade (C).
(b) 98.6 degrees Fahrenheit (F).
(c) 44.5 degrees centigrade (C).
(d) Both answers (a) and (b).
(e) Both answers (b) and (c).

70. When measured rectally, the normal body temperature is

(a) 99.6 degrees F.
(b) 98.6 degrees F.
(c) 97.6 degrees F.
(d) 37 degrees C.
(e) 44.5 degrees C.

71. The medical term for a pathogenic elevation of normal body temperature (fever) is

(a) hyperthermia.
(b) Fahrenheit.
(c) pyrexia.
(d) hypertemperature.
(e) hyproxia.

72. Mild hypothermia is defined as a core temperature of _____ degrees Fahrenheit.

(a) 95 to 98
(b) 90 to 95
(c) 86 to 96
(d) less than 90
(e) less than 86

73. When a centigrade thermometer is used, mild hypothermia is defined as a core temperature of

(a) less than 30 degrees.
(b) less than 32 degrees.
(c) 30 to 36 degrees.
(d) 32 to 35 degrees.
(e) 35 to 37 degrees.

74. Severe hypothermia is defined as a core temperature of _____ degrees Fahrenheit.

(a) 95 to 98
(b) 90 to 95
(c) 86 to 96
(d) less than 90
(e) less than 86

75. When a centigrade thermometer is used, severe hypothermia is defined as a core temperature of

(a) less than 30 degrees.
(b) less than 32 degrees.

(c) 30 to 36 degrees.
(d) 32 to 35 degrees.
(e) 35 to 37 degrees.

76. The J wave (Osborn wave) occurs in many hypothermia victims. The wave size increases as the body core temperature decreases. The J wave can be observed on the ECG

(a) between the P wave and the QRS.
(b) on the upslope of the R wave.
(c) on the downslope of the R wave, or at the junction of the QRS and S–T segment.
(d) on the downslope of the T wave.
(e) at the junction of the T wave's downslope and the isoelectric line.

77. Treatment of mild hypothermia includes

(a) gentle handling and insulation from cold after removal of all wet clothing.
(b) addition of heat to the patient's head, neck, chest, and groin (respiratory warmers may also be used).
(c) oral administration of warm fluids and sugar sources after shivering has stopped.
(d) Answers (a) and (b) only.
(e) Answers (a), (b), and (c).

78. Treatment of severe hypothermia, when vital signs are present, includes all of the following, except

(a) gentle handling and insulation from cold after removal of all wet clothing.
(b) addition of heat to the patient's head, neck, chest, and groin (respiratory warmers may also be used).
(c) cardiac monitoring.
(d) establishment of IV access and administration of prophylactic lidocaine 1 mg/kg IV.
(e) withholding oxygen supplementation unless the oxygen is heated to above 99 degrees F.

79. Treatment of severe hypothermia, when vital signs are absent, includes all of the following, except

(a) assessment of cardiopulmonary status for up to two minutes prior to initiation of CPR.
(b) initial defibrillation of V-fib at 400 ws; if unsuccessful, repeat defibrillation only if core temperature is 85 degrees F or greater.
(c) administration of lidocaine 1 mg/kg IV after successful defibrillation, followed by 0.5 mg/kg IV in 15 minutes.
(d) intubation and ventilation with warm oxygen.
(e) addition of heat to the patient's head, neck, chest, and groin regardless of proximity to ER.

80. Hypothermia may also be precipitated by metabolic factors that include

(a) hypoglycemia or hypothyroidism.
(b) brain tumor or head injuries.
(c) malnutrition or old age.
(d) Answers (a) and (b) only.
(e) Answers (a), (b), and (c).

81. Which of the following statements regarding the pathophysiology of frostbite is false?

(a) Exposure to subfreezing temperatures results in vasoconstriction, producing increased circulation and accelerated heat loss.
(b) Ice crystals form in extra- and intracellular fluid.
(c) Expansion of forming ice crystals results in structural damage of the cells.
(d) Biochemical changes occur that result in structural damage of the cells.
(e) Frostbite results in anoxia/ischemia of the frozen tissues and may result in necrosis.

82. Which of the following statements regarding treatment of frostbite is true?

(a) Rubbing or massaging the frozen area is extremely painful but enhances rapid thawing and prevents further crystallization damage.
(b) Keep the thawed part dependent (below heart level) to enhance circulation to the thawed tissues.
(c) Do not thaw the frozen part if there is any possibility that it will be frozen again.

(d) Do not administer analgesia to a patient with frostbite.
(e) Using a sterile needle, puncture and allow drainage of frostbite blisters to allow for external expansion of ice crystals (thus minimizing further internal damage).

83. Compared to the osmotic pressure of human blood, freshwater is
(a) hypertonic.
(b) isotonic.
(c) hypotonic.
(d) atonic.
(e) None of the above.

84. Compared to the osmotic pressure of human blood, saltwater is
(a) hypertonic.
(b) isotonic.
(c) hypotonic.
(d) atonic.
(e) None of the above.

85. Freshwater aspiration produces
(a) possible hemodilution.
(b) surfactant destruction with subsequent pulmonary edema and atelectasis.
(c) drowning without pulmonary edema.
(d) Both answers (a) and (b).
(e) Both answers (a) and (c).

86. Saltwater aspiration produces pulmonary edema by
(a) filling the alveoli with saltwater, which prevents communication with the capillaries for exchange of gas or fluid.
(b) drawing fluid from the circulatory system into the alveoli.
(c) "washing away" the alveolar surfactant as it is drawn into the circulatory system by osmotic pressure, creating a generalized fluid overload.
(d) obstructing alveoli with sea matter and causing an antigen-antibody reaction.
(e) None of the above.

87. Which of the following statements regarding factors that influence success of resuscitation and near-drowning survival is false?

(a) The colder the water, the greater the probability of successful resuscitation and survival.
(b) Beyond 60 minutes of submersion, resuscitation is unlikely.
(c) Children survive longer submersion times with a greater probability of successful resuscitation and survival.
(d) Cleanliness of the water affects probability of survival.
(e) The mammalian diving reflex improves the probability of successful resuscitation and survival by increasing pulse and cardiac output to all vital organs.

88. Which of the following statements regarding treatment of near-drowning victims is false?

(a) Only a trained rescue swimmer, secured by a safety line, should be allowed to enter the water.
(b) Suspect a head or neck injury if there is any indication of trauma.
(c) Mouth-to-mouth resuscitation should be immediately initiated, even while still in the water.
(d) The Heimlich maneuver should be performed 4 to 6 times prior to intubation to facilitate draining of the lungs prior to positive pressure ventilation.
(e) All near-drowning victims require hospital admission for observation.

89. Which of the following statements regarding injuries associated with fresh or saltwater diving is true?

(a) Injuries may occur on the water surface secondary to direct trauma or entanglement in lines or aquatic matter.
(b) Barotrauma to the middle ear may occur during descent to depth, especially if the diver has an upper respiratory tract infection.
(c) At the bottom of the dive, nitrogen narcosis may result in trauma secondary to euphoria, impaired judgment, or diminished motor coordination.
(d) Barotrauma to the middle ear or lung parenchyma may occur during ascent.
(e) All of the above are true.

90. A diver who holds her/his breath when ascending from depth may suffer from
(a) lung trauma caused by expansion of trapped air as atmospheric pressure decreases.
(b) air embolism secondary to ruptured alveoli.
(c) subcutaneous or mediastinal emphysema due to gas diffusion from the lungs.
(d) a pneumothorax.
(e) Any of the above.

91. Signs and symptoms of decompression sickness Type I ("the bends") include
(a) localized musculoskeletal pain, particularly in the joints.
(b) skin changes: puritis erythema, spotted pallor or cyanosis, pitting edema.
(c) dyspnea, chest pain, headache, nausea/vomiting, paresthesis, or paralysis.
(d) Answers (a) and (b) only.
(e) Answers (a), (b), and (c).

92. Signs and symptoms of decompression sickness Type II include
(a) localized musculoskeletal pain, particularly in the joints.
(b) skin changes: puritis erythema, spotted pallor or cyanosis, pitting edema.
(c) dyspnea, chest pain, headache, nausea/vomiting, paresthesis, or paralysis.
(d) Answers (a) and (b) only.
(e) Answers (a), (b), and (c).

93. Management of decompression sickness includes all of the following, except
(a) ABC assessment with CPR as needed.
(b) oxygen administration with nonrebreather mask at 15 LPM for the conscious patient.
(c) IV access with D_5W TKO.
(d) administration of dexamethasone, heparin, Dextran, or diazepam as ordered by the base physician.
(e) deferral of air evacuation unless sea level cabin pressure can be maintained or low altitude flight safely possible.

94. Air embolism secondary to pulmonary overpressure accidents may result in

(a) AMI.
(b) CVA.
(c) tension pneumothorax.
(d) Answers (a) and (b) only.
(e) Answers (a), (b), and (c).

95. Management of pulmonary overpressure accidents includes all of the following, except

(a) ABC assessment with CPR as needed.
(b) oxygen administration with a nonrebreather mask at 15 LPM for the conscious patient.
(c) right lateral Trendelenburg positioning.
(d) administration of dexamethasone as ordered by the base physician.
(e) consideration of transport to a hospital with a hyperbaric oxygen chamber.

96. Signs and symptoms of a pneumomediastinum include all of the following, except

(a) profound cyanosis in all cases.
(b) substernal chest pain.
(c) irregular pulse.
(d) hypotension and/or narrow pulse pressure.
(e) voice changes.

97. Treatment of pneumomediastinum includes all of the following, except

(a) NRB oxygenation at 15 LPM.
(b) IV access of D_5W TKO.
(c) cardiac monitoring.
(d) treatment of acute symptoms per related protocols.
(e) transportation to the hospital.

98. Which of the following statements regarding radiation exposure is true?

(a) Alpha and beta particles are the most penetrating and dangerous.
(b) Gamma rays are dangerous only if the particles they irradiate are swallowed or inhaled.
(c) X-ray machines do not irradiate body cells.
(d) All of the above are true.
(e) None of the above is true.

99. Sources of radiation that provide potential for radiation injuries include all of the following, except

(a) natural cosmic radiation from terrestrial radionucleides.
(b) diagnostic X-rays and radiotherapy.
(c) radiopharmaceuticals.
(d) atomic energy industry and laboratories.
(e) transportation of radioactive (nuclear) waste.

100. Which of the following statements regarding exposure factors is true?

(a) The longer the exposure duration, the greater the radiation received.
(b) The greater the distance from the source, the smaller the amount of radiation received.
(c) The more material located between the source and the victim (shielding), the less radiation received.
(d) the stronger the radiation source, the more distance and/or shielding required for safety.
(e) All of the above are true.

101. Which of the following statements regarding the geriatric patient is false?

(a) Normal aging results in a general decline of all organ systems.
(b) Normal aging results in significantly slower metabolic activity.
(c) The geriatric patient may have difficulty separating the consequences of aging from the effects of disease, and may fail to report important symptoms.
(d) Chronic problems may make assessment of acute signs and symptoms difficult and confusing.
(e) Because pain perception is diminished or absent, the patient or responder may underestimate the severity of the patient's condition.

102. Although not limited to the geriatric populace, normal aging and/or chronic illness leads to a thickening of the walls of the heart (especially the left ventricle) without a corresponding increase in the size of the cavities. This contributes to a decline in stroke volume and is called cardiac

(a) osteoporosis.
(b) kyphosis.
(c) hypertrophy.
(d) fibrosis.
(e) spondylolysis.

103. The elderly musculoskeletal structure becomes more porous, softer, more brittle, and less flexible. This condition is referred to as

(a) osteoporosis.
(b) kyphosis.
(c) hypertrophy.
(d) fibrosis.
(e) spondylolysis.

104. A breakdown or degeneration of vertebrae occurs with age. Sudden neck movement, even without fracture, may cause spinal cord injury. This vertebral degeneration is called

(a) osteoporosis.
(b) kyphosis.
(c) hypertrophy.
(d) fibrosis.
(e) spondylolysis.

105. Many elderly people develop a "hunchback" appearance from rheumatoid arthritis, vertebral degeneration, and/or poor posture. This exaggeration or angulation of the normal posterior curve of the thoracic spine is called

(a) osteoporosis.
(b) kyphosis.
(c) hypertrophy.
(d) fibrosis.
(e) spondylolysis.

106. Scar tissue formation within the connective tissue framework of aging organs (and peripheral vascular system) contributes to the decline of their function. This formation is called

(a) osteoporosis.
(b) kyphosis.
(c) hypertrophy.
(d) fibrosis.
(e) spondylolysis.

107. Communication problems are common when one is dealing with the elderly. Social and emotional factors are often compounded by diminished hearing and sight. A cloudlike opacity of the eye's lens is produced by

(a) cataracts.
(b) dysphagia.

(c) glaucoma.
(d) intraocular hydrocephalopathy.
(e) chronic hyphema.

108. A disease of the eye that produces increased intraocular pressure, and results in atrophy of the optic nerve and blindness, is called

(a) cataracts.
(b) dysphagia.
(c) glaucoma.
(d) intraocular hydrocephalopathy.
(e) chronic hyphema.

109. Which of the following statements regarding mental status considerations when one is assessing the geriatric patient is false?

(a) The majority of geriatric patients are either senile or simply imagining physical ailments for attention.
(b) Alcoholism is more common in the elderly than is generally realized, and may interfere with obtaining an accurate history.
(c) Depression is common and may mimic senility or organic brain syndrome.
(d) Suicide is the fourth leading cause of death among the elderly population of the United States.
(e) None of the above is false.

110. Which of the following statements regarding trauma and the elderly is false?

(a) Elderly people are biologically more at risk for trauma (especially falls) because of slower reflexes, failing eyesight/hearing, arthritis, less elastic blood vessels, and more fragile tissues and bones.
(b) Elderly people are at high risk for trauma from criminal assault.
(c) The elderly are more prone to head injury, even from minor mechanisms.
(d) Elderly patients are slower to develop signs and symptoms from brain trauma (sometimes requiring days or weeks) and may have forgotten the original injury.
(e) None of the above is false.

111. Priorities of care when one is treating the geriatric trauma patient are similar to those for all trauma patients. However, consideration of factors specific to the elderly is important. Which of the following statements regarding these factors is false?

(a) There may be decreased response of the heart to hypovolemia in terms of rate and stroke volume adjustment.
(b) The elderly patient requires a greater amount of IV fluid infusion to support the higher arterial pressures required for perfusion of vital organs because of chronically increased peripheral vascular resistance and history of hypertension.
(c) Positive pressure ventilation may rupture pulmonary parenchyma.
(d) In the geriatric patient, all organs have less tolerance for anoxia.
(e) Physical deformities (arthritis, spinal abnormalities, atrophied limbs) will require modification of standard immobilization techniques.

112. Which of the following statements regarding medical emergencies and the geriatric patient is false?

(a) Dyspnea may be the only symptom when AMI occurs.
(b) Congestive heart failure may be acute or chronic.
(c) Syncope is frequently experienced by the elderly and is rarely significant.
(d) Occlusive stroke is statistically more common in the elderly.
(e) Seizures may be mistaken for CVA in the elderly patient; conversely, a first-time seizure may result from previous CVA and is likely to affect the partially paralyzed limb.

113. Loss of neurons begins slowly in people in their forties and fifties, continuing at a progressive rate into the later decades. Which of the following statements regarding this loss is false?

(a) True psychiatric disorders are uncommon in the elderly.
(b) "Senility," "dementia," and "organic brain syndrome" (OBS) are pathologically and clinically alike.

(c) Alzheimer's is a progressive disease that produces memory impairment, cognitive degeneration, and abnormal behaviors.
(d) Etiologies of chronic organic dementia include aging, CVA, TIA, cerebral atherosclerosis, neurologic and hereditary diseases.
(e) None of the above is false.

114. It is vitally important to determine whether altered mentation is chronic, acute, or acutely different. Acute etiologies that may be mistaken for "senile dementia" include
(a) subdural hematoma, brain lesions, or tumors.
(b) drug or alcohol intoxication, excessive drug doses, or medication interaction reactions.
(c) CNS infections or cardiac failure.
(d) nutritional deficiencies or electrolyte abnormalities.
(e) All of the above.

115. Which of the following statements regarding other disorders/emergencies and the geriatric patient is false?
(a) True psychiatric disorders are uncommon in the elderly.
(b) The elderly are more susceptible to hypothermia.
(c) The elderly are more susceptible to hyperthermia from both environmental and metabolic causes.
(d) GI bleed is the most common geriatric gastrointestinal disorder.
(e) Adverse drug interactions frequently occur secondary to multiple prescriptions. Also, forgetfulness and/or confusion often results in overdose or underdose of medications.

116. Geriatric abuse is defined as a syndrome in which elderly persons have received serious physical (or psychological) injury from their children or care providers. Which of the following statements regarding geriatric abuse is false?
(a) The average age of the abused elder is older than 80.
(b) Elderly abuse is generally limited to low-income socioeconomic situations.
(c) The older person is no longer able to be independent, and the family has difficulty upholding commitment to care of this person.
(d) Unexplained trauma is the primary finding.
(e) Inconsistencies between the patient-provided and family-provided histories may be important clues to report to authorities.

117. Physical examination of the neonate or infant includes observation of the anterior fontanelle. In a normal state, the fontanelle

(a) feels tight and may bulge.
(b) is level with the skull or slightly sunken.
(c) falls below the level of the skull surface and appears sunken.
(d) Any of the above.
(e) None of the above.

118. With increased intracranial pressure, the fontanelle

(a) feels tight and may bulge.
(b) is level with the skull or slightly sunken.
(c) falls below the level of the skull surface and appears sunken.
(d) Any of the above.
(e) None of the above.

119. With dehydration, the fontanelle

(a) feels tight and may bulge.
(b) is level with the skull or slightly sunken.
(c) falls below the level of the skull surface and appears sunken.
(d) Any of the above.
(e) None of the above.

120. The generally accepted definition of a neonate is the infant from birth to the age of

(a) 1 week old.
(b) 2 weeks old.
(c) 1 month old.
(d) 2 months old.
(e) 6 months old.

121. Sudden infant death syndrome (SIDS) is defined as the sudden death of an infant between the ages of 1 week and 1 year, which is

(a) the result of regurgitation and aspiration of vomitus.
(b) unexpected by history and in which a thorough postmortem examination fails to reveal an adequate cause of death.
(c) the result of external suffocation (for example, from a pillow or blanket).
(d) the result of child abuse.
(e) Any of the above.

122. Which of the following statements regarding SIDS is true?
- (a) The incidence of SIDS is greatest during the hot months of summer.
- (b) SIDS is more common in female infants than males.
- (c) SIDS is more prevalent among families from middle and upper socioeconomic groups.
- (d) Infants with low birth weight are at greatest risk for SIDS.
- (e) All of the above are true.

123. A SIDS infant may exhibit any of the following, except
- (a) a normal state of nutrition and hydration.
- (b) frothy fluids in and around the mouth and nostrils which may be blood tinged.
- (c) the presence of vomitus.
- (d) signs of repeated trauma (such as multiple bruises of different ages).
- (e) unusual positioning at time of death.

124. When presented with an apneic and pulseless infant exhibiting rigor mortis or dependent lividity,
- (a) follow your local protocol regarding performance of BLS alone, BLS and ACLS, or deferral of resuscitation efforts.
- (b) never attempt resuscitation; protect the crime scene until the police arrive.
- (c) provide unconditional support for the parents.
- (d) Both answers (a) and (c).
- (e) Both answers (b) and (c).

125. Which of the following characteristics of the abused child is false?
- (a) Girls are more often abused than boys.
- (b) Handicapped or frequently ill children, or those with special needs are at great risk for abuse.
- (c) Illegitimate or unwanted children are at high risk for abuse.
- (d) Uncommunicative (autistic) children are at great risk.
- (e) Premature infants or twins are at high risk.

126. Which of the following characteristics of the child abuser is false?

(a) The abuser is usually a parent or individual in the role of a parent.
(b) The abuser may come from any geographic, religious, ethnic, or occupational group.
(c) When the mother is the parent who spends the most time with the child, she is frequently identified as the abuser.
(d) The abuser often was the victim of physical or emotional abuse as a child.
(e) The vast majority of abusers are either high-school dropouts or of a low-income socioeconomic group.

127. Physical findings that should prompt suspicion of child abuse include all of the following, except

(a) any obvious or suspected fractures in a child less than 2 years old.
(b) more injuries than usually seen in children of the same age, or multiple injuries of various ages.
(c) any infant less than 1 year old who appears healthy and atraumatic, but who "died in his sleep."
(d) bruises or burns in patterns that suggest intentional infliction.
(e) any injury that does not fit with the provided description of cause.

128. Historical points that should prompt suspicion of child abuse include all of the following, except

(a) the parent's account is vague, inconsistent, or does not account for the nature or severity of injury.
(b) history of a previous SIDS death (SIDS does not occur twice in the same family).
(c) accusation or inference that the child injured herself/himself intentionally.
(d) an inappropriate delay in seeking help occurred.
(e) the child is dressed inappropriately for the situation.

129. Which of the following statements regarding physical neglect and sexual abuse is false?

(a) Extreme malnutrition suggests neglect.
(b) Genital injury is required to have occurred before sexual abuse can be suspected.

(c) Multiple insect bites are suggestive of neglect.
(d) Extreme lack of cleanliness is suggestive of neglect.
(e) None of the above is false.

130. Causes of seizures in the pediatric patient include
(a) fever, infection, or electrolyte abnormalities.
(b) hypoxia or hypoglycemia.
(c) head trauma, tumors, CNS malformations, or idiopathic epilepsy.
(d) toxic ingestions or exposure.
(e) All of the above.

131. Which of the following statements regarding pediatric febrile seizures is false?
(a) Seizures are normal for any febrile child, easily diagnosed in the field, and do not constitute a significant illness.
(b) Febrile seizures commonly occur between the ages of 6 months and 6 years and are caused by a rapid rise in temperature.
(c) Suspect fever as the cause of seizure if the temperature is above 103 degrees F (39.2 C).
(d) All pediatric patients who experience a seizure must be transported for evaluation to rule out a significant underlying illness or injury.
(e) None of the above is false.

132. The pediatric dose of diazepam for infants and children up to the age of 5 years is
(a) 0.2–0.5 mg (slowly) every 2–5 minutes, up to a maximum of 2.5 mg.
(b) 1 mg every 2–5 minutes, up to a maximum of 5 mg.
(c) 2–5 mg every 5 minutes, up to a maximum of 20 mg.
(d) 0.001 mg every 5 minutes, up to a maximum of 1 mg.
(e) 0.2–0.5 mg/kg every 5 minutes, up to a maximum of 10 mg.

133. The pediatric dose of diazepam for children 5 years old or older is
(a) 0.2–0.5 mg (slowly) every 2–5 minutes, up to a maximum of 2.5 mg.
(b) 1 mg every 2–5 minutes, up to a maximum of 5 mg.
(c) 2–5 mg every 5 minutes, up to a maximum of 20 mg.
(d) 0.001 mg every 5 minutes, up to a maximum of 1 mg.
(e) 0.2–0.5 mg/kg every 5 minutes, up to a maximum of 10 mg.

134. Which of the following statements regarding dehydration and the pediatric patient is true?

(a) Because of their youth and resilient systems, children rarely suffer from dehydration.
(b) Suspicion of neglect should be prompted by signs and symptoms of dehydration in any pediatric patient, and the proper authorities should be notified immediately.
(c) Because of the danger of fluid overload in the pediatric patient, dehydration should be treated with D_5W on a microdrip infusion set only.
(d) The initial IV fluid bolus for the dehydrated pediatric patient exhibiting signs and symptoms of shock is 20 ml/kg.
(e) All of the above are true.

135. A bacterial infection of the bloodstream is called

(a) meningitis.
(b) sepsis.
(c) Reyes syndrome.
(d) Both answers (a) and (b).
(e) Answers (a), (b), and (c).

136. Aspirin administration to pediatric patients is contra-indicated because of its correlation to the incidence of

(a) meningitis.
(b) sepsis.
(c) Reyes syndrome.
(d) Both answers (a) and (b).
(e) Answers (a), (b), and (c).

137. Fever, lethargy, and irritability are often associated with

(a) meningitis.
(b) sepsis.
(c) Reyes syndrome.
(d) Both answers (a) and (b).
(e) Answers (a), (b), and (c).

138. A full or bulging fontanelle may be indicative of

(a) meningitis.
(b) sepsis.
(c) measles.
(d) Both answers (a) and (b).
(e) None of the above.

139. Complaints of severe headache and stiff neck are indicative of

(a) meningitis.
(b) sepsis.
(c) Reyes syndrome.
(d) Both answers (a) and (b).
(e) Answers (a), (b), and (c).

140. A recent history of chicken pox is present in 10 to 20 percent of the cases of

(a) meningitis.
(b) sepsis.
(c) Reyes syndrome.
(d) Both answers (a) and (b).
(e) Answers (a), (b), and (c).

141. A recent history of upper respiratory tract infection is present in many cases of

(a) meningitis.
(b) sepsis.
(c) Reyes syndrome.
(d) Both answers (a) and (b).
(e) Answers (a), (b), and (c).

142. Laryngotracheobronchitis is more commonly referred to as

(a) bronchiolitis.
(b) epiglottitis.
(c) croup.
(d) All of the above.
(e) None of the above.

143. Prominent expiratory wheezing is characteristic of

(a) bronchiolitis.
(b) epiglottitis.
(c) croup.
(d) Answers (a) and (b) only.
(e) Answers (a), (b), and (c).

144. Fever often accompanies

(a) bronchiolitis.
(b) epiglottitis.
(c) croup.
(d) All of the above.
(e) None of the above.

145. A harsh, barking cough is characteristic of
(a) bronchiolitis.
(b) epiglottitis.
(c) croup.
(d) Either answer (a) or (b).
(e) Either answer (b) or (c).

146. Resistance to being cradled or placed supine, difficulty in swallowing, and drooling is characteristic of
(a) bronchiolitis.
(b) epiglottitis.
(c) croup.
(d) Both answers (a) and b).
(e) Both answers (b) and (c).

147. Stridor may accompany
(a) bronchiolitis.
(b) epiglottitis.
(c) croup.
(d) Either answer (a) or (b).
(e) Either answer (b) or (c).

148. Attempted visualization of the posterior oropharynx is contraindicated in
(a) bronchiolitis.
(b) epiglottitis.
(c) croup.
(d) Both answers (a) and (b).
(e) Both answers (b) and (c).

149. Humidified oxygen should be administered in all cases of
(a) bronchiolitis.
(b) epiglottitis.
(c) croup.
(d) Answers (a) and (b) only.
(e) Answers (a), (b), and (c).

150. When one is defibrillating a child, the correct initial energy setting is
(a) 200 ws (joules).
(b) 1 ws (joule) per kilogram.
(c) 2 ws (joules) per kilogram.
(d) 4 ws (joules) per kilogram.
(e) 10 ws (joules) per kilogram.

151. If the initial defibrillation is unsuccessful, the subsequent energy setting is

(a) 200 ws (joules).
(b) 1 ws (joule) per kilogram.
(c) 2 ws (joules) per kilogram.
(d) 4 ws (joules) per kilogram.
(e) 10 ws (joules) per kilogram.

152. Which of the following statements regarding pediatric endotracheal intubation is false?

(a) The technique for pediatric endotracheal intubation is similar to that of the adult.
(b) The pediatric tongue is larger and the glottis is found higher than that of an adult.
(c) Nasotracheal intubation is contraindicated in pediatric patients.
(d) Pediatric laryngoscope blades will fit any adult-sized laryngoscope handle.
(e) None of the above is false.

153. Pelvic inflammatory disease (PID)

(a) is an acute infection of the uterus, ovaries, and/or fallopian tubes.
(b) is a chronic infection of the uterus, ovaries, and/or fallopian tubes.
(c) may involve the peritoneum and intestines.
(d) Answers (b) and (c) only.
(e) Answers (a), (b), and (c).

154. Causes of PID include

(a) gonorrhea.
(b) staph infection.
(c) strep infection.
(d) All of the above.
(e) None of the above.

155. PID may produce any of the following signs and symptoms, except

(a) a rigid abdomen and rebound tenderness.
(b) severe postural hypotension.
(c) fever, chills, and nausea/vomiting.
(d) tachycardia.
(e) vaginal discharge and erratic menstrual periods.

156. Which of the following statements regarding sexual assault is false?

(a) The rapist is primarily motivated by sexual desire.
(b) There is no "typical victim" profile for sexual assault.
(c) Sexual assault is not limited to vaginal or anal penetration.
(d) The rapist is frequently motivated by aggression and the desire to inflict pain.
(e) The rapist is frequently motivated by the desire to control and humiliate the victim.

157. Management of the sexual assault victim includes assessment of and treatment for any life-threatening injuries. Preservation of evidence is also an important concern. Which of the following statements regarding preservation of evidence is false?

(a) Do not allow the victim to change clothing, shower, or even wash the face or hands.
(b) Handle clothing as little as possible.
(c) Place any blood- or fluid-stained articles that must be removed in separate plastic bags.
(d) Avoid cleansing of any wounds.
(e) Do not disturb the crime scene.

158. The placenta performs all of the following functions, except

(a) providing an extra cushion effect to protect the fetus from trauma.
(b) transfer of gases (fetal respiration).
(c) transport of nutrients and excretion of wastes.
(d) transfer of heat.
(e) hormone production.

159. A woman who is pregnant for the first time is referred to as a

(a) primagravida.
(b) primapara.
(c) multigravida.
(d) multipara.
(e) None of the above.

160. A woman who has given birth to her first child is referred to as a

(a) primagravida.
(b) primapara.

(c) multigravida.
(d) multipara.
(e) None of the above.

161. A woman who has borne more than one child is referred to as a

(a) primagravida.
(b) primapara.
(c) multigravida.
(d) multipara.
(e) None of the above.

162. A woman who has been pregnant several times is referred to as a

(a) primagravida.
(b) primapara.
(c) multigravida.
(d) multipara.
(e) None of the above.

163. Your patient is ready to deliver her second child. She has a two-year-old son, and has had one previous miscarriage (no abortions or adoptions). You would describe her as being

(a) para 1, gravida 2.
(b) gravida 2, para 1.
(c) para 2, gravida 3.
(d) gravida 3, para 1.
(e) para 3, gravida 1.

164. Which of the following statements regarding pregnancy and medical disorders is false?

(a) Pregnancy may cause previously diagnosed diabetes to become unstable.
(b) Pregnancy may cause the onset of diabetes in a previously nondiabetic patient.
(c) Pregnancy may produce hypertension in the previously normotensive patient.
(d) Pregnancy may precipitate CHF in patients who had previously asymptomatic heart disease.
(e) None of the above is false.

165. A spontaneous abortion is commonly called a miscarriage, and frequently occurs

(a) when the embryo or fetus is malformed.
(b) as a result of maternal infection.
(c) during the third trimester of pregnancy.
(d) Answers (a) and (b) only.
(e) Answers (a), (b), and (c).

166. An ectopic pregnancy occurs when a fertilized ovum is implanted

(a) anywhere outside of the uterine cavity.
(b) in a fallopian tube only.
(c) in the abdominal cavity only.
(d) at the mouth of the cervix.
(e) within the uterus.

167. The most common site of ectopic implantation is

(a) on an ovary.
(b) in a fallopian tube.
(c) in the abdominal cavity.
(d) at the mouth of the cervix.
(e) within the uterus.

168. Any female patient of childbearing age complaining of abdominal pain, with or without vaginal bleeding, is suffering from ______ until proven otherwise.

(a) spontaneous abortion
(b) ectopic pregnancy
(c) criminal abortion
(d) appendicitis
(e) Any of the above.

169. The premature separation of the placenta from the uterine wall is called

(a) spontaneous abortion.
(b) traumatic abortion.
(c) placenta previa.
(d) abruptio placenta.
(e) ectopic placenta.

170. Occasionally a placenta develops too low in the uterus and partially or completely covers the opening of the cervix. This condition is called

(a) spontaneous abortion.
(b) traumatic abortion.
(c) placenta previa.
(d) abruptio placenta.
(e) ectopic placenta.

171. Painless vaginal bleeding late in pregnancy is considered to be

(a) spontaneous abortion.
(b) traumatic abortion.
(c) placenta previa.
(d) abruptio placenta.
(e) ectopic placenta.

172. Treatment of spontaneous abortion, placenta previa, abruptio placenta, and ectopic pregnancy includes all of the following, except

(a) high-flow oxygen administration.
(b) one or more IV lines of NS or LR.
(c) pneumatic antishock garment with inflation of the abdominal compartment only (tamponade effect).
(d) left laterally recumbent positioning for second- and third-trimester patients.
(e) emergency transport if signs and symptoms of shock are present.

173. Preeclampsia is characterized by

(a) pregnancy-induced hypertension.
(b) excessive weight gain with edema (most pronounced in the hands and face).
(c) seizures.
(d) Answers (a) and (b) only.
(e) Answers (a), (b), and (c).

174. Eclampsia is characterized by

(a) pregnancy-induced hypertension.
(b) excessive weight gain with edema (most pronounced in the hands and face).
(c) seizures.
(d) Answers (a) and (b) only.
(e) Answers (a), (b), and (c).

175. Management of the patient with preeclampsia includes
- (a) oxygen administration and IV access.
- (b) left laterally recumbent positioning.
- (c) IV administration of magnesium sulfate and/or diazepam.
- (d) Answers (a) and (b) only.
- (e) Answers (a), (b), and (c).

176. Management of the patient with eclampsia includes
- (a) oxygen administration and IV access.
- (b) left laterally recumbent positioning.
- (c) IV administration of magnesium sulfate and/or diazepam.
- (d) Answers (a) and (b) only.
- (e) Answers (a), (b), and (c).

177. Supine-hypotensive syndrome occurs when the abdominal mass of the gravid uterus is large and
- (a) demands more of the maternal blood supply than can be compensated for by the already increased cardiac output.
- (b) compresses the inferior vena cava, thereby reducing venous return, and cardiac output falls.
- (c) compresses the abdominal aorta, thereby obstructing cardiac output.
- (d) Both answers (a) and (b).
- (e) Both answers (a) and (c).

178. In the absence of any signs and symptoms of volume depletion, management of supine-hypotensive syndrome includes
- (a) left laterally recumbent positioning.
- (b) Trendelenburg positioning.
- (c) semi-Fowler's positioning.
- (d) right laterally recumbent positioning.
- (e) oxygen and two large-bore IVs.

179. Full dilation of the cervix signifies
- (a) the beginning of the fourth stage of labor.
- (b) the end of the second stage of labor.
- (c) the end of the first stage of labor.
- (d) the beginning of the first stage of labor.
- (e) the end of the third stage of labor.

180. The birth of the baby signifies
(a) the beginning of the fourth stage of labor.
(b) the end of the second stage of labor.
(c) the end of the first stage of labor.
(d) the beginning of the first stage of labor.
(e) the end of the third stage of labor.

181. The beginning of regular uterine contractions signifies
(a) the beginning of the fourth stage of labor.
(b) the end of the second stage of labor.
(c) the end of the first stage of labor.
(d) the beginning of the first stage of labor.
(e) the end of the third stage of labor.

182. The delivery of the placenta signifies
(a) the beginning of the fourth stage of labor.
(b) the end of the second stage of labor.
(c) the end of the first stage of labor.
(d) the beginning of the first stage of labor.
(e) the end of the third stage of labor.

183. The classic sign/symptom that delivery is imminent is
(a) the uncontrollable maternal urge to push or the crowning of the fetus.
(b) contractions only 5 minutes apart.
(c) the rupture of the amniotic sac.
(d) uterine contractions only 15 to 30 seconds in duration.
(e) Any of the above.

184. As the delivery begins, if the baby's head is the first structure to be seen it is called
(a) a cephalic presentation.
(b) a breech presentation.
(c) an occipital presentation.
(d) a head presentation.
(e) a posterior presentation.

185. As the delivery begins, if the baby's buttocks are the first structure to be seen it is called
(a) a cephalic presentation.
(b) a breech presentation.
(c) an occipital presentation.
(d) a head presentation.
(e) a posterior presentation.

186. Preparation for an imminent delivery includes all of the following, except

(a) maternal oxygen administration.
(b) IV access with NS or LR.
(c) use of drapes and sterile gloves.
(d) obtaining a receiving blanket and towels.
(e) performing a vaginal exam to determine presenting part prior to attempted field delivery.

187. If the amniotic sac has not ruptured, as the baby's membrane-enclosed head emerges

(a) use your fingers to pinch and puncture the sac, and then push the sac away from the baby's nose and mouth for suctioning.
(b) continue with the delivery; the sac will break on its own once the baby is completely delivered.
(c) use a sterile scalpel to carefully open the sac and provide suction before the baby takes its first breath.
(d) leave the sac alone unless the baby appears to be breathing.
(e) delay further delivery and transport immediately for an emergency cesarean section.

188. When the head emerges, if the umbilical cord is wrapped around the baby's neck and cannot be gently loosened,

(a) continue with the delivery; the cord will unwind naturally as the baby delivers.
(b) delay further delivery and transport immediately for an emergency cesarean section.
(c) clamp or tie the cord in two places and cut the cord between the clamps/ties.
(d) forcefully pull the cord over the baby's head before the baby is strangled.
(e) do not delay for clamping or tying; cut the cord and remove it before the baby is strangled.

189. The proximal clamp is placed on the umbilical cord at approximately

(a) 4 inches (10 centimeters) from the baby.
(b) 6 inches (15 centimeters) from the baby.
(c) 8 inches (20 centimeters) from the baby.
(d) 10 inches (26 centimeters) from the baby.
(e) 12 inches (31 centimeters) from the baby.

190. The distal clamp is placed on the umbilical cord at approximately

(a) 4 inches (10 centimeters) from the baby.
(b) 6 inches (15 centimeters) from the baby.
(c) 8 inches (20 centimeters) from the baby.
(d) 10 inches (26 centimeters) from the baby.
(e) 12 inches (31 centimeters) from the baby.

191. Given the fact that delivery of the placenta is normally accompanied by bleeding, the paramedic should

(a) gently pack the vagina with a sterile sanitary napkin.
(b) expect excessive blood loss, which will rarely lead to shock.
(c) gently massage the abdomen (uterine fundus) to assist in uterine contraction and reduction of hemorrhage.
(d) All of the above.
(e) None of the above.

192. Cephalopelvic disproportion

(a) results in the baby's head not passing through the maternal pelvis.
(b) is the result of a contracted pelvis, oversized baby, or fetal abnormalities.
(c) results in strong, frequent contractions for a prolonged period of time without delivery of the baby.
(d) will result in fetal demise and/or uterine rupture if a cesarean section is not performed.
(e) All of the above.

193. Which of the following statements regarding abnormal deliveries is false?

(a) If a prolapsed cord becomes cold, the oxygen supply to the fetus will cease and the infant will die.
(b) The insertion of a gloved hand into the mother's vagina may be necessary, to gently push the baby's head off of the prolapsed cord.
(c) The insertion of a gloved hand into the mother's vagina may be necessary to gently replace a single-limb presentation prior to field delivery.
(d) The insertion of a gloved hand into the mother's vagina may be necessary to provide an airway for the breech-birth baby whose head will not deliver.
(e) If both feet present first, the baby may still be delivered in the field.

194. When a delivery becomes a multiple-birth situation, it is important to

(a) be on your way to the hospital before the second or third baby delivers.
(b) provide suction, clamp and cut the umbilical cord, and warm one infant before delivering the next.
(c) wait for delivery of the first baby's placenta prior to allowing the next baby to deliver.
(d) All of the above.
(e) None of the above.

195. A ruptured uterus

(a) frequently occurs spontaneously and is a life-threatening emergency.
(b) can occur because of abdominal trauma or during prolonged labor.
(c) causes extreme pain and will always present with excessive external bleeding.
(d) All of the above.
(e) None of the above.

196. Uterine inversion is when the uterus literally turns inside out upon delivery of the infant and/or placenta. Uterine inversion

(a) produces profound, life-threatening shock.
(b) can be caused by pulling on the umbilical cord.
(c) can be caused by attempts to express the placenta when the uterus is relaxed.
(d) Answers (a) and (b) only.
(e) Answers (a), (b), and (c).

197. Management of uterine inversion includes all of the following, except

(a) high-flow oxygen and supine patient positioning.
(b) two or more IVs of NS or LR.
(c) one attempt to replace the uterus manually, by exerting pressure on the area surrounding the cervix.
(d) one attempt to detach the placenta from the extruded uterus.
(e) moist towel packing of all protruding tissue and an emergency return to the ER.

198. Pulmonary embolism (PE) is one of the most common causes of maternal death. Which of the following statements regarding PE is true?

(a) PE may occur at any time during pregnancy.
(b) PE may occur during labor.
(c) PE may occur in the postpartum period.
(d) PE most commonly follows a cesarean section.
(e) All of the above are true.

199. Suctioning of the newborn's mouth and nose should be performed

(a) immediately upon emergence of the head.
(b) immediately after birth (prior to severing the umbilical cord) and as often as needed to ensure a clear airway (but without deprivation of oxygen).
(c) only after the umbilical cord is cut.
(d) Both answers (a) and (b).
(e) Any of the above (depending upon local protocols).

200. As soon as the baby is born, place it at the level of the mother's vagina,

(a) on its side with the head elevated to allow for adequate suction of its mouth and nose.
(b) providing stimulation by exposure to room temperature air.
(c) on its back with the head slightly lower than its body to allow for adequate drainage of the airway.
(d) wrapped in a warm blanket and on its side with the head elevated to allow for suctioning.
(e) on its side with the head slightly lower to allow for drainage and suction of fluids.

201. APGAR assessment of the newborn should be performed

(a) only in the ambulance, on the way to the hospital.
(b) at 1 and 5 minutes after birth.
(c) at 1 and 5 minutes after the placenta delivers.
(d) at 5 and 10 minutes after the placenta delivers.
(e) at 5-minute intervals after birth.

202. When the APGAR mnemonic and scoring system is used to assess the newborn, the two "A"s stand for

(a) appearance (skin color) and apnea (absent respirations).
(b) activity (extremity movement) and altered mentation (level of consciousness).
(c) appearance (skin color) and activity (extremity movement).
(d) appearance (skin color) and activity (heart rate).
(e) activity (grimace) and appearance (extremity movement).

203. When the APGAR mnemonic and scoring system is used to assess the newborn, the "P" stands for

(a) pulse (heart rate).
(b) pupils (equal or unequal).
(c) pink (skin color).
(d) pallor (skin color).
(e) purple (cyanosis).

204. When the APGAR mnemonic and scoring system is used to assess the newborn, the "G" stands for

(a) gasping (poor respiratory effort).
(b) gurgling (indication of aspiration).
(c) gross deformity (infant abnormalities).
(d) grimace (irritability/crying response to stimulus).
(e) groaning (poor ability of cry).

205. When the APGAR mnemonic and scoring system is used to assess the newborn, the "R" stands for

(a) rigor (stillborn infant).
(b) rapid (heart rate or respiratory rate).
(c) rate (of pulse or respirations).
(d) respirations (rate and effort of respirations).
(e) robust (healthy, crying infant).

206. When caring for the premature newborn, the paramedic must pay even stricter attention to

(a) prevention of heat loss by thoroughly drying the baby and wrapping it in a dry, foil-lined blanket.
(b) provision of oxygen and suctioning of the airway.
(c) protection from external sources of contamination.
(d) Answers (a) and (b) only.
(e) Answers (a), (b), and (c).

207. Fetal feces is greenish-black to brown and may be present in the amniotic fluid. Presence of this fecal matter is produced by a period of fetal distress and can cause severe lung infection if aspirated. This material is called

(a) meconium.
(b) mercronial feces.
(c) melena.
(d) fetal diarrhea.
(e) mucous stools.

208. CPR is required for a newborn when

(a) respirations are absent.
(b) the pulse rate is less than 60 per minute.
(c) the pulse is absent.
(d) Answers (a) and (c) only.
(e) Answers (a), (b), and (c).

209. Pathologies that may mimic behavioral disorders include

(a) drug or alcohol overdose/abuse.
(b) head trauma, hypoxia, or hypovolemia.
(c) diabetes or electrolyte imbalance.
(d) dementia (organic brain syndrome).
(e) All of the above.

210. All of the following are examples of "open-ended" questions, except

(a) What kinds of medications did you take?
(b) How often has this happened before?
(c) Have you done anything else to hurt yourself?
(d) When did you begin feeling like you wanted to die?
(e) Whom have you turned to for help in the past?

211. Which of the following statements regarding risk factors for suicide is false?

(a) The more specific and detailed the plan, the greater the suicide potential.
(b) Women are more successful at committing suicide than men.
(c) Men use more violent methods of suicide than women.
(d) A history of prior attempts increases the potential for successful suicide.
(e) The presence of depression increases the potential for successful suicide.

212. A suicidal patient who is in possession of a weapon should be

(a) approached calmly and supportively, without a show of fear.
(b) threatened with arrest if she/he does not relinquish the weapon immediately.
(c) approached rapidly and in force to prevent the use of the weapon for self harm.
(d) considered homicidal as well as suicidal.
(e) None of the above.

213. Which of the following statements regarding management of behavioral or psychiatric disorders is false?

(a) Routine use of restraint devices is recommended for the safety of the paramedic, regardless of the amount of "cooperation" exhibited by the patient.
(b) The paramedic should intervene in the situation to the extent that she/he feels capable, being aware of personal and professional limitations.
(c) The paramedic should seek professional assistance in dealing with any situation that is beyond the scope of her/his training or capabilities.
(d) Overreaction to the patient's behavior or emotional condition will interfere with the paramedic's ability to assess and address the needs of the patient.
(e) None of the above is false.

The answer key to Section Seven begins on page 399.

Appendix

EMT-Paramedic National Standards Review Self Test, 3rd Edition

TEST SECTION ONE ANSWER SHEET Page one of 3 pages

No.					
1.	ⓐ	ⓑ	ⓒ	ⓓ	ⓔ
2.	ⓐ	ⓑ	ⓒ	ⓓ	ⓔ
3.	ⓐ	ⓑ	ⓒ	ⓓ	ⓔ
4.	ⓐ	ⓑ	ⓒ	ⓓ	ⓔ
5.	ⓐ	ⓑ	ⓒ	ⓓ	ⓔ
6.	ⓐ	ⓑ	ⓒ	ⓓ	ⓔ
7.	ⓐ	ⓑ	ⓒ	ⓓ	ⓔ
8.	ⓐ	ⓑ	ⓒ	ⓓ	ⓔ
9.	ⓐ	ⓑ	ⓒ	ⓓ	ⓔ
10.	ⓐ	ⓑ	ⓒ	ⓓ	ⓔ
11.	ⓐ	ⓑ	ⓒ	ⓓ	ⓔ
12.	ⓐ	ⓑ	ⓒ	ⓓ	ⓔ
13.	ⓐ	ⓑ	ⓒ	ⓓ	ⓔ
14.	ⓐ	ⓑ	ⓒ	ⓓ	ⓔ
15.	ⓐ	ⓑ	ⓒ	ⓓ	ⓔ
16.	ⓐ	ⓑ	ⓒ	ⓓ	ⓔ
17.	ⓐ	ⓑ	ⓒ	ⓓ	ⓔ
18.	ⓐ	ⓑ	ⓒ	ⓓ	ⓔ
19.	ⓐ	ⓑ	ⓒ	ⓓ	ⓔ
20.	ⓐ	ⓑ	ⓒ	ⓓ	ⓔ
21.	ⓐ	ⓑ	ⓒ	ⓓ	ⓔ
22.	ⓐ	ⓑ	ⓒ	ⓓ	ⓔ
23.	ⓐ	ⓑ	ⓒ	ⓓ	ⓔ
24.	ⓐ	ⓑ	ⓒ	ⓓ	ⓔ
25.	ⓐ	ⓑ	ⓒ	ⓓ	ⓔ
26.	ⓐ	ⓑ	ⓒ	ⓓ	ⓔ
27.	ⓐ	ⓑ	ⓒ	ⓓ	ⓔ
28.	ⓐ	ⓑ	ⓒ	ⓓ	ⓔ
29.	ⓐ	ⓑ	ⓒ	ⓓ	ⓔ
30.	ⓐ	ⓑ	ⓒ	ⓓ	ⓔ
31.	ⓐ	ⓑ	ⓒ	ⓓ	ⓔ
32.	ⓐ	ⓑ	ⓒ	ⓓ	ⓔ
33.	ⓐ	ⓑ	ⓒ	ⓓ	ⓔ
34.	ⓐ	ⓑ	ⓒ	ⓓ	ⓔ
35.	ⓐ	ⓑ	ⓒ	ⓓ	ⓔ
36.	ⓐ	ⓑ	ⓒ	ⓓ	ⓔ
37.	ⓐ	ⓑ	ⓒ	ⓓ	ⓔ
38.	ⓐ	ⓑ	ⓒ	ⓓ	ⓔ
39.	ⓐ	ⓑ	ⓒ	ⓓ	ⓔ
40.	ⓐ	ⓑ	ⓒ	ⓓ	ⓔ
41.	ⓐ	ⓑ	ⓒ	ⓓ	ⓔ
42.	ⓐ	ⓑ	ⓒ	ⓓ	ⓔ
43.	ⓐ	ⓑ	ⓒ	ⓓ	ⓔ
44.	ⓐ	ⓑ	ⓒ	ⓓ	ⓔ
45.	ⓐ	ⓑ	ⓒ	ⓓ	ⓔ
46.	ⓐ	ⓑ	ⓒ	ⓓ	ⓔ
47.	ⓐ	ⓑ	ⓒ	ⓓ	ⓔ
48.	ⓐ	ⓑ	ⓒ	ⓓ	ⓔ
49.	ⓐ	ⓑ	ⓒ	ⓓ	ⓔ
50.	ⓐ	ⓑ	ⓒ	ⓓ	ⓔ
51.	ⓐ	ⓑ	ⓒ	ⓓ	ⓔ
52.	ⓐ	ⓑ	ⓒ	ⓓ	ⓔ
53.	ⓐ	ⓑ	ⓒ	ⓓ	ⓔ
54.	ⓐ	ⓑ	ⓒ	ⓓ	ⓔ
55.	ⓐ	ⓑ	ⓒ	ⓓ	ⓔ
56.	ⓐ	ⓑ	ⓒ	ⓓ	ⓔ
57.	ⓐ	ⓑ	ⓒ	ⓓ	ⓔ
58.	ⓐ	ⓑ	ⓒ	ⓓ	ⓔ
59.	ⓐ	ⓑ	ⓒ	ⓓ	ⓔ
60.	ⓐ	ⓑ	ⓒ	ⓓ	ⓔ
61.	ⓐ	ⓑ	ⓒ	ⓓ	ⓔ
62.	ⓐ	ⓑ	ⓒ	ⓓ	ⓔ
63.	ⓐ	ⓑ	ⓒ	ⓓ	ⓔ
64.	ⓐ	ⓑ	ⓒ	ⓓ	ⓔ
65.	ⓐ	ⓑ	ⓒ	ⓓ	ⓔ
66.	ⓐ	ⓑ	ⓒ	ⓓ	ⓔ
67.	ⓐ	ⓑ	ⓒ	ⓓ	ⓔ
68.	ⓐ	ⓑ	ⓒ	ⓓ	ⓔ
69.	ⓐ	ⓑ	ⓒ	ⓓ	ⓔ
70.	ⓐ	ⓑ	ⓒ	ⓓ	ⓔ
71.	ⓐ	ⓑ	ⓒ	ⓓ	ⓔ
72.	ⓐ	ⓑ	ⓒ	ⓓ	ⓔ
73.	ⓐ	ⓑ	ⓒ	ⓓ	ⓔ
74.	ⓐ	ⓑ	ⓒ	ⓓ	ⓔ
75.	ⓐ	ⓑ	ⓒ	ⓓ	ⓔ
76.	ⓐ	ⓑ	ⓒ	ⓓ	ⓔ
77.	ⓐ	ⓑ	ⓒ	ⓓ	ⓔ
78.	ⓐ	ⓑ	ⓒ	ⓓ	ⓔ
79.	ⓐ	ⓑ	ⓒ	ⓓ	ⓔ
80.	ⓐ	ⓑ	ⓒ	ⓓ	ⓔ
81.	ⓐ	ⓑ	ⓒ	ⓓ	ⓔ
82.	ⓐ	ⓑ	ⓒ	ⓓ	ⓔ
83.	ⓐ	ⓑ	ⓒ	ⓓ	ⓔ
84.	ⓐ	ⓑ	ⓒ	ⓓ	ⓔ

TEST SECTION ONE ANSWER SHEET Page two of _3_ pages

No.					
85.	ⓐ	ⓑ	ⓒ	ⓓ	ⓔ
86.	ⓐ	ⓑ	ⓒ	ⓓ	ⓔ
87.	ⓐ	ⓑ	ⓒ	ⓓ	ⓔ
88.	ⓐ	ⓑ	ⓒ	ⓓ	ⓔ
89.	ⓐ	ⓑ	ⓒ	ⓓ	ⓔ
90.	ⓐ	ⓑ	ⓒ	ⓓ	ⓔ
91.	ⓐ	ⓑ	ⓒ	ⓓ	ⓔ
92.	ⓐ	ⓑ	ⓒ	ⓓ	ⓔ
93.	ⓐ	ⓑ	ⓒ	ⓓ	ⓔ
94.	ⓐ	ⓑ	ⓒ	ⓓ	ⓔ
95.	ⓐ	ⓑ	ⓒ	ⓓ	ⓔ
96.	ⓐ	ⓑ	ⓒ	ⓓ	ⓔ
97.	ⓐ	ⓑ	ⓒ	ⓓ	ⓔ
98.	ⓐ	ⓑ	ⓒ	ⓓ	ⓔ
99.	ⓐ	ⓑ	ⓒ	ⓓ	ⓔ
100.	ⓐ	ⓑ	ⓒ	ⓓ	ⓔ
101.	ⓐ	ⓑ	ⓒ	ⓓ	ⓔ
102.	ⓐ	ⓑ	ⓒ	ⓓ	ⓔ
103.	ⓐ	ⓑ	ⓒ	ⓓ	ⓔ
104.	ⓐ	ⓑ	ⓒ	ⓓ	ⓔ
105.	ⓐ	ⓑ	ⓒ	ⓓ	ⓔ
106.	ⓐ	ⓑ	ⓒ	ⓓ	ⓔ
107.	ⓐ	ⓑ	ⓒ	ⓓ	ⓔ
108.	ⓐ	ⓑ	ⓒ	ⓓ	ⓔ
109.	ⓐ	ⓑ	ⓒ	ⓓ	ⓔ
110.	ⓐ	ⓑ	ⓒ	ⓓ	ⓔ
111.	ⓐ	ⓑ	ⓒ	ⓓ	ⓔ
112.	ⓐ	ⓑ	ⓒ	ⓓ	ⓔ
113.	ⓐ	ⓑ	ⓒ	ⓓ	ⓔ
114.	ⓐ	ⓑ	ⓒ	ⓓ	ⓔ
115.	ⓐ	ⓑ	ⓒ	ⓓ	ⓔ
116.	ⓐ	ⓑ	ⓒ	ⓓ	ⓔ
117.	ⓐ	ⓑ	ⓒ	ⓓ	ⓔ
118.	ⓐ	ⓑ	ⓒ	ⓓ	ⓔ
119.	ⓐ	ⓑ	ⓒ	ⓓ	ⓔ
120.	ⓐ	ⓑ	ⓒ	ⓓ	ⓔ
121.	ⓐ	ⓑ	ⓒ	ⓓ	ⓔ
122.	ⓐ	ⓑ	ⓒ	ⓓ	ⓔ
123.	ⓐ	ⓑ	ⓒ	ⓓ	ⓔ
124.	ⓐ	ⓑ	ⓒ	ⓓ	ⓔ
125.	ⓐ	ⓑ	ⓒ	ⓓ	ⓔ
126.	ⓐ	ⓑ	ⓒ	ⓓ	ⓔ
127.	ⓐ	ⓑ	ⓒ	ⓓ	ⓔ
128.	ⓐ	ⓑ	ⓒ	ⓓ	ⓔ
129.	ⓐ	ⓑ	ⓒ	ⓓ	ⓔ
130.	ⓐ	ⓑ	ⓒ	ⓓ	ⓔ
131.	ⓐ	ⓑ	ⓒ	ⓓ	ⓔ
132.	ⓐ	ⓑ	ⓒ	ⓓ	ⓔ
133.	ⓐ	ⓑ	ⓒ	ⓓ	ⓔ
134.	ⓐ	ⓑ	ⓒ	ⓓ	ⓔ
135.	ⓐ	ⓑ	ⓒ	ⓓ	ⓔ
136.	ⓐ	ⓑ	ⓒ	ⓓ	ⓔ
137.	ⓐ	ⓑ	ⓒ	ⓓ	ⓔ
138.	ⓐ	ⓑ	ⓒ	ⓓ	ⓔ
139.	ⓐ	ⓑ	ⓒ	ⓓ	ⓔ
140.	ⓐ	ⓑ	ⓒ	ⓓ	ⓔ
141.	ⓐ	ⓑ	ⓒ	ⓓ	ⓔ
142.	ⓐ	ⓑ	ⓒ	ⓓ	ⓔ
143.	ⓐ	ⓑ	ⓒ	ⓓ	ⓔ
144.	ⓐ	ⓑ	ⓒ	ⓓ	ⓔ
145.	ⓐ	ⓑ	ⓒ	ⓓ	ⓔ
146.	ⓐ	ⓑ	ⓒ	ⓓ	ⓔ
147.	ⓐ	ⓑ	ⓒ	ⓓ	ⓔ
148.	ⓐ	ⓑ	ⓒ	ⓓ	ⓔ
149.	ⓐ	ⓑ	ⓒ	ⓓ	ⓔ
150.	ⓐ	ⓑ	ⓒ	ⓓ	ⓔ
151.	ⓐ	ⓑ	ⓒ	ⓓ	ⓔ
152.	ⓐ	ⓑ	ⓒ	ⓓ	ⓔ
153.	ⓐ	ⓑ	ⓒ	ⓓ	ⓔ
154.	ⓐ	ⓑ	ⓒ	ⓓ	ⓔ
155.	ⓐ	ⓑ	ⓒ	ⓓ	ⓔ
156.	ⓐ	ⓑ	ⓒ	ⓓ	ⓔ
157.	ⓐ	ⓑ	ⓒ	ⓓ	ⓔ
158.	ⓐ	ⓑ	ⓒ	ⓓ	ⓔ
159.	ⓐ	ⓑ	ⓒ	ⓓ	ⓔ
160.	ⓐ	ⓑ	ⓒ	ⓓ	ⓔ
161.	ⓐ	ⓑ	ⓒ	ⓓ	ⓔ
162.	ⓐ	ⓑ	ⓒ	ⓓ	ⓔ
163.	ⓐ	ⓑ	ⓒ	ⓓ	ⓔ
164.	ⓐ	ⓑ	ⓒ	ⓓ	ⓔ
165.	ⓐ	ⓑ	ⓒ	ⓓ	ⓔ
166.	ⓐ	ⓑ	ⓒ	ⓓ	ⓔ
167.	ⓐ	ⓑ	ⓒ	ⓓ	ⓔ
168.	ⓐ	ⓑ	ⓒ	ⓓ	ⓔ
169.	ⓐ	ⓑ	ⓒ	ⓓ	ⓔ
170.	ⓐ	ⓑ	ⓒ	ⓓ	ⓔ

TEST SECTION ONE ANSWER SHEET Page three of 3 pages

171.	ⓐ	ⓑ	ⓒ	ⓓ	ⓔ
172.	ⓐ	ⓑ	ⓒ	ⓓ	ⓔ
173.	ⓐ	ⓑ	ⓒ	ⓓ	ⓔ
174.	ⓐ	ⓑ	ⓒ	ⓓ	ⓔ
175.	ⓐ	ⓑ	ⓒ	ⓓ	ⓔ
176.	ⓐ	ⓑ	ⓒ	ⓓ	ⓔ
177.	ⓐ	ⓑ	ⓒ	ⓓ	ⓔ
178.	ⓐ	ⓑ	ⓒ	ⓓ	ⓔ
179.	ⓐ	ⓑ	ⓒ	ⓓ	ⓔ
180.	ⓐ	ⓑ	ⓒ	ⓓ	ⓔ
181.	ⓐ	ⓑ	ⓒ	ⓓ	ⓔ
182.	ⓐ	ⓑ	ⓒ	ⓓ	ⓔ
183.	ⓐ	ⓑ	ⓒ	ⓓ	ⓔ
184.	ⓐ	ⓑ	ⓒ	ⓓ	ⓔ
185.	ⓐ	ⓑ	ⓒ	ⓓ	ⓔ
186.	ⓐ	ⓑ	ⓒ	ⓓ	ⓔ
187.	ⓐ	ⓑ	ⓒ	ⓓ	ⓔ
188.	ⓐ	ⓑ	ⓒ	ⓓ	ⓔ
189.	ⓐ	ⓑ	ⓒ	ⓓ	ⓔ
190.	ⓐ	ⓑ	ⓒ	ⓓ	ⓔ
191.	ⓐ	ⓑ	ⓒ	ⓓ	ⓔ
192.	ⓐ	ⓑ	ⓒ	ⓓ	ⓔ
193.	ⓐ	ⓑ	ⓒ	ⓓ	ⓔ
194.	ⓐ	ⓑ	ⓒ	ⓓ	ⓔ
195.	ⓐ	ⓑ	ⓒ	ⓓ	ⓔ
196.	ⓐ	ⓑ	ⓒ	ⓓ	ⓔ
197.	ⓐ	ⓑ	ⓒ	ⓓ	ⓔ
198.	ⓐ	ⓑ	ⓒ	ⓓ	ⓔ
199.	ⓐ	ⓑ	ⓒ	ⓓ	ⓔ
200.	ⓐ	ⓑ	ⓒ	ⓓ	ⓔ

EMT-Paramedic National Standards Review Self Test, 3rd Edition

TEST SECTION TWO ANSWER SHEET Page one of 3 pages

No.					
1.	a	b	c	d	e
2.	a	b	c	d	e
3.	a	b	c	d	e
4.	a	b	c	d	e
5.	a	b	c	d	e
6.	a	b	c	d	e
7.	a	b	c	d	e
8.	a	b	c	d	e
9.	a	b	c	d	e
10.	a	b	c	d	e
11.	a	b	c	d	e
12.	a	b	c	d	e
13.	a	b	c	d	e
14.	a	b	c	d	e
15.	a	b	c	d	e
16.	a	b	c	d	e
17.	a	b	c	d	e
18.	a	b	c	d	e
19.	a	b	c	d	e
20.	a	b	c	d	e
21.	a	b	c	d	e
22.	a	b	c	d	e
23.	a	b	c	d	e
24.	a	b	c	d	e
25.	a	b	c	d	e
26.	a	b	c	d	e
27.	a	b	c	d	e
28.	a	b	c	d	e
29.	a	b	c	d	e
30.	a	b	c	d	e
31.	a	b	c	d	e
32.	a	b	c	d	e
33.	a	b	c	d	e
34.	a	b	c	d	e
35.	a	b	c	d	e
36.	a	b	c	d	e
37.	a	b	c	d	e
38.	a	b	c	d	e
39.	a	b	c	d	e
40.	a	b	c	d	e
41.	a	b	c	d	e
42.	a	b	c	d	e
43.	a	b	c	d	e
44.	a	b	c	d	e
45.	a	b	c	d	e
46.	a	b	c	d	e
47.	a	b	c	d	e
48.	a	b	c	d	e
49.	a	b	c	d	e
50.	a	b	c	d	e
51.	a	b	c	d	e
52.	a	b	c	d	e
53.	a	b	c	d	e
54.	a	b	c	d	e
55.	a	b	c	d	e
56.	a	b	c	d	e
57.	a	b	c	d	e
58.	a	b	c	d	e
59.	a	b	c	d	e
60.	a	b	c	d	e
61.	a	b	c	d	e
62.	a	b	c	d	e
63.	a	b	c	d	e
64.	a	b	c	d	e
65.	a	b	c	d	e
66.	a	b	c	d	e
67.	a	b	c	d	e
68.	a	b	c	d	e
69.	a	b	c	d	e
70.	a	b	c	d	e
71.	a	b	c	d	e
72.	a	b	c	d	e
73.	a	b	c	d	e
74.	a	b	c	d	e
75.	a	b	c	d	e
76.	a	b	c	d	e
77.	a	b	c	d	e
78.	a	b	c	d	e
79.	a	b	c	d	e
80.	a	b	c	d	e
81.	a	b	c	d	e
82.	a	b	c	d	e
83.	a	b	c	d	e
84.	a	b	c	d	e

TEST SECTION TWO ANSWER SHEET Page two of 3 pages

85.	ⓐ	ⓑ	ⓒ	ⓓ	ⓔ
86.	ⓐ	ⓑ	ⓒ	ⓓ	ⓔ
87.	ⓐ	ⓑ	ⓒ	ⓓ	ⓔ
88.	ⓐ	ⓑ	ⓒ	ⓓ	ⓔ
89.	ⓐ	ⓑ	ⓒ	ⓓ	ⓔ
90.	ⓐ	ⓑ	ⓒ	ⓓ	ⓔ
91.	ⓐ	ⓑ	ⓒ	ⓓ	ⓔ
92.	ⓐ	ⓑ	ⓒ	ⓓ	ⓔ
93.	ⓐ	ⓑ	ⓒ	ⓓ	ⓔ
94.	ⓐ	ⓑ	ⓒ	ⓓ	ⓔ
95.	ⓐ	ⓑ	ⓒ	ⓓ	ⓔ
96.	ⓐ	ⓑ	ⓒ	ⓓ	ⓔ
97.	ⓐ	ⓑ	ⓒ	ⓓ	ⓔ
98.	ⓐ	ⓑ	ⓒ	ⓓ	ⓔ
99.	ⓐ	ⓑ	ⓒ	ⓓ	ⓔ
100.	ⓐ	ⓑ	ⓒ	ⓓ	ⓔ
101.	ⓐ	ⓑ	ⓒ	ⓓ	ⓔ
102.	ⓐ	ⓑ	ⓒ	ⓓ	ⓔ
103.	ⓐ	ⓑ	ⓒ	ⓓ	ⓔ
104.	ⓐ	ⓑ	ⓒ	ⓓ	ⓔ
105.	ⓐ	ⓑ	ⓒ	ⓓ	ⓔ
106.	ⓐ	ⓑ	ⓒ	ⓓ	ⓔ
107.	ⓐ	ⓑ	ⓒ	ⓓ	ⓔ
108.	ⓐ	ⓑ	ⓒ	ⓓ	ⓔ
109.	ⓐ	ⓑ	ⓒ	ⓓ	ⓔ
110.	ⓐ	ⓑ	ⓒ	ⓓ	ⓔ
111.	ⓐ	ⓑ	ⓒ	ⓓ	ⓔ
112.	ⓐ	ⓑ	ⓒ	ⓓ	ⓔ
113.	ⓐ	ⓑ	ⓒ	ⓓ	ⓔ
114.	ⓐ	ⓑ	ⓒ	ⓓ	ⓔ
115.	ⓐ	ⓑ	ⓒ	ⓓ	ⓔ
116.	ⓐ	ⓑ	ⓒ	ⓓ	ⓔ
117.	ⓐ	ⓑ	ⓒ	ⓓ	ⓔ
118.	ⓐ	ⓑ	ⓒ	ⓓ	ⓔ
119.	ⓐ	ⓑ	ⓒ	ⓓ	ⓔ
120.	ⓐ	ⓑ	ⓒ	ⓓ	ⓔ
121.	ⓐ	ⓑ	ⓒ	ⓓ	ⓔ
122.	ⓐ	ⓑ	ⓒ	ⓓ	ⓔ
123.	ⓐ	ⓑ	ⓒ	ⓓ	ⓔ
124.	ⓐ	ⓑ	ⓒ	ⓓ	ⓔ
125.	ⓐ	ⓑ	ⓒ	ⓓ	ⓔ
126.	ⓐ	ⓑ	ⓒ	ⓓ	ⓔ
127.	ⓐ	ⓑ	ⓒ	ⓓ	ⓔ
128.	ⓐ	ⓑ	ⓒ	ⓓ	ⓔ
129.	ⓐ	ⓑ	ⓒ	ⓓ	ⓔ
130.	ⓐ	ⓑ	ⓒ	ⓓ	ⓔ
131.	ⓐ	ⓑ	ⓒ	ⓓ	ⓔ
132.	ⓐ	ⓑ	ⓒ	ⓓ	ⓔ
133.	ⓐ	ⓑ	ⓒ	ⓓ	ⓔ
134.	ⓐ	ⓑ	ⓒ	ⓓ	ⓔ
135.	ⓐ	ⓑ	ⓒ	ⓓ	ⓔ
136.	ⓐ	ⓑ	ⓒ	ⓓ	ⓔ
137.	ⓐ	ⓑ	ⓒ	ⓓ	ⓔ
138.	ⓐ	ⓑ	ⓒ	ⓓ	ⓔ
139.	ⓐ	ⓑ	ⓒ	ⓓ	ⓔ
140.	ⓐ	ⓑ	ⓒ	ⓓ	ⓔ
141.	ⓐ	ⓑ	ⓒ	ⓓ	ⓔ
142.	ⓐ	ⓑ	ⓒ	ⓓ	ⓔ
143.	ⓐ	ⓑ	ⓒ	ⓓ	ⓔ
144.	ⓐ	ⓑ	ⓒ	ⓓ	ⓔ
145.	ⓐ	ⓑ	ⓒ	ⓓ	ⓔ
146.	ⓐ	ⓑ	ⓒ	ⓓ	ⓔ
147.	ⓐ	ⓑ	ⓒ	ⓓ	ⓔ
148.	ⓐ	ⓑ	ⓒ	ⓓ	ⓔ
149.	ⓐ	ⓑ	ⓒ	ⓓ	ⓔ
150.	ⓐ	ⓑ	ⓒ	ⓓ	ⓔ
151.	ⓐ	ⓑ	ⓒ	ⓓ	ⓔ
152.	ⓐ	ⓑ	ⓒ	ⓓ	ⓔ
153.	ⓐ	ⓑ	ⓒ	ⓓ	ⓔ
154.	ⓐ	ⓑ	ⓒ	ⓓ	ⓔ
155.	ⓐ	ⓑ	ⓒ	ⓓ	ⓔ
156.	ⓐ	ⓑ	ⓒ	ⓓ	ⓔ
157.	ⓐ	ⓑ	ⓒ	ⓓ	ⓔ
158.	ⓐ	ⓑ	ⓒ	ⓓ	ⓔ
159.	ⓐ	ⓑ	ⓒ	ⓓ	ⓔ
160.	ⓐ	ⓑ	ⓒ	ⓓ	ⓔ
161.	ⓐ	ⓑ	ⓒ	ⓓ	ⓔ
162.	ⓐ	ⓑ	ⓒ	ⓓ	ⓔ
163.	ⓐ	ⓑ	ⓒ	ⓓ	ⓔ
164.	ⓐ	ⓑ	ⓒ	ⓓ	ⓔ
165.	ⓐ	ⓑ	ⓒ	ⓓ	ⓔ
166.	ⓐ	ⓑ	ⓒ	ⓓ	ⓔ
167.	ⓐ	ⓑ	ⓒ	ⓓ	ⓔ
168.	ⓐ	ⓑ	ⓒ	ⓓ	ⓔ
169.	ⓐ	ⓑ	ⓒ	ⓓ	ⓔ
170.	ⓐ	ⓑ	ⓒ	ⓓ	ⓔ

TEST SECTION TWO ANSWER SHEET Page three of 3 pages

171.	ⓐ	ⓑ	ⓒ	ⓓ	ⓔ
172.	ⓐ	ⓑ	ⓒ	ⓓ	ⓔ
173.	ⓐ	ⓑ	ⓒ	ⓓ	ⓔ
174.	ⓐ	ⓑ	ⓒ	ⓓ	ⓔ
175.	ⓐ	ⓑ	ⓒ	ⓓ	ⓔ
176.	ⓐ	ⓑ	ⓒ	ⓓ	ⓔ
177.	ⓐ	ⓑ	ⓒ	ⓓ	ⓔ
178.	ⓐ	ⓑ	ⓒ	ⓓ	ⓔ
179.	ⓐ	ⓑ	ⓒ	ⓓ	ⓔ
180.	ⓐ	ⓑ	ⓒ	ⓓ	ⓔ
181.	ⓐ	ⓑ	ⓒ	ⓓ	ⓔ
182.	ⓐ	ⓑ	ⓒ	ⓓ	ⓔ
183.	ⓐ	ⓑ	ⓒ	ⓓ	ⓔ
184.	ⓐ	ⓑ	ⓒ	ⓓ	ⓔ
185.	ⓐ	ⓑ	ⓒ	ⓓ	ⓔ
186.	ⓐ	ⓑ	ⓒ	ⓓ	ⓔ
187.	ⓐ	ⓑ	ⓒ	ⓓ	ⓔ
188.	ⓐ	ⓑ	ⓒ	ⓓ	ⓔ
189.	ⓐ	ⓑ	ⓒ	ⓓ	ⓔ
190.	ⓐ	ⓑ	ⓒ	ⓓ	ⓔ
191.	ⓐ	ⓑ	ⓒ	ⓓ	ⓔ
192.	ⓐ	ⓑ	ⓒ	ⓓ	ⓔ
193.	ⓐ	ⓑ	ⓒ	ⓓ	ⓔ
194.	ⓐ	ⓑ	ⓒ	ⓓ	ⓔ
195.	ⓐ	ⓑ	ⓒ	ⓓ	ⓔ
196.	ⓐ	ⓑ	ⓒ	ⓓ	ⓔ
197.	ⓐ	ⓑ	ⓒ	ⓓ	ⓔ
198.	ⓐ	ⓑ	ⓒ	ⓓ	ⓔ
199.	ⓐ	ⓑ	ⓒ	ⓓ	ⓔ
200.	ⓐ	ⓑ	ⓒ	ⓓ	ⓔ
201.	ⓐ	ⓑ	ⓒ	ⓓ	ⓔ
202.	ⓐ	ⓑ	ⓒ	ⓓ	ⓔ
203.	ⓐ	ⓑ	ⓒ	ⓓ	ⓔ
204.	ⓐ	ⓑ	ⓒ	ⓓ	ⓔ
205.	ⓐ	ⓑ	ⓒ	ⓓ	ⓔ
206.	ⓐ	ⓑ	ⓒ	ⓓ	ⓔ
207.	ⓐ	ⓑ	ⓒ	ⓓ	ⓔ
208.	ⓐ	ⓑ	ⓒ	ⓓ	ⓔ
209.	ⓐ	ⓑ	ⓒ	ⓓ	ⓔ
210.	ⓐ	ⓑ	ⓒ	ⓓ	ⓔ
211.	ⓐ	ⓑ	ⓒ	ⓓ	ⓔ
212.	ⓐ	ⓑ	ⓒ	ⓓ	ⓔ
213.	ⓐ	ⓑ	ⓒ	ⓓ	ⓔ
214.	ⓐ	ⓑ	ⓒ	ⓓ	ⓔ
215.	ⓐ	ⓑ	ⓒ	ⓓ	ⓔ
216.	ⓐ	ⓑ	ⓒ	ⓓ	ⓔ
217.	ⓐ	ⓑ	ⓒ	ⓓ	ⓔ
218.	ⓐ	ⓑ	ⓒ	ⓓ	ⓔ
219.	ⓐ	ⓑ	ⓒ	ⓓ	ⓔ
220.	ⓐ	ⓑ	ⓒ	ⓓ	ⓔ
221.	ⓐ	ⓑ	ⓒ	ⓓ	ⓔ
222.	ⓐ	ⓑ	ⓒ	ⓓ	ⓔ
223.	ⓐ	ⓑ	ⓒ	ⓓ	ⓔ
224.	ⓐ	ⓑ	ⓒ	ⓓ	ⓔ
225.	ⓐ	ⓑ	ⓒ	ⓓ	ⓔ
226.	ⓐ	ⓑ	ⓒ	ⓓ	ⓔ
227.	ⓐ	ⓑ	ⓒ	ⓓ	ⓔ
228.	ⓐ	ⓑ	ⓒ	ⓓ	ⓔ
229.	ⓐ	ⓑ	ⓒ	ⓓ	ⓔ
230.	ⓐ	ⓑ	ⓒ	ⓓ	ⓔ
231.	ⓐ	ⓑ	ⓒ	ⓓ	ⓔ
232.	ⓐ	ⓑ	ⓒ	ⓓ	ⓔ
233.	ⓐ	ⓑ	ⓒ	ⓓ	ⓔ
234.	ⓐ	ⓑ	ⓒ	ⓓ	ⓔ
235.	ⓐ	ⓑ	ⓒ	ⓓ	ⓔ
236.	ⓐ	ⓑ	ⓒ	ⓓ	ⓔ
237.	ⓐ	ⓑ	ⓒ	ⓓ	ⓔ
238.	ⓐ	ⓑ	ⓒ	ⓓ	ⓔ
239.	ⓐ	ⓑ	ⓒ	ⓓ	ⓔ
240.	ⓐ	ⓑ	ⓒ	ⓓ	ⓔ
241.	ⓐ	ⓑ	ⓒ	ⓓ	ⓔ
242.	ⓐ	ⓑ	ⓒ	ⓓ	ⓔ
243.	ⓐ	ⓑ	ⓒ	ⓓ	ⓔ
244.	ⓐ	ⓑ	ⓒ	ⓓ	ⓔ
245.	ⓐ	ⓑ	ⓒ	ⓓ	ⓔ
246.	ⓐ	ⓑ	ⓒ	ⓓ	ⓔ
247.	ⓐ	ⓑ	ⓒ	ⓓ	ⓔ
248.	ⓐ	ⓑ	ⓒ	ⓓ	ⓔ
249.	ⓐ	ⓑ	ⓒ	ⓓ	ⓔ
250.	ⓐ	ⓑ	ⓒ	ⓓ	ⓔ
251.	ⓐ	ⓑ	ⓒ	ⓓ	ⓔ
252.	ⓐ	ⓑ	ⓒ	ⓓ	ⓔ
253.	ⓐ	ⓑ	ⓒ	ⓓ	ⓔ

EMT-Paramedic National Standards Review Self Test, 3rd Edition

TEST SECTION THREE ANSWER SHEET Page one of 2 pages

1.	a	b	c	d	e	43.	a	b	c	d	e
2.	a	b	c	d	e	44.	a	b	c	d	e
3.	a	b	c	d	e	45.	a	b	c	d	e
4.	a	b	c	d	e	46.	a	b	c	d	e
5.	a	b	c	d	e	47.	a	b	c	d	e
6.	a	b	c	d	e	48.	a	b	c	d	e
7.	a	b	c	d	e	49.	a	b	c	d	e
8.	a	b	c	d	e	50.	a	b	c	d	e
9.	a	b	c	d	e	51.	a	b	c	d	e
10.	a	b	c	d	e	52.	a	b	c	d	e
11.	a	b	c	d	e	53.	a	b	c	d	e
12.	a	b	c	d	e	54.	a	b	c	d	e
13.	a	b	c	d	e	55.	a	b	c	d	e
14.	a	b	c	d	e	56.	a	b	c	d	e
15.	a	b	c	d	e	57.	a	b	c	d	e
16.	a	b	c	d	e	58.	a	b	c	d	e
17.	a	b	c	d	e	59.	a	b	c	d	e
18.	a	b	c	d	e	60.	a	b	c	d	e
19.	a	b	c	d	e	61.	a	b	c	d	e
20.	a	b	c	d	e	62.	a	b	c	d	e
21.	a	b	c	d	e	63.	a	b	c	d	e
22.	a	b	c	d	e	64.	a	b	c	d	e
23.	a	b	c	d	e	65.	a	b	c	d	e
24.	a	b	c	d	e	66.	a	b	c	d	e
25.	a	b	c	d	e	67.	a	b	c	d	e
26.	a	b	c	d	e	68.	a	b	c	d	e
27.	a	b	c	d	e	69.	a	b	c	d	e
28.	a	b	c	d	e	70.	a	b	c	d	e
29.	a	b	c	d	e	71.	a	b	c	d	e
30.	a	b	c	d	e	72.	a	b	c	d	e
31.	a	b	c	d	e	73.	a	b	c	d	e
32.	a	b	c	d	e	74.	a	b	c	d	e
33.	a	b	c	d	e	75.	a	b	c	d	e
34.	a	b	c	d	e	76.	a	b	c	d	e
35.	a	b	c	d	e	77.	a	b	c	d	e
36.	a	b	c	d	e	78.	a	b	c	d	e
37.	a	b	c	d	e	79.	a	b	c	d	e
38.	a	b	c	d	e	80.	a	b	c	d	e
39.	a	b	c	d	e	81.	a	b	c	d	e
40.	a	b	c	d	e	82.	a	b	c	d	e
41.	a	b	c	d	e	83.	a	b	c	d	e
42.	a	b	c	d	e	84.	a	b	c	d	e

TEST SECTION THREE ANSWER SHEET Page two of 2 pages

85.	ⓐ	ⓑ	ⓒ	ⓓ	ⓔ
86.	ⓐ	ⓑ	ⓒ	ⓓ	ⓔ
87.	ⓐ	ⓑ	ⓒ	ⓓ	ⓔ
88.	ⓐ	ⓑ	ⓒ	ⓓ	ⓔ
89.	ⓐ	ⓑ	ⓒ	ⓓ	ⓔ
90.	ⓐ	ⓑ	ⓒ	ⓓ	ⓔ
91.	ⓐ	ⓑ	ⓒ	ⓓ	ⓔ
92.	ⓐ	ⓑ	ⓒ	ⓓ	ⓔ
93.	ⓐ	ⓑ	ⓒ	ⓓ	ⓔ
94.	ⓐ	ⓑ	ⓒ	ⓓ	ⓔ
95.	ⓐ	ⓑ	ⓒ	ⓓ	ⓔ
96.	ⓐ	ⓑ	ⓒ	ⓓ	ⓔ
97.	ⓐ	ⓑ	ⓒ	ⓓ	ⓔ
98.	ⓐ	ⓑ	ⓒ	ⓓ	ⓔ
99.	ⓐ	ⓑ	ⓒ	ⓓ	ⓔ
100.	ⓐ	ⓑ	ⓒ	ⓓ	ⓔ
101.	ⓐ	ⓑ	ⓒ	ⓓ	ⓔ
102.	ⓐ	ⓑ	ⓒ	ⓓ	ⓔ
103.	ⓐ	ⓑ	ⓒ	ⓓ	ⓔ
104.	ⓐ	ⓑ	ⓒ	ⓓ	ⓔ
105.	ⓐ	ⓑ	ⓒ	ⓓ	ⓔ
106.	ⓐ	ⓑ	ⓒ	ⓓ	ⓔ
107.	ⓐ	ⓑ	ⓒ	ⓓ	ⓔ
108.	ⓐ	ⓑ	ⓒ	ⓓ	ⓔ
109.	ⓐ	ⓑ	ⓒ	ⓓ	ⓔ
110.	ⓐ	ⓑ	ⓒ	ⓓ	ⓔ
111.	ⓐ	ⓑ	ⓒ	ⓓ	ⓔ
112.	ⓐ	ⓑ	ⓒ	ⓓ	ⓔ
113.	ⓐ	ⓑ	ⓒ	ⓓ	ⓔ
114.	ⓐ	ⓑ	ⓒ	ⓓ	ⓔ
115.	ⓐ	ⓑ	ⓒ	ⓓ	ⓔ
116.	ⓐ	ⓑ	ⓒ	ⓓ	ⓔ
117.	ⓐ	ⓑ	ⓒ	ⓓ	ⓔ
118.	ⓐ	ⓑ	ⓒ	ⓓ	ⓔ
119.	ⓐ	ⓑ	ⓒ	ⓓ	ⓔ
120.	ⓐ	ⓑ	ⓒ	ⓓ	ⓔ
121.	ⓐ	ⓑ	ⓒ	ⓓ	ⓔ
122.	ⓐ	ⓑ	ⓒ	ⓓ	ⓔ
123.	ⓐ	ⓑ	ⓒ	ⓓ	ⓔ
124.	ⓐ	ⓑ	ⓒ	ⓓ	ⓔ
125.	ⓐ	ⓑ	ⓒ	ⓓ	ⓔ
126.	ⓐ	ⓑ	ⓒ	ⓓ	ⓔ
127.	ⓐ	ⓑ	ⓒ	ⓓ	ⓔ
128.	ⓐ	ⓑ	ⓒ	ⓓ	ⓔ
129.	ⓐ	ⓑ	ⓒ	ⓓ	ⓔ
130.	ⓐ	ⓑ	ⓒ	ⓓ	ⓔ
131.	ⓐ	ⓑ	ⓒ	ⓓ	ⓔ
132.	ⓐ	ⓑ	ⓒ	ⓓ	ⓔ
133.	ⓐ	ⓑ	ⓒ	ⓓ	ⓔ
134.	ⓐ	ⓑ	ⓒ	ⓓ	ⓔ
135.	ⓐ	ⓑ	ⓒ	ⓓ	ⓔ
136.	ⓐ	ⓑ	ⓒ	ⓓ	ⓔ
137.	ⓐ	ⓑ	ⓒ	ⓓ	ⓔ
138.	ⓐ	ⓑ	ⓒ	ⓓ	ⓔ
139.	ⓐ	ⓑ	ⓒ	ⓓ	ⓔ
140.	ⓐ	ⓑ	ⓒ	ⓓ	ⓔ
141.	ⓐ	ⓑ	ⓒ	ⓓ	ⓔ
142.	ⓐ	ⓑ	ⓒ	ⓓ	ⓔ
143.	ⓐ	ⓑ	ⓒ	ⓓ	ⓔ
144.	ⓐ	ⓑ	ⓒ	ⓓ	ⓔ
145.	ⓐ	ⓑ	ⓒ	ⓓ	ⓔ
146.	ⓐ	ⓑ	ⓒ	ⓓ	ⓔ
147.	ⓐ	ⓑ	ⓒ	ⓓ	ⓔ
148.	ⓐ	ⓑ	ⓒ	ⓓ	ⓔ
149.	ⓐ	ⓑ	ⓒ	ⓓ	ⓔ
150.	ⓐ	ⓑ	ⓒ	ⓓ	ⓔ
151.	ⓐ	ⓑ	ⓒ	ⓓ	ⓔ
152.	ⓐ	ⓑ	ⓒ	ⓓ	ⓔ
153.	ⓐ	ⓑ	ⓒ	ⓓ	ⓔ
154.	ⓐ	ⓑ	ⓒ	ⓓ	ⓔ
155.	ⓐ	ⓑ	ⓒ	ⓓ	ⓔ
156.	ⓐ	ⓑ	ⓒ	ⓓ	ⓔ
157.	ⓐ	ⓑ	ⓒ	ⓓ	ⓔ
158.	ⓐ	ⓑ	ⓒ	ⓓ	ⓔ
159.	ⓐ	ⓑ	ⓒ	ⓓ	ⓔ
160.	ⓐ	ⓑ	ⓒ	ⓓ	ⓔ
161.	ⓐ	ⓑ	ⓒ	ⓓ	ⓔ
162.	ⓐ	ⓑ	ⓒ	ⓓ	ⓔ
163.	ⓐ	ⓑ	ⓒ	ⓓ	ⓔ
164.	ⓐ	ⓑ	ⓒ	ⓓ	ⓔ
165.	ⓐ	ⓑ	ⓒ	ⓓ	ⓔ
166.	ⓐ	ⓑ	ⓒ	ⓓ	ⓔ
167.	ⓐ	ⓑ	ⓒ	ⓓ	ⓔ
168.	ⓐ	ⓑ	ⓒ	ⓓ	ⓔ
169.	ⓐ	ⓑ	ⓒ	ⓓ	ⓔ
170.	ⓐ	ⓑ	ⓒ	ⓓ	ⓔ

EMT-Paramedic National Standards Review Self Test, 3rd Edition

TEST SECTION FOUR ANSWER SHEET Page one of 4 pages

1.	ⓐ	ⓑ	ⓒ	ⓓ	ⓔ
2.	ⓐ	ⓑ	ⓒ	ⓓ	ⓔ
3.	ⓐ	ⓑ	ⓒ	ⓓ	ⓔ
4.	ⓐ	ⓑ	ⓒ	ⓓ	ⓔ
5.	ⓐ	ⓑ	ⓒ	ⓓ	ⓔ
6.	ⓐ	ⓑ	ⓒ	ⓓ	ⓔ
7.	ⓐ	ⓑ	ⓒ	ⓓ	ⓔ
8.	ⓐ	ⓑ	ⓒ	ⓓ	ⓔ
9.	ⓐ	ⓑ	ⓒ	ⓓ	ⓔ
10.	ⓐ	ⓑ	ⓒ	ⓓ	ⓔ
11.	ⓐ	ⓑ	ⓒ	ⓓ	ⓔ
12.	ⓐ	ⓑ	ⓒ	ⓓ	ⓔ
13.	ⓐ	ⓑ	ⓒ	ⓓ	ⓔ
14.	ⓐ	ⓑ	ⓒ	ⓓ	ⓔ
15.	ⓐ	ⓑ	ⓒ	ⓓ	ⓔ
16.	ⓐ	ⓑ	ⓒ	ⓓ	ⓔ
17.	ⓐ	ⓑ	ⓒ	ⓓ	ⓔ
18.	ⓐ	ⓑ	ⓒ	ⓓ	ⓔ
19.	ⓐ	ⓑ	ⓒ	ⓓ	ⓔ
20.	ⓐ	ⓑ	ⓒ	ⓓ	ⓔ
21.	ⓐ	ⓑ	ⓒ	ⓓ	ⓔ
22.	ⓐ	ⓑ	ⓒ	ⓓ	ⓔ
23.	ⓐ	ⓑ	ⓒ	ⓓ	ⓔ
24.	ⓐ	ⓑ	ⓒ	ⓓ	ⓔ
25.	ⓐ	ⓑ	ⓒ	ⓓ	ⓔ
26.	ⓐ	ⓑ	ⓒ	ⓓ	ⓔ
27.	ⓐ	ⓑ	ⓒ	ⓓ	ⓔ
28.	ⓐ	ⓑ	ⓒ	ⓓ	ⓔ
29.	ⓐ	ⓑ	ⓒ	ⓓ	ⓔ
30.	ⓐ	ⓑ	ⓒ	ⓓ	ⓔ
31.	ⓐ	ⓑ	ⓒ	ⓓ	ⓔ
32.	ⓐ	ⓑ	ⓒ	ⓓ	ⓔ
33.	ⓐ	ⓑ	ⓒ	ⓓ	ⓔ
34.	ⓐ	ⓑ	ⓒ	ⓓ	ⓔ
35.	ⓐ	ⓑ	ⓒ	ⓓ	ⓔ
36.	ⓐ	ⓑ	ⓒ	ⓓ	ⓔ
37.	ⓐ	ⓑ	ⓒ	ⓓ	ⓔ
38.	ⓐ	ⓑ	ⓒ	ⓓ	ⓔ
39.	ⓐ	ⓑ	ⓒ	ⓓ	ⓔ
40.	ⓐ	ⓑ	ⓒ	ⓓ	ⓔ
41.	ⓐ	ⓑ	ⓒ	ⓓ	ⓔ
42.	ⓐ	ⓑ	ⓒ	ⓓ	ⓔ
43.	ⓐ	ⓑ	ⓒ	ⓓ	ⓔ
44.	ⓐ	ⓑ	ⓒ	ⓓ	ⓔ
45.	ⓐ	ⓑ	ⓒ	ⓓ	ⓔ
46.	ⓐ	ⓑ	ⓒ	ⓓ	ⓔ
47.	ⓐ	ⓑ	ⓒ	ⓓ	ⓔ
48.	ⓐ	ⓑ	ⓒ	ⓓ	ⓔ
49.	ⓐ	ⓑ	ⓒ	ⓓ	ⓔ
50.	ⓐ	ⓑ	ⓒ	ⓓ	ⓔ
51.	ⓐ	ⓑ	ⓒ	ⓓ	ⓔ
52.	ⓐ	ⓑ	ⓒ	ⓓ	ⓔ
53.	ⓐ	ⓑ	ⓒ	ⓓ	ⓔ
54.	ⓐ	ⓑ	ⓒ	ⓓ	ⓔ
55.	ⓐ	ⓑ	ⓒ	ⓓ	ⓔ
56.	ⓐ	ⓑ	ⓒ	ⓓ	ⓔ
57.	ⓐ	ⓑ	ⓒ	ⓓ	ⓔ
58.	ⓐ	ⓑ	ⓒ	ⓓ	ⓔ
59.	ⓐ	ⓑ	ⓒ	ⓓ	ⓔ
60.	ⓐ	ⓑ	ⓒ	ⓓ	ⓔ
61.	ⓐ	ⓑ	ⓒ	ⓓ	ⓔ
62.	ⓐ	ⓑ	ⓒ	ⓓ	ⓔ
63.	ⓐ	ⓑ	ⓒ	ⓓ	ⓔ
64.	ⓐ	ⓑ	ⓒ	ⓓ	ⓔ
65.	ⓐ	ⓑ	ⓒ	ⓓ	ⓔ
66.	ⓐ	ⓑ	ⓒ	ⓓ	ⓔ
67.	ⓐ	ⓑ	ⓒ	ⓓ	ⓔ
68.	ⓐ	ⓑ	ⓒ	ⓓ	ⓔ
69.	ⓐ	ⓑ	ⓒ	ⓓ	ⓔ
70.	ⓐ	ⓑ	ⓒ	ⓓ	ⓔ
71.	ⓐ	ⓑ	ⓒ	ⓓ	ⓔ
72.	ⓐ	ⓑ	ⓒ	ⓓ	ⓔ
73.	ⓐ	ⓑ	ⓒ	ⓓ	ⓔ
74.	ⓐ	ⓑ	ⓒ	ⓓ	ⓔ
75.	ⓐ	ⓑ	ⓒ	ⓓ	ⓔ
76.	ⓐ	ⓑ	ⓒ	ⓓ	ⓔ
77.	ⓐ	ⓑ	ⓒ	ⓓ	ⓔ
78.	ⓐ	ⓑ	ⓒ	ⓓ	ⓔ
79.	ⓐ	ⓑ	ⓒ	ⓓ	ⓔ
80.	ⓐ	ⓑ	ⓒ	ⓓ	ⓔ
81.	ⓐ	ⓑ	ⓒ	ⓓ	ⓔ
82.	ⓐ	ⓑ	ⓒ	ⓓ	ⓔ
83.	ⓐ	ⓑ	ⓒ	ⓓ	ⓔ
84.	ⓐ	ⓑ	ⓒ	ⓓ	ⓔ

TEST SECTION FOUR ANSWER SHEET Page two of 4 pages

85.	ⓐ	ⓑ	ⓒ	ⓓ	ⓔ
86.	ⓐ	ⓑ	ⓒ	ⓓ	ⓔ
87.	ⓐ	ⓑ	ⓒ	ⓓ	ⓔ
88.	ⓐ	ⓑ	ⓒ	ⓓ	ⓔ
89.	ⓐ	ⓑ	ⓒ	ⓓ	ⓔ
90.	ⓐ	ⓑ	ⓒ	ⓓ	ⓔ
91.	ⓐ	ⓑ	ⓒ	ⓓ	ⓔ
92.	ⓐ	ⓑ	ⓒ	ⓓ	ⓔ
93.	ⓐ	ⓑ	ⓒ	ⓓ	ⓔ
94.	ⓐ	ⓑ	ⓒ	ⓓ	ⓔ
95.	ⓐ	ⓑ	ⓒ	ⓓ	ⓔ
96.	ⓐ	ⓑ	ⓒ	ⓓ	ⓔ
97.	ⓐ	ⓑ	ⓒ	ⓓ	ⓔ
98.	ⓐ	ⓑ	ⓒ	ⓓ	ⓔ
99.	ⓐ	ⓑ	ⓒ	ⓓ	ⓔ
100.	ⓐ	ⓑ	ⓒ	ⓓ	ⓔ
101.	ⓐ	ⓑ	ⓒ	ⓓ	ⓔ
102.	ⓐ	ⓑ	ⓒ	ⓓ	ⓔ
103.	ⓐ	ⓑ	ⓒ	ⓓ	ⓔ
104.	ⓐ	ⓑ	ⓒ	ⓓ	ⓔ
105.	ⓐ	ⓑ	ⓒ	ⓓ	ⓔ
106.	ⓐ	ⓑ	ⓒ	ⓓ	ⓔ
107.	ⓐ	ⓑ	ⓒ	ⓓ	ⓔ
108.	ⓐ	ⓑ	ⓒ	ⓓ	ⓔ
109.	ⓐ	ⓑ	ⓒ	ⓓ	ⓔ
110.	ⓐ	ⓑ	ⓒ	ⓓ	ⓔ
111.	ⓐ	ⓑ	ⓒ	ⓓ	ⓔ
112.	ⓐ	ⓑ	ⓒ	ⓓ	ⓔ
113.	ⓐ	ⓑ	ⓒ	ⓓ	ⓔ
114.	ⓐ	ⓑ	ⓒ	ⓓ	ⓔ
115.	ⓐ	ⓑ	ⓒ	ⓓ	ⓔ
116.	ⓐ	ⓑ	ⓒ	ⓓ	ⓔ
117.	ⓐ	ⓑ	ⓒ	ⓓ	ⓔ
118.	ⓐ	ⓑ	ⓒ	ⓓ	ⓔ
119.	ⓐ	ⓑ	ⓒ	ⓓ	ⓔ
120.	ⓐ	ⓑ	ⓒ	ⓓ	ⓔ
121.	ⓐ	ⓑ	ⓒ	ⓓ	ⓔ
122.	ⓐ	ⓑ	ⓒ	ⓓ	ⓔ
123.	ⓐ	ⓑ	ⓒ	ⓓ	ⓔ
124.	ⓐ	ⓑ	ⓒ	ⓓ	ⓔ
125.	ⓐ	ⓑ	ⓒ	ⓓ	ⓔ
126.	ⓐ	ⓑ	ⓒ	ⓓ	ⓔ
127.	ⓐ	ⓑ	ⓒ	ⓓ	ⓔ
128.	ⓐ	ⓑ	ⓒ	ⓓ	ⓔ
129.	ⓐ	ⓑ	ⓒ	ⓓ	ⓔ
130.	ⓐ	ⓑ	ⓒ	ⓓ	ⓔ
131.	ⓐ	ⓑ	ⓒ	ⓓ	ⓔ
132.	ⓐ	ⓑ	ⓒ	ⓓ	ⓔ
133.	ⓐ	ⓑ	ⓒ	ⓓ	ⓔ
134.	ⓐ	ⓑ	ⓒ	ⓓ	ⓔ
135.	ⓐ	ⓑ	ⓒ	ⓓ	ⓔ
136.	ⓐ	ⓑ	ⓒ	ⓓ	ⓔ
137.	ⓐ	ⓑ	ⓒ	ⓓ	ⓔ
138.	ⓐ	ⓑ	ⓒ	ⓓ	ⓔ
139.	ⓐ	ⓑ	ⓒ	ⓓ	ⓔ
140.	ⓐ	ⓑ	ⓒ	ⓓ	ⓔ
141.	ⓐ	ⓑ	ⓒ	ⓓ	ⓔ
142.	ⓐ	ⓑ	ⓒ	ⓓ	ⓔ
143.	ⓐ	ⓑ	ⓒ	ⓓ	ⓔ
144.	ⓐ	ⓑ	ⓒ	ⓓ	ⓔ
145.	ⓐ	ⓑ	ⓒ	ⓓ	ⓔ
146.	ⓐ	ⓑ	ⓒ	ⓓ	ⓔ
147.	ⓐ	ⓑ	ⓒ	ⓓ	ⓔ
148.	ⓐ	ⓑ	ⓒ	ⓓ	ⓔ
149.	ⓐ	ⓑ	ⓒ	ⓓ	ⓔ
150.	ⓐ	ⓑ	ⓒ	ⓓ	ⓔ
151.	ⓐ	ⓑ	ⓒ	ⓓ	ⓔ
152.	ⓐ	ⓑ	ⓒ	ⓓ	ⓔ
153.	ⓐ	ⓑ	ⓒ	ⓓ	ⓔ
154.	ⓐ	ⓑ	ⓒ	ⓓ	ⓔ
155.	ⓐ	ⓑ	ⓒ	ⓓ	ⓔ
156.	ⓐ	ⓑ	ⓒ	ⓓ	ⓔ
157.	ⓐ	ⓑ	ⓒ	ⓓ	ⓔ
158.	ⓐ	ⓑ	ⓒ	ⓓ	ⓔ
159.	ⓐ	ⓑ	ⓒ	ⓓ	ⓔ
160.	ⓐ	ⓑ	ⓒ	ⓓ	ⓔ
161.	ⓐ	ⓑ	ⓒ	ⓓ	ⓔ
162.	ⓐ	ⓑ	ⓒ	ⓓ	ⓔ
163.	ⓐ	ⓑ	ⓒ	ⓓ	ⓔ
164.	ⓐ	ⓑ	ⓒ	ⓓ	ⓔ
165.	ⓐ	ⓑ	ⓒ	ⓓ	ⓔ
166.	ⓐ	ⓑ	ⓒ	ⓓ	ⓔ
167.	ⓐ	ⓑ	ⓒ	ⓓ	ⓔ
168.	ⓐ	ⓑ	ⓒ	ⓓ	ⓔ
169.	ⓐ	ⓑ	ⓒ	ⓓ	ⓔ
170.	ⓐ	ⓑ	ⓒ	ⓓ	ⓔ

TEST SECTION FOUR ANSWER SHEET Page three of 4 pages

171.	ⓐ	ⓑ	ⓒ	ⓓ	ⓔ
172.	ⓐ	ⓑ	ⓒ	ⓓ	ⓔ
173.	ⓐ	ⓑ	ⓒ	ⓓ	ⓔ
174.	ⓐ	ⓑ	ⓒ	ⓓ	ⓔ
175.	ⓐ	ⓑ	ⓒ	ⓓ	ⓔ
176.	ⓐ	ⓑ	ⓒ	ⓓ	ⓔ
177.	ⓐ	ⓑ	ⓒ	ⓓ	ⓔ
178.	ⓐ	ⓑ	ⓒ	ⓓ	ⓔ
179.	ⓐ	ⓑ	ⓒ	ⓓ	ⓔ
180.	ⓐ	ⓑ	ⓒ	ⓓ	ⓔ
181.	ⓐ	ⓑ	ⓒ	ⓓ	ⓔ
182.	ⓐ	ⓑ	ⓒ	ⓓ	ⓔ
183.	ⓐ	ⓑ	ⓒ	ⓓ	ⓔ
184.	ⓐ	ⓑ	ⓒ	ⓓ	ⓔ
185.	ⓐ	ⓑ	ⓒ	ⓓ	ⓔ
186.	ⓐ	ⓑ	ⓒ	ⓓ	ⓔ
187.	ⓐ	ⓑ	ⓒ	ⓓ	ⓔ
188.	ⓐ	ⓑ	ⓒ	ⓓ	ⓔ
189.	ⓐ	ⓑ	ⓒ	ⓓ	ⓔ
190.	ⓐ	ⓑ	ⓒ	ⓓ	ⓔ
191.	ⓐ	ⓑ	ⓒ	ⓓ	ⓔ
192.	ⓐ	ⓑ	ⓒ	ⓓ	ⓔ
193.	ⓐ	ⓑ	ⓒ	ⓓ	ⓔ
194.	ⓐ	ⓑ	ⓒ	ⓓ	ⓔ
195.	ⓐ	ⓑ	ⓒ	ⓓ	ⓔ
196.	ⓐ	ⓑ	ⓒ	ⓓ	ⓔ
197.	ⓐ	ⓑ	ⓒ	ⓓ	ⓔ
198.	ⓐ	ⓑ	ⓒ	ⓓ	ⓔ
199.	ⓐ	ⓑ	ⓒ	ⓓ	ⓔ
200.	ⓐ	ⓑ	ⓒ	ⓓ	ⓔ
201.	ⓐ	ⓑ	ⓒ	ⓓ	ⓔ
202.	ⓐ	ⓑ	ⓒ	ⓓ	ⓔ
203.	ⓐ	ⓑ	ⓒ	ⓓ	ⓔ
204.	ⓐ	ⓑ	ⓒ	ⓓ	ⓔ
205.	ⓐ	ⓑ	ⓒ	ⓓ	ⓔ
206.	ⓐ	ⓑ	ⓒ	ⓓ	ⓔ
207.	ⓐ	ⓑ	ⓒ	ⓓ	ⓔ
208.	ⓐ	ⓑ	ⓒ	ⓓ	ⓔ
209.	ⓐ	ⓑ	ⓒ	ⓓ	ⓔ
210.	ⓐ	ⓑ	ⓒ	ⓓ	ⓔ
211.	ⓐ	ⓑ	ⓒ	ⓓ	ⓔ
212.	ⓐ	ⓑ	ⓒ	ⓓ	ⓔ
213.	ⓐ	ⓑ	ⓒ	ⓓ	ⓔ

214.	ⓐ	ⓑ	ⓒ	ⓓ	ⓔ
215.	ⓐ	ⓑ	ⓒ	ⓓ	ⓔ
216.	ⓐ	ⓑ	ⓒ	ⓓ	ⓔ
217.	ⓐ	ⓑ	ⓒ	ⓓ	ⓔ
218.	ⓐ	ⓑ	ⓒ	ⓓ	ⓔ
219.	ⓐ	ⓑ	ⓒ	ⓓ	ⓔ
220.	ⓐ	ⓑ	ⓒ	ⓓ	ⓔ
221.	ⓐ	ⓑ	ⓒ	ⓓ	ⓔ
222.	ⓐ	ⓑ	ⓒ	ⓓ	ⓔ
223.	ⓐ	ⓑ	ⓒ	ⓓ	ⓔ
224.	ⓐ	ⓑ	ⓒ	ⓓ	ⓔ
225.	ⓐ	ⓑ	ⓒ	ⓓ	ⓔ
226.	ⓐ	ⓑ	ⓒ	ⓓ	ⓔ
227.	ⓐ	ⓑ	ⓒ	ⓓ	ⓔ
228.	ⓐ	ⓑ	ⓒ	ⓓ	ⓔ
229.	ⓐ	ⓑ	ⓒ	ⓓ	ⓔ
230.	ⓐ	ⓑ	ⓒ	ⓓ	ⓔ
231.	ⓐ	ⓑ	ⓒ	ⓓ	ⓔ
232.	ⓐ	ⓑ	ⓒ	ⓓ	ⓔ
233.	ⓐ	ⓑ	ⓒ	ⓓ	ⓔ
234.	ⓐ	ⓑ	ⓒ	ⓓ	ⓔ
235.	ⓐ	ⓑ	ⓒ	ⓓ	ⓔ
236.	ⓐ	ⓑ	ⓒ	ⓓ	ⓔ
237.	ⓐ	ⓑ	ⓒ	ⓓ	ⓔ
238.	ⓐ	ⓑ	ⓒ	ⓓ	ⓔ
239.	ⓐ	ⓑ	ⓒ	ⓓ	ⓔ
240.	ⓐ	ⓑ	ⓒ	ⓓ	ⓔ
241.	ⓐ	ⓑ	ⓒ	ⓓ	ⓔ
242.	ⓐ	ⓑ	ⓒ	ⓓ	ⓔ
243.	ⓐ	ⓑ	ⓒ	ⓓ	ⓔ
244.	ⓐ	ⓑ	ⓒ	ⓓ	ⓔ
245.	ⓐ	ⓑ	ⓒ	ⓓ	ⓔ
246.	ⓐ	ⓑ	ⓒ	ⓓ	ⓔ
247.	ⓐ	ⓑ	ⓒ	ⓓ	ⓔ
248.	ⓐ	ⓑ	ⓒ	ⓓ	ⓔ
249.	ⓐ	ⓑ	ⓒ	ⓓ	ⓔ
250.	ⓐ	ⓑ	ⓒ	ⓓ	ⓔ
251.	ⓐ	ⓑ	ⓒ	ⓓ	ⓔ
252.	ⓐ	ⓑ	ⓒ	ⓓ	ⓔ
253.	ⓐ	ⓑ	ⓒ	ⓓ	ⓔ
254.	ⓐ	ⓑ	ⓒ	ⓓ	ⓔ
255.	ⓐ	ⓑ	ⓒ	ⓓ	ⓔ
256.	ⓐ	ⓑ	ⓒ	ⓓ	ⓔ

TEST SECTION FOUR ANSWER SHEET Page four of 4 pages

257.	ⓐ	ⓑ	ⓒ	ⓓ	ⓔ
258.	ⓐ	ⓑ	ⓒ	ⓓ	ⓔ
259.	ⓐ	ⓑ	ⓒ	ⓓ	ⓔ
260.	ⓐ	ⓑ	ⓒ	ⓓ	ⓔ
261.	ⓐ	ⓑ	ⓒ	ⓓ	ⓔ
262.	ⓐ	ⓑ	ⓒ	ⓓ	ⓔ
263.	ⓐ	ⓑ	ⓒ	ⓓ	ⓔ
264.	ⓐ	ⓑ	ⓒ	ⓓ	ⓔ
265.	ⓐ	ⓑ	ⓒ	ⓓ	ⓔ
266.	ⓐ	ⓑ	ⓒ	ⓓ	ⓔ
267.	ⓐ	ⓑ	ⓒ	ⓓ	ⓔ
268.	ⓐ	ⓑ	ⓒ	ⓓ	ⓔ
269.	ⓐ	ⓑ	ⓒ	ⓓ	ⓔ
270.	ⓐ	ⓑ	ⓒ	ⓓ	ⓔ
271.	ⓐ	ⓑ	ⓒ	ⓓ	ⓔ
272.	ⓐ	ⓑ	ⓒ	ⓓ	ⓔ
273.	ⓐ	ⓑ	ⓒ	ⓓ	ⓔ

EMT-Paramedic National Standards Review Self Test, 3rd Edition

TEST SECTION FIVE ANSWER SHEET Page one of 3 pages

1.	ⓐ	ⓑ	ⓒ	ⓓ	ⓔ
2.	ⓐ	ⓑ	ⓒ	ⓓ	ⓔ
3.	ⓐ	ⓑ	ⓒ	ⓓ	ⓔ
4.	ⓐ	ⓑ	ⓒ	ⓓ	ⓔ
5.	ⓐ	ⓑ	ⓒ	ⓓ	ⓔ
6.	ⓐ	ⓑ	ⓒ	ⓓ	ⓔ
7.	ⓐ	ⓑ	ⓒ	ⓓ	ⓔ
8.	ⓐ	ⓑ	ⓒ	ⓓ	ⓔ
9.	ⓐ	ⓑ	ⓒ	ⓓ	ⓔ
10.	ⓐ	ⓑ	ⓒ	ⓓ	ⓔ
11.	ⓐ	ⓑ	ⓒ	ⓓ	ⓔ
12.	ⓐ	ⓑ	ⓒ	ⓓ	ⓔ
13.	ⓐ	ⓑ	ⓒ	ⓓ	ⓔ
14.	ⓐ	ⓑ	ⓒ	ⓓ	ⓔ
15.	ⓐ	ⓑ	ⓒ	ⓓ	ⓔ
16.	ⓐ	ⓑ	ⓒ	ⓓ	ⓔ
17.	ⓐ	ⓑ	ⓒ	ⓓ	ⓔ
18.	ⓐ	ⓑ	ⓒ	ⓓ	ⓔ
19.	ⓐ	ⓑ	ⓒ	ⓓ	ⓔ
20.	ⓐ	ⓑ	ⓒ	ⓓ	ⓔ
21.	ⓐ	ⓑ	ⓒ	ⓓ	ⓔ
22.	ⓐ	ⓑ	ⓒ	ⓓ	ⓔ
23.	ⓐ	ⓑ	ⓒ	ⓓ	ⓔ
24.	ⓐ	ⓑ	ⓒ	ⓓ	ⓔ
25.	ⓐ	ⓑ	ⓒ	ⓓ	ⓔ
26.	ⓐ	ⓑ	ⓒ	ⓓ	ⓔ
27.	ⓐ	ⓑ	ⓒ	ⓓ	ⓔ
28.	ⓐ	ⓑ	ⓒ	ⓓ	ⓔ
29.	ⓐ	ⓑ	ⓒ	ⓓ	ⓔ
30.	ⓐ	ⓑ	ⓒ	ⓓ	ⓔ
31.	ⓐ	ⓑ	ⓒ	ⓓ	ⓔ
32.	ⓐ	ⓑ	ⓒ	ⓓ	ⓔ
33.	ⓐ	ⓑ	ⓒ	ⓓ	ⓔ
34.	ⓐ	ⓑ	ⓒ	ⓓ	ⓔ
35.	ⓐ	ⓑ	ⓒ	ⓓ	ⓔ
36.	ⓐ	ⓑ	ⓒ	ⓓ	ⓔ
37.	ⓐ	ⓑ	ⓒ	ⓓ	ⓔ
38.	ⓐ	ⓑ	ⓒ	ⓓ	ⓔ
39.	ⓐ	ⓑ	ⓒ	ⓓ	ⓔ
40.	ⓐ	ⓑ	ⓒ	ⓓ	ⓔ
41.	ⓐ	ⓑ	ⓒ	ⓓ	ⓔ
42.	ⓐ	ⓑ	ⓒ	ⓓ	ⓔ
43.	ⓐ	ⓑ	ⓒ	ⓓ	ⓔ
44.	ⓐ	ⓑ	ⓒ	ⓓ	ⓔ
45.	ⓐ	ⓑ	ⓒ	ⓓ	ⓔ
46.	ⓐ	ⓑ	ⓒ	ⓓ	ⓔ
47.	ⓐ	ⓑ	ⓒ	ⓓ	ⓔ
48.	ⓐ	ⓑ	ⓒ	ⓓ	ⓔ
49.	ⓐ	ⓑ	ⓒ	ⓓ	ⓔ
50.	ⓐ	ⓑ	ⓒ	ⓓ	ⓔ
51.	ⓐ	ⓑ	ⓒ	ⓓ	ⓔ
52.	ⓐ	ⓑ	ⓒ	ⓓ	ⓔ
53.	ⓐ	ⓑ	ⓒ	ⓓ	ⓔ
54.	ⓐ	ⓑ	ⓒ	ⓓ	ⓔ
55.	ⓐ	ⓑ	ⓒ	ⓓ	ⓔ
56.	ⓐ	ⓑ	ⓒ	ⓓ	ⓔ
57.	ⓐ	ⓑ	ⓒ	ⓓ	ⓔ
58.	ⓐ	ⓑ	ⓒ	ⓓ	ⓔ
59.	ⓐ	ⓑ	ⓒ	ⓓ	ⓔ
60.	ⓐ	ⓑ	ⓒ	ⓓ	ⓔ
61.	ⓐ	ⓑ	ⓒ	ⓓ	ⓔ
62.	ⓐ	ⓑ	ⓒ	ⓓ	ⓔ
63.	ⓐ	ⓑ	ⓒ	ⓓ	ⓔ
64.	ⓐ	ⓑ	ⓒ	ⓓ	ⓔ
65.	ⓐ	ⓑ	ⓒ	ⓓ	ⓔ
66.	ⓐ	ⓑ	ⓒ	ⓓ	ⓔ
67.	ⓐ	ⓑ	ⓒ	ⓓ	ⓔ
68.	ⓐ	ⓑ	ⓒ	ⓓ	ⓔ
69.	ⓐ	ⓑ	ⓒ	ⓓ	ⓔ
70.	ⓐ	ⓑ	ⓒ	ⓓ	ⓔ
71.	ⓐ	ⓑ	ⓒ	ⓓ	ⓔ
72.	ⓐ	ⓑ	ⓒ	ⓓ	ⓔ
73.	ⓐ	ⓑ	ⓒ	ⓓ	ⓔ
74.	ⓐ	ⓑ	ⓒ	ⓓ	ⓔ
75.	ⓐ	ⓑ	ⓒ	ⓓ	ⓔ
76.	ⓐ	ⓑ	ⓒ	ⓓ	ⓔ
77.	ⓐ	ⓑ	ⓒ	ⓓ	ⓔ
78.	ⓐ	ⓑ	ⓒ	ⓓ	ⓔ
79.	ⓐ	ⓑ	ⓒ	ⓓ	ⓔ
80.	ⓐ	ⓑ	ⓒ	ⓓ	ⓔ
81.	ⓐ	ⓑ	ⓒ	ⓓ	ⓔ
82.	ⓐ	ⓑ	ⓒ	ⓓ	ⓔ
83.	ⓐ	ⓑ	ⓒ	ⓓ	ⓔ
84.	ⓐ	ⓑ	ⓒ	ⓓ	ⓔ

TEST SECTION FIVE ANSWER SHEET Page two of 3 pages

85.	ⓐ	ⓑ	ⓒ	ⓓ	ⓔ
86.	ⓐ	ⓑ	ⓒ	ⓓ	ⓔ
87.	ⓐ	ⓑ	ⓒ	ⓓ	ⓔ
88.	ⓐ	ⓑ	ⓒ	ⓓ	ⓔ
89.	ⓐ	ⓑ	ⓒ	ⓓ	ⓔ
90.	ⓐ	ⓑ	ⓒ	ⓓ	ⓔ
91.	ⓐ	ⓑ	ⓒ	ⓓ	ⓔ
92.	ⓐ	ⓑ	ⓒ	ⓓ	ⓔ
93.	ⓐ	ⓑ	ⓒ	ⓓ	ⓔ
94.	ⓐ	ⓑ	ⓒ	ⓓ	ⓔ
95.	ⓐ	ⓑ	ⓒ	ⓓ	ⓔ
96.	ⓐ	ⓑ	ⓒ	ⓓ	ⓔ
97.	ⓐ	ⓑ	ⓒ	ⓓ	ⓔ
98.	ⓐ	ⓑ	ⓒ	ⓓ	ⓔ
99.	ⓐ	ⓑ	ⓒ	ⓓ	ⓔ
100.	ⓐ	ⓑ	ⓒ	ⓓ	ⓔ
101.	ⓐ	ⓑ	ⓒ	ⓓ	ⓔ
102.	ⓐ	ⓑ	ⓒ	ⓓ	ⓔ
103.	ⓐ	ⓑ	ⓒ	ⓓ	ⓔ
104.	ⓐ	ⓑ	ⓒ	ⓓ	ⓔ
105.	ⓐ	ⓑ	ⓒ	ⓓ	ⓔ
106.	ⓐ	ⓑ	ⓒ	ⓓ	ⓔ
107.	ⓐ	ⓑ	ⓒ	ⓓ	ⓔ
108.	ⓐ	ⓑ	ⓒ	ⓓ	ⓔ
109.	ⓐ	ⓑ	ⓒ	ⓓ	ⓔ
110.	ⓐ	ⓑ	ⓒ	ⓓ	ⓔ
111.	ⓐ	ⓑ	ⓒ	ⓓ	ⓔ
112.	ⓐ	ⓑ	ⓒ	ⓓ	ⓔ
113.	ⓐ	ⓑ	ⓒ	ⓓ	ⓔ
114.	ⓐ	ⓑ	ⓒ	ⓓ	ⓔ
115.	ⓐ	ⓑ	ⓒ	ⓓ	ⓔ
116.	ⓐ	ⓑ	ⓒ	ⓓ	ⓔ
117.	ⓐ	ⓑ	ⓒ	ⓓ	ⓔ
118.	ⓐ	ⓑ	ⓒ	ⓓ	ⓔ
119.	ⓐ	ⓑ	ⓒ	ⓓ	ⓔ
120.	ⓐ	ⓑ	ⓒ	ⓓ	ⓔ
121.	ⓐ	ⓑ	ⓒ	ⓓ	ⓔ
122.	ⓐ	ⓑ	ⓒ	ⓓ	ⓔ
123.	ⓐ	ⓑ	ⓒ	ⓓ	ⓔ
124.	ⓐ	ⓑ	ⓒ	ⓓ	ⓔ
125.	ⓐ	ⓑ	ⓒ	ⓓ	ⓔ
126.	ⓐ	ⓑ	ⓒ	ⓓ	ⓔ
127.	ⓐ	ⓑ	ⓒ	ⓓ	ⓔ
128.	ⓐ	ⓑ	ⓒ	ⓓ	ⓔ
129.	ⓐ	ⓑ	ⓒ	ⓓ	ⓔ
130.	ⓐ	ⓑ	ⓒ	ⓓ	ⓔ
131.	ⓐ	ⓑ	ⓒ	ⓓ	ⓔ
132.	ⓐ	ⓑ	ⓒ	ⓓ	ⓔ
133.	ⓐ	ⓑ	ⓒ	ⓓ	ⓔ
134.	ⓐ	ⓑ	ⓒ	ⓓ	ⓔ
135.	ⓐ	ⓑ	ⓒ	ⓓ	ⓔ
136.	ⓐ	ⓑ	ⓒ	ⓓ	ⓔ
137.	ⓐ	ⓑ	ⓒ	ⓓ	ⓔ
138.	ⓐ	ⓑ	ⓒ	ⓓ	ⓔ
139.	ⓐ	ⓑ	ⓒ	ⓓ	ⓔ
140.	ⓐ	ⓑ	ⓒ	ⓓ	ⓔ
141.	ⓐ	ⓑ	ⓒ	ⓓ	ⓔ
142.	ⓐ	ⓑ	ⓒ	ⓓ	ⓔ
143.	ⓐ	ⓑ	ⓒ	ⓓ	ⓔ
144.	ⓐ	ⓑ	ⓒ	ⓓ	ⓔ
145.	ⓐ	ⓑ	ⓒ	ⓓ	ⓔ
146.	ⓐ	ⓑ	ⓒ	ⓓ	ⓔ
147.	ⓐ	ⓑ	ⓒ	ⓓ	ⓔ
148.	ⓐ	ⓑ	ⓒ	ⓓ	ⓔ
149.	ⓐ	ⓑ	ⓒ	ⓓ	ⓔ
150.	ⓐ	ⓑ	ⓒ	ⓓ	ⓔ
151.	ⓐ	ⓑ	ⓒ	ⓓ	ⓔ
152.	ⓐ	ⓑ	ⓒ	ⓓ	ⓔ
153.	ⓐ	ⓑ	ⓒ	ⓓ	ⓔ
154.	ⓐ	ⓑ	ⓒ	ⓓ	ⓔ
155.	ⓐ	ⓑ	ⓒ	ⓓ	ⓔ
156.	ⓐ	ⓑ	ⓒ	ⓓ	ⓔ
157.	ⓐ	ⓑ	ⓒ	ⓓ	ⓔ
158.	ⓐ	ⓑ	ⓒ	ⓓ	ⓔ
159.	ⓐ	ⓑ	ⓒ	ⓓ	ⓔ
160.	ⓐ	ⓑ	ⓒ	ⓓ	ⓔ
161.	ⓐ	ⓑ	ⓒ	ⓓ	ⓔ
162.	ⓐ	ⓑ	ⓒ	ⓓ	ⓔ
163.	ⓐ	ⓑ	ⓒ	ⓓ	ⓔ
164.	ⓐ	ⓑ	ⓒ	ⓓ	ⓔ
165.	ⓐ	ⓑ	ⓒ	ⓓ	ⓔ
166.	ⓐ	ⓑ	ⓒ	ⓓ	ⓔ
167.	ⓐ	ⓑ	ⓒ	ⓓ	ⓔ
168.	ⓐ	ⓑ	ⓒ	ⓓ	ⓔ
169.	ⓐ	ⓑ	ⓒ	ⓓ	ⓔ
170.	ⓐ	ⓑ	ⓒ	ⓓ	ⓔ

TEST SECTION FIVE ANSWER SHEET Page three of 3 pages

171.	ⓐ	ⓑ	ⓒ	ⓓ	ⓔ
172.	ⓐ	ⓑ	ⓒ	ⓓ	ⓔ
173.	ⓐ	ⓑ	ⓒ	ⓓ	ⓔ
174.	ⓐ	ⓑ	ⓒ	ⓓ	ⓔ
175.	ⓐ	ⓑ	ⓒ	ⓓ	ⓔ
176.	ⓐ	ⓑ	ⓒ	ⓓ	ⓔ
177.	ⓐ	ⓑ	ⓒ	ⓓ	ⓔ
178.	ⓐ	ⓑ	ⓒ	ⓓ	ⓔ
179.	ⓐ	ⓑ	ⓒ	ⓓ	ⓔ
180.	ⓐ	ⓑ	ⓒ	ⓓ	ⓔ
181.	ⓐ	ⓑ	ⓒ	ⓓ	ⓔ
182.	ⓐ	ⓑ	ⓒ	ⓓ	ⓔ
183.	ⓐ	ⓑ	ⓒ	ⓓ	ⓔ
184.	ⓐ	ⓑ	ⓒ	ⓓ	ⓔ
185.	ⓐ	ⓑ	ⓒ	ⓓ	ⓔ
186.	ⓐ	ⓑ	ⓒ	ⓓ	ⓔ
187.	ⓐ	ⓑ	ⓒ	ⓓ	ⓔ
188.	ⓐ	ⓑ	ⓒ	ⓓ	ⓔ
189.	ⓐ	ⓑ	ⓒ	ⓓ	ⓔ
190.	ⓐ	ⓑ	ⓒ	ⓓ	ⓔ
191.	ⓐ	ⓑ	ⓒ	ⓓ	ⓔ
192.	ⓐ	ⓑ	ⓒ	ⓓ	ⓔ
193.	ⓐ	ⓑ	ⓒ	ⓓ	ⓔ
194.	ⓐ	ⓑ	ⓒ	ⓓ	ⓔ
195.	ⓐ	ⓑ	ⓒ	ⓓ	ⓔ
196.	ⓐ	ⓑ	ⓒ	ⓓ	ⓔ

EMT-Paramedic National Standards Review Self Test, 3rd Edition

TEST SECTION SIX ANSWER SHEET Page one of 6 pages

1.	ⓐ	ⓑ	ⓒ	ⓓ	ⓔ
2.	ⓐ	ⓑ	ⓒ	ⓓ	ⓔ
3.	ⓐ	ⓑ	ⓒ	ⓓ	ⓔ
4.	ⓐ	ⓑ	ⓒ	ⓓ	ⓔ
5.	ⓐ	ⓑ	ⓒ	ⓓ	ⓔ
6.	ⓐ	ⓑ	ⓒ	ⓓ	ⓔ
7.	ⓐ	ⓑ	ⓒ	ⓓ	ⓔ
8.	ⓐ	ⓑ	ⓒ	ⓓ	ⓔ
9.	ⓐ	ⓑ	ⓒ	ⓓ	ⓔ
10.	ⓐ	ⓑ	ⓒ	ⓓ	ⓔ
11.	ⓐ	ⓑ	ⓒ	ⓓ	ⓔ
12.	ⓐ	ⓑ	ⓒ	ⓓ	ⓔ
13.	ⓐ	ⓑ	ⓒ	ⓓ	ⓔ
14.	ⓐ	ⓑ	ⓒ	ⓓ	ⓔ
15.	ⓐ	ⓑ	ⓒ	ⓓ	ⓔ
16.	ⓐ	ⓑ	ⓒ	ⓓ	ⓔ
17.	ⓐ	ⓑ	ⓒ	ⓓ	ⓔ
18.	ⓐ	ⓑ	ⓒ	ⓓ	ⓔ
19.	ⓐ	ⓑ	ⓒ	ⓓ	ⓔ
20.	ⓐ	ⓑ	ⓒ	ⓓ	ⓔ
21.	ⓐ	ⓑ	ⓒ	ⓓ	ⓔ
22.	ⓐ	ⓑ	ⓒ	ⓓ	ⓔ
23.	ⓐ	ⓑ	ⓒ	ⓓ	ⓔ
24.	ⓐ	ⓑ	ⓒ	ⓓ	ⓔ
25.	ⓐ	ⓑ	ⓒ	ⓓ	ⓔ
26.	ⓐ	ⓑ	ⓒ	ⓓ	ⓔ
27.	ⓐ	ⓑ	ⓒ	ⓓ	ⓔ
28.	ⓐ	ⓑ	ⓒ	ⓓ	ⓔ
29.	ⓐ	ⓑ	ⓒ	ⓓ	ⓔ
30.	ⓐ	ⓑ	ⓒ	ⓓ	ⓔ
31.	ⓐ	ⓑ	ⓒ	ⓓ	ⓔ
32.	ⓐ	ⓑ	ⓒ	ⓓ	ⓔ
33.	ⓐ	ⓑ	ⓒ	ⓓ	ⓔ
34.	ⓐ	ⓑ	ⓒ	ⓓ	ⓔ
35.	ⓐ	ⓑ	ⓒ	ⓓ	ⓔ
36.	ⓐ	ⓑ	ⓒ	ⓓ	ⓔ
37.	ⓐ	ⓑ	ⓒ	ⓓ	ⓔ
38.	ⓐ	ⓑ	ⓒ	ⓓ	ⓔ
39.	ⓐ	ⓑ	ⓒ	ⓓ	ⓔ
40.	ⓐ	ⓑ	ⓒ	ⓓ	ⓔ
41.	ⓐ	ⓑ	ⓒ	ⓓ	ⓔ
42.	ⓐ	ⓑ	ⓒ	ⓓ	ⓔ
43.	ⓐ	ⓑ	ⓒ	ⓓ	ⓔ
44.	ⓐ	ⓑ	ⓒ	ⓓ	ⓔ
45.	ⓐ	ⓑ	ⓒ	ⓓ	ⓔ
46.	ⓐ	ⓑ	ⓒ	ⓓ	ⓔ
47.	ⓐ	ⓑ	ⓒ	ⓓ	ⓔ
48.	ⓐ	ⓑ	ⓒ	ⓓ	ⓔ
49.	ⓐ	ⓑ	ⓒ	ⓓ	ⓔ
50.	ⓐ	ⓑ	ⓒ	ⓓ	ⓔ
51.	ⓐ	ⓑ	ⓒ	ⓓ	ⓔ
52.	ⓐ	ⓑ	ⓒ	ⓓ	ⓔ
53.	ⓐ	ⓑ	ⓒ	ⓓ	ⓔ
54.	ⓐ	ⓑ	ⓒ	ⓓ	ⓔ
55.	ⓐ	ⓑ	ⓒ	ⓓ	ⓔ
56.	ⓐ	ⓑ	ⓒ	ⓓ	ⓔ
57.	ⓐ	ⓑ	ⓒ	ⓓ	ⓔ
58.	ⓐ	ⓑ	ⓒ	ⓓ	ⓔ
59.	ⓐ	ⓑ	ⓒ	ⓓ	ⓔ
60.	ⓐ	ⓑ	ⓒ	ⓓ	ⓔ
61.	ⓐ	ⓑ	ⓒ	ⓓ	ⓔ
62.	ⓐ	ⓑ	ⓒ	ⓓ	ⓔ
63.	ⓐ	ⓑ	ⓒ	ⓓ	ⓔ
64.	ⓐ	ⓑ	ⓒ	ⓓ	ⓔ
65.	ⓐ	ⓑ	ⓒ	ⓓ	ⓔ
66.	ⓐ	ⓑ	ⓒ	ⓓ	ⓔ
67.	ⓐ	ⓑ	ⓒ	ⓓ	ⓔ
68.	ⓐ	ⓑ	ⓒ	ⓓ	ⓔ
69.	ⓐ	ⓑ	ⓒ	ⓓ	ⓔ
70.	ⓐ	ⓑ	ⓒ	ⓓ	ⓔ
71.	ⓐ	ⓑ	ⓒ	ⓓ	ⓔ
72.	ⓐ	ⓑ	ⓒ	ⓓ	ⓔ
73.	ⓐ	ⓑ	ⓒ	ⓓ	ⓔ
74.	ⓐ	ⓑ	ⓒ	ⓓ	ⓔ
75.	ⓐ	ⓑ	ⓒ	ⓓ	ⓔ
76.	ⓐ	ⓑ	ⓒ	ⓓ	ⓔ
77.	ⓐ	ⓑ	ⓒ	ⓓ	ⓔ
78.	ⓐ	ⓑ	ⓒ	ⓓ	ⓔ
79.	ⓐ	ⓑ	ⓒ	ⓓ	ⓔ
80.	ⓐ	ⓑ	ⓒ	ⓓ	ⓔ
81.	ⓐ	ⓑ	ⓒ	ⓓ	ⓔ
82.	ⓐ	ⓑ	ⓒ	ⓓ	ⓔ
83.	ⓐ	ⓑ	ⓒ	ⓓ	ⓔ
84.	ⓐ	ⓑ	ⓒ	ⓓ	ⓔ

TEST SECTION SIX ANSWER SHEET Page of two of _6_ pages

85.	a	b	c	d	e	106.	a	b	c	d	e
86.	a	b	c	d	e	107.	a	b	c	d	e
87.	a	b	c	d	e	108.	a	b	c	d	e
88.	a	b	c	d	e	109.	a	b	c	d	e
89.	a	b	c	d	e	110.	a	b	c	d	e
90.	a	b	c	d	e	111.	a	b	c	d	e
91.	a	b	c	d	e	112.	a	b	c	d	e
92.	a	b	c	d	e	113.	a	b	c	d	e
93.	a	b	c	d	e	114.	a	b	c	d	e
94.	a	b	c	d	e	115.	a	b	c	d	e
95.	a	b	c	d	e	116.	a	b	c	d	e
96.	a	b	c	d	e	117.	a	b	c	d	e
97.	a	b	c	d	e	118.	a	b	c	d	e
98.	a	b	c	d	e	119.	a	b	c	d	e
99.	a	b	c	d	e	120.	a	b	c	d	e
100.	a	b	c	d	e	121.	a	b	c	d	e
101.	a	b	c	d	e	122.	a	b	c	d	e
102.	a	b	c	d	e	123.	a	b	c	d	e
103.	a	b	c	d	e	124.	a	b	c	d	e
104.	a	b	c	d	e	125.	a	b	c	d	e
105.	a	b	c	d	e	126.	a	b	c	d	e

Figure 6-15____________________

Figure 6-16____________________

Figure 6-17____________________

Figure 6-18____________________

Figure 6-19____________________

Figure 6-20____________________

TEST SECTION SIX ANSWER SHEET Page three of _6_ pages

Figure 6-21 ______________________________

____

Figure 6-22 ______________________________

____

Figure 6-23 ______________________________

____

Figure 6-24 ______________________________

____

Figure 6-25 ______________________________

____

Figure 6-26 ______________________________

____

Figure 6-27 ______________________________

____

Figure 6-28 ______________________________

____

Figure 6-29 ______________________________

____

Figure 6-30 ______________________________

____

Figure 6-31 ______________________________

____

Figure 6-32 ______________________________

____

Figure 6-33 ______________________________

____

TEST SECTION SIX ANSWER SHEET Page four of 6 pages

Figure 6-34 ____________________

____

Figure 6-35 ____________________

____

Figure 6-36 ____________________

____

Figure 6-37 ____________________

____

Figure 6-38 ____________________

____

Figure 6-39 ____________________

____

Figure 6-40 ____________________

____

Figure 6-41 ____________________

____

Figure 6-42 ____________________

____

Figure 6-43 ____________________

____

Figure 6-44 ____________________

____

Figure 6-45 ____________________

____

Figure 6-46 ____________________

____

TEST SECTION SIX ANSWER SHEET Page five of _6_ pages

Figure 6-47____

____

Figure 6-48____

____

Figure 6-49____

____

Figure 6-50____

____

Figure 6-51____

____

Figure 6-52____

____

Figure 6-53____

____

Figure 6-54____

____

Figure 6-55____

____

Figure 6-56____

____

Figure 6-57____

____

Figure 6-58____

____

Figure 6-59____

____

TEST SECTION SIX ANSWER SHEET Page six of 6 pages

Figure 6-60 ______________________________

____

Figure 6-61 ______________________________

____

Figure 6-62 ______________________________

____

Figure 6-63 ______________________________

____

Figure 6-64 ______________________________

____

Figure 6-65 ______________________________

____

EMT-Paramedic National Standards Review Self Test, 3rd Edition

TEST SECTION SEVEN ANSWER SHEET Page one of 3 pages

1.	a	b	c	d	e	43.	a	b	c	d	e
2.	a	b	c	d	e	44.	a	b	c	d	e
3.	a	b	c	d	e	45.	a	b	c	d	e
4.	a	b	c	d	e	46.	a	b	c	d	e
5.	a	b	c	d	e	47.	a	b	c	d	e
6.	a	b	c	d	e	48.	a	b	c	d	e
7.	a	b	c	d	e	49.	a	b	c	d	e
8.	a	b	c	d	e	50.	a	b	c	d	e
9.	a	b	c	d	e	51.	a	b	c	d	e
10.	a	b	c	d	e	52.	a	b	c	d	e
11.	a	b	c	d	e	53.	a	b	c	d	e
12.	a	b	c	d	e	54.	a	b	c	d	e
13.	a	b	c	d	e	55.	a	b	c	d	e
14.	a	b	c	d	e	56.	a	b	c	d	e
15.	a	b	c	d	e	57.	a	b	c	d	e
16.	a	b	c	d	e	58.	a	b	c	d	e
17.	a	b	c	d	e	59.	a	b	c	d	e
18.	a	b	c	d	e	60.	a	b	c	d	e
19.	a	b	c	d	e	61.	a	b	c	d	e
20.	a	b	c	d	e	62.	a	b	c	d	e
21.	a	b	c	d	e	63.	a	b	c	d	e
22.	a	b	c	d	e	64.	a	b	c	d	e
23.	a	b	c	d	e	65.	a	b	c	d	e
24.	a	b	c	d	e	66.	a	b	c	d	e
25.	a	b	c	d	e	67.	a	b	c	d	e
26.	a	b	c	d	e	68.	a	b	c	d	e
27.	a	b	c	d	e	69.	a	b	c	d	e
28.	a	b	c	d	e	70.	a	b	c	d	e
29.	a	b	c	d	e	71.	a	b	c	d	e
30.	a	b	c	d	e	72.	a	b	c	d	e
31.	a	b	c	d	e	73.	a	b	c	d	e
32.	a	b	c	d	e	74.	a	b	c	d	e
33.	a	b	c	d	e	75.	a	b	c	d	e
34.	a	b	c	d	e	76.	a	b	c	d	e
35.	a	b	c	d	e	77.	a	b	c	d	e
36.	a	b	c	d	e	78.	a	b	c	d	e
37.	a	b	c	d	e	79.	a	b	c	d	e
38.	a	b	c	d	e	80.	a	b	c	d	e
39.	a	b	c	d	e	81.	a	b	c	d	e
40.	a	b	c	d	e	82.	a	b	c	d	e
41.	a	b	c	d	e	83.	a	b	c	d	e
42.	a	b	c	d	e	84.	a	b	c	d	e

TEST SECTION SEVEN ANSWER SHEET Page two of 3 pages

85.	a	b	c	d	e
86.	a	b	c	d	e
87.	a	b	c	d	e
88.	a	b	c	d	e
89.	a	b	c	d	e
90.	a	b	c	d	e
91.	a	b	c	d	e
92.	a	b	c	d	e
93.	a	b	c	d	e
94.	a	b	c	d	e
95.	a	b	c	d	e
96.	a	b	c	d	e
97.	a	b	c	d	e
98.	a	b	c	d	e
99.	a	b	c	d	e
100.	a	b	c	d	e
101.	a	b	c	d	e
102.	a	b	c	d	e
103.	a	b	c	d	e
104.	a	b	c	d	e
105.	a	b	c	d	e
106.	a	b	c	d	e
107.	a	b	c	d	e
108.	a	b	c	d	e
109.	a	b	c	d	e
110.	a	b	c	d	e
111.	a	b	c	d	e
112.	a	b	c	d	e
113.	a	b	c	d	e
114.	a	b	c	d	e
115.	a	b	c	d	e
116.	a	b	c	d	e
117.	a	b	c	d	e
118.	a	b	c	d	e
119.	a	b	c	d	e
120.	a	b	c	d	e
121.	a	b	c	d	e
122.	a	b	c	d	e
123.	a	b	c	d	e
124.	a	b	c	d	e
125.	a	b	c	d	e
126.	a	b	c	d	e
127.	a	b	c	d	e
128.	a	b	c	d	e
129.	a	b	c	d	e
130.	a	b	c	d	e
131.	a	b	c	d	e
132.	a	b	c	d	e
133.	a	b	c	d	e
134.	a	b	c	d	e
135.	a	b	c	d	e
136.	a	b	c	d	e
137.	a	b	c	d	e
138.	a	b	c	d	e
139.	a	b	c	d	e
140.	a	b	c	d	e
141.	a	b	c	d	e
142.	a	b	c	d	e
143.	a	b	c	d	e
144.	a	b	c	d	e
145.	a	b	c	d	e
146.	a	b	c	d	e
147.	a	b	c	d	e
148.	a	b	c	d	e
149.	a	b	c	d	e
150.	a	b	c	d	e
151.	a	b	c	d	e
152.	a	b	c	d	e
153.	a	b	c	d	e
154.	a	b	c	d	e
155.	a	b	c	d	e
156.	a	b	c	d	e
157.	a	b	c	d	e
158.	a	b	c	d	e
159.	a	b	c	d	e
160.	a	b	c	d	e
161.	a	b	c	d	e
162.	a	b	c	d	e
163.	a	b	c	d	e
164.	a	b	c	d	e
165.	a	b	c	d	e
166.	a	b	c	d	e
167.	a	b	c	d	e
168.	a	b	c	d	e
169.	a	b	c	d	e
170.	a	b	c	d	e

TEST SECTION SEVEN ANSWER SHEET Page three of 3 pages

171.	ⓐ	ⓑ	ⓒ	ⓓ	ⓔ
172.	ⓐ	ⓑ	ⓒ	ⓓ	ⓔ
173.	ⓐ	ⓑ	ⓒ	ⓓ	ⓔ
174.	ⓐ	ⓑ	ⓒ	ⓓ	ⓔ
175.	ⓐ	ⓑ	ⓒ	ⓓ	ⓔ
176.	ⓐ	ⓑ	ⓒ	ⓓ	ⓔ
177.	ⓐ	ⓑ	ⓒ	ⓓ	ⓔ
178.	ⓐ	ⓑ	ⓒ	ⓓ	ⓔ
179.	ⓐ	ⓑ	ⓒ	ⓓ	ⓔ
180.	ⓐ	ⓑ	ⓒ	ⓓ	ⓔ
181.	ⓐ	ⓑ	ⓒ	ⓓ	ⓔ
182.	ⓐ	ⓑ	ⓒ	ⓓ	ⓔ
183.	ⓐ	ⓑ	ⓒ	ⓓ	ⓔ
184.	ⓐ	ⓑ	ⓒ	ⓓ	ⓔ
185.	ⓐ	ⓑ	ⓒ	ⓓ	ⓔ
186.	ⓐ	ⓑ	ⓒ	ⓓ	ⓔ
187.	ⓐ	ⓑ	ⓒ	ⓓ	ⓔ
188.	ⓐ	ⓑ	ⓒ	ⓓ	ⓔ
189.	ⓐ	ⓑ	ⓒ	ⓓ	ⓔ
190.	ⓐ	ⓑ	ⓒ	ⓓ	ⓔ
191.	ⓐ	ⓑ	ⓒ	ⓓ	ⓔ
192.	ⓐ	ⓑ	ⓒ	ⓓ	ⓔ
193.	ⓐ	ⓑ	ⓒ	ⓓ	ⓔ
194.	ⓐ	ⓑ	ⓒ	ⓓ	ⓔ
195.	ⓐ	ⓑ	ⓒ	ⓓ	ⓔ
196.	ⓐ	ⓑ	ⓒ	ⓓ	ⓔ
197.	ⓐ	ⓑ	ⓒ	ⓓ	ⓔ
198.	ⓐ	ⓑ	ⓒ	ⓓ	ⓔ
199.	ⓐ	ⓑ	ⓒ	ⓓ	ⓔ
200.	ⓐ	ⓑ	ⓒ	ⓓ	ⓔ
201.	ⓐ	ⓑ	ⓒ	ⓓ	ⓔ
202.	ⓐ	ⓑ	ⓒ	ⓓ	ⓔ
203.	ⓐ	ⓑ	ⓒ	ⓓ	ⓔ
204.	ⓐ	ⓑ	ⓒ	ⓓ	ⓔ
205.	ⓐ	ⓑ	ⓒ	ⓓ	ⓔ
206.	ⓐ	ⓑ	ⓒ	ⓓ	ⓔ
207.	ⓐ	ⓑ	ⓒ	ⓓ	ⓔ
208.	ⓐ	ⓑ	ⓒ	ⓓ	ⓔ
209.	ⓐ	ⓑ	ⓒ	ⓓ	ⓔ
210.	ⓐ	ⓑ	ⓒ	ⓓ	ⓔ
211.	ⓐ	ⓑ	ⓒ	ⓓ	ⓔ
212.	ⓐ	ⓑ	ⓒ	ⓓ	ⓔ
213.	ⓐ	ⓑ	ⓒ	ⓓ	ⓔ

EMT-PARAMEDIC NATIONAL STANDARDS REVIEW SELF TEST, 3/E

TEST SECTION ONE ANSWER KEY

The page numbers following each answer indicate that subject's reference page within Brady's *Paramedic Emergency Care*, 3rd edition. If you do not have access to Brady's text, use the index of the text you do have to obtain information for review of that subject.

Following some answers is "g/d." This indicates that you should consult the glossary of your paramedic text or a medical dictionary.

QUESTION	ANSWER	PAGE REFERENCE *Paramedic Emergency Care, 3/e*	
1.	(e)	pages 6-7	ethics
2.	(a)	page 9	professionalism
3.	(c)	pages 10-11	roles of the EMT-P
4.	(b)	page 12	certification
5.	(c)	page 13	licensure
6.	(a)	page 13	reciprocity
7.	(d)	page 13	continuing education
8.	(a)	page 14	professional journals
9.	(e)	page 18	EMS
10.	(b)	page 23	indirect medical control
11.	(a)	page 22	direct medical control
12.	(b)	page 23	indirect medical control
13.	(b)	page 23	indirect medical control
14.	(c)	page 23	scene control
15.	(c)	page 41	civil (tort) law
16.	(a)	page 41	criminal law
17.	(c)	page 44	malpractice
18.	(a)	page 41	unlicensed practice of medicine
19.	(e)	page 41	medical practice act
20.	(a)	page 41	the Good Samaritan act
21.	(d)	page 41	motor vehicle laws
22.	(d)	page 42	paramedic obligations
23.	(b)	page 42	living will
24.	(e)	page 42	terminal illness
25.	(d)	page 44	standard of care
26.	(b)	page 44	negligence
27.	(a)	page 44	malpractice
28.	(c)	page 45	implied consent
29.	(d)	page 45	informed/expressed consent
30.	(b)	page 45	expressed consent
31.	(c)	page 46	abandonment
32.	(a)	page 46	refusal of treatment
33.	(d)	page 46	assault, battery, false imprisonment
34.	(d)	page 46	assault, battery, false imprisonment

QUESTION	ANSWER	PAGE REFERENCE *Paramedic Emergency Care, 3/e*	
35.	(b)	page 46	battery
36.	(a)	page 46	assault
37.	(c)	page 46	false imprisonment
38.	(d)	page 47	slander
39.	(d)	page 47	slander
40.	(b)	page 47	libel
41.	(b)	page 48	documentation
42.	(d)	page 48	malpractice insurance
43.	(c)	page 55	portable radio
44.	(b)	page 55	mobile two-way radio
45.	(a)	page 54	base station
46.	(c)	page 55	repeater
47.	(d)	page 56	encoder
48.	(e)	page 57	decoder
49.	(b)	page 58	Hertz
50.	(a)	page 58	kilohertz
51.	(e)	page 58	megahertz
52.	(c)	page 58	gigahertz
53.	(d)	page 58	VHF
54.	(e)	page 58	UHF
55.	(e)	page 58	FM/AM
56.	(b)	page 59	biotelemetry
57.	(a)	page 60	simplex transmission systems
58.	(b)	page 60	duplex transmission systems
59.	(c)	page 61	multiplex transmission systems
60.	(c)	page 61	EMS dispatcher
61.	(c)	page 63	radio codes
62.	(c)	page 63	radio techniques
63.	(b)	page 65	radio reports
64.	(a)	page 65	radio reports
65.	(d)	page 65	radio reply to orders
66.	(c)	page 66	biotelemetry
67.	(e)	page 66	EMS forms
68.	(c)	page 70	rescue
69.	(d)	page 70	rescue safety
70.	(c)	page 72	patient safety
71.	(b)	page 74	rescue phases
72.	(e)	page 75	scene hazards
73.	(e)	pages 88-90	multiple-casualty incidents
74.	(d)	page 91	incident command system
75.	(b)	page 91	incident commander
76.	(c)	page 94	incident commander
77.	(d)	page 96	transfer of command
78.	(d)	page 99	triage/treatment sector officer
79.	(a)	page 97	extrication sector
80.	(b)	page 98	treatment sector
81.	(c)	page 101	staging sector

QUESTION	ANSWER	PAGE REFERENCE *Paramedic Emergency Care, 3/e*	
82.	(c)	page 112	stress
83.	(e)	page 112	stress
84.	(e)	page 112	stress
85.	(b)	page 113	stages of stress
86.	(e)	page 113	stages of stress
87.	(a)	page 113	stages of stress
88.	(d)	page 116	defense mechanisms
89.	(e)	g/d	compensation
90.	(b)	g/d	regression
91.	(a)	g/d	repression
92.	(c)	g/d	projection
93.	(b)	g/d	regression
94.	(c)	g/d	projection
95.	(d)	g/d	rationalization
96.	(d)	g/d	rationalization
97.	(d)	g/d	substitution
98.	(d)	g/d	substitution
99.	(b)	g/d	sublimation
100.	(e)	g/d	isolation
101.	(c)	g/d	denial
102.	(a)	g/d	reaction formation
103.	(a)	page 116	anxiety
104.	(e)	page 116	anxiety
105.	(d)	page 116	anxiety
106.	(b)	page 116	anxiety
107.	(e)	page 113	anxiety
108.	(c)	page 117	anxiety
109.	(d)	page 119	stress management
110.	(e)	page 120	death and dying
111.	(a)	page 120	death and dying
112.	(d)	page 121	death and dying
113.	(c)	page 121	death and dying
114.	(b)	page 121	death and dying
115.	(d)	page 121	death and dying
116.	(e)	page 121	death and dying
117.	(c)	page 121	death and dying
118.	(a)	page 132	trans-
119.	(c)	page 132	supra-
120.	(e)	page 132	para-
121.	(d)	page 131	intra-
122.	(b)	page 131	inter-
123.	(c)	pages 125–133	neuralgia
124.	(b)	pages 125–133	leukocyte
125.	(d)	pages 125–133	anemia
126.	(a)	pages 125–133	neuroma
127.	(e)	pages 125–133	neuropathy
128.	(e)	page 130	arthro-

QUESTION	ANSWER	PAGE REFERENCE *Paramedic Emergency Care, 3/e*	
129.	(a)	page 130	angio-
130.	(c)	page 130	anti-
131.	(d)	page 133	-centesis
132.	(b)	page 133	-ectomy
133.	(a)	page 133	-ostomy
134.	(e)	page 133	-plasty
135.	(c)	page 133	-lysis
136.	(a)	page 130	chole-
137.	(c)	page 127	hepat-
138.	(e)	page 131	my-
139.	(d)	page 131	nephr-
140.	(b)	page 126	chondr-
141.	(c)	page 130	a-, an-
142.	(b)	page 130	cephal-
143.	(d)	page 131	cyst-
144.	(a)	page 131	infra-
145.	(e)	page 132	olig-
146.	(c)	page 133	-algia
147.	(b)	page 126	-asthenia
148.	(d)	page 133	-emia
149.	(e)	page 133	-itis
150.	(a)	page 133	-osis
151.	(b)	pages 125-133	retroflexion
152.	(a)	pages 125-133	unilateral
153.	(d)	pages 125-133	rhinitis
154.	(e)	pages 125-133	circumoral
155.	(c)	pages 125-133	intercostal
156.	(e)	page 131	ecto-
157.	(d)	page 131	endo-
158.	(a)	page 131	entero-, enter-
159.	(b)	page 131	epi-
160.	(b)	g/d	-paresis
161.	(d)	page 133	-phagia
162.	(a)	g/d	-esthesia
163.	(e)	page 133	-phasia
164.	(c)	g/d	-plegia
165.	(d)	page 130	ab-
166.	(e)	page 130	bi-
167.	(a)	page 131	contra-
168.	(b)	page 131	hemi-
169.	(c)	page 132	peri-
170.	(c)	page 137	$\overline{p}$
171.	(a)	page 134	$\overline{a}$
172.	(b)	page 134	$\overline{c}$
173.	(d)	page 138	$\overline{s}$
174.	(e)	page 137	$\overline{q}$
175.	(e)	page 134	b.i.d.

QUESTION	ANSWER	PAGE REFERENCE *Paramedic Emergency Care 3/e*	
176.	(c)	page 137	q.i.d.
177.	(b)	page 138	t.i.d.
178.	(e)	pages 166-195	general assessment
179.	(c)	page 170	primary survey
180.	(c)	g/d	paraplegia
181.	(b)	g/d	quadriplegia
182.	(a)	g/d	hemiplegia
183.	(d)	pages 170-172	primary survey
184.	(d)	page 172	pulses and blood pressure
185.	(c)	page 172	pulses and blood pressure
186.	(b)	page 172	pulses and blood pressure
187.	(a)	page 173	level of consciousness
188.	(e)	page 173	AVPU
189.	(b)	page 173	AVPU
190.	(a)	page 173	AVPU
191.	(d)	page 173	AVPU
192.	(d)	page 177	inspection
193.	(c)	page 177	auscultation
194.	(a)	page 178	percussion
195.	(e)	page 177	palpation
196.	(e)	page 197	pain assessment
197.	(c)	page 198	pertinent negatives
198.	(c)	page 198	AMPLE
199.	(b)	page 198	AMPLE
200.	(d)	page 199	AMPLE

TEST SECTION TWO ANSWER KEY

The page numbers following each answer indicate that subject's reference page within Brady's *Paramedic Emergency Care*, 3rd edition. If you do not have access to Brady's text, use the index of the text you do have to obtain information for review of that subject.

QUESTION	ANSWER	PAGE REFERENCE *Paramedic Emergency Care, 3/e*	
1.	(d)	page 142	anatomy
2.	(b)	page 142	physiology
3.	(b)	page 144	mitochondria
4.	(a)	page 144	nucleus
5.	(e)	page 144	tissue
6.	(c)	page 144	cell membranes
7.	(b)	page 144	epithelial tissue
8.	(d)	page 144	connective tissue
9.	(d)	page 144	connective tissue
10.	(c)	page 144	nerve tissue
11.	(a)	page 144	muscle tissue
12.	(d)	page 144	connective tissue
13.	(b)	page 144	epithelial tissue
14.	(c)	page 144	nerve tissue
15.	(e)	page 144	muscle tissues
16.	(b)	page 144	cardiac muscle
17.	(c)	page 144	smooth muscle
18.	(e)	page 144	organ
19.	(d)	page 146	muscular system
20.	(e)	page 146	skeletal system
21.	(a)	page 145	cardiovascular system
22.	(c)	page 146	nervous system
23.	(b)	page 145	respiratory system
24.	(d)	page 146	lymphatic system
25.	(e)	page 146	endocrine system
26.	(e)	page 146	endocrine system
27.	(a)	page 145	gastrointestinal system
28.	(c)	page 146	reproductive system
29.	(b)	page 146	urinary system
30.	(d)	page 146	homeostasis
31.	(d)	page 159	medial
32.	(c)	page 159	lateral
33.	(a)	page 159	ventral
34.	(b)	page 159	dorsal
35.	(b)	page 159	posterior

QUESTION	ANSWER	PAGE REFERENCE *Paramedic Emergency Care, 3/e*	
36.	(a)	page 159	anterior
37.	(e)	page 159	midline
38.	(d)	page 159	superior
39.	(c)	page 159	inferior
40.	(d)	page 159	proximal
41.	(c)	page 159	distal
42.	(a)	page 159	superficial
43.	(e)	page 159	abduction
44.	(b)	page 159	flexion
45.	(a)	page 159	extension
46.	(d)	page 159	adduction
47.	(b)	page 159	prone position
48.	(a)	page 159	Trendelenburg position
49.	(c)	page 159	laterally recumbent position
50.	(d)	page 159	semi-Fowler's position
51.	(e)	page 159	supine position
52.	(c)	page 159	laterally recumbent position
53.	(c)	page 208	hypopharynx
54.	(a)	page 208	nasopharynx
55.	(b)	page 208	oropharynx
56.	(c)	page 208	laryngopharynx
57.	(e)	page 207	septum
58.	(d)	page 208	cilia
59.	(d)	page 208	tongue
60.	(e)	page 208	epiglottis
61.	(d)	page 210	stimulation of larynx during intubation
62.	(c)	page 210	carina
63.	(d)	pages 210, 211	trachea and bronchi
64.	(d)	pages 210, 211	trachea and bronchi
65.	(b)	page 211	right mainstem bronchus
66.	(c)	page 211	left mainstem bronchus
67.	(c)	page 211	secondary bronchi/bronchioles
68.	(a)	page 211	alveoli
69.	(a)	page 211	alveoli
70.	(d)	page 217	tidal volume
71.	(c)	page 217	alveolar air
72.	(a)	page 217	dead space air
73.	(c)	page 211	parietal pleura
74.	(b)	page 211	visceral pleura
75.	(c)	page 215	respiratory center
76.	(d)	page 212	respiration
77.	(c)	page 217	(DOT standard) adult respiratory rate (some systems recognize 12 to 24)

QUESTION	ANSWER	PAGE REFERENCE *Paramedic Emergency Care, 3/e*	
78.	(d)	page 217	child respiratory rate
79.	(e)	page 217	infant respiratory rate
80.	(c)	page 213	room air oxygen percent
81.	(a)	page 214	oxygen and carbon dioxide exchange
82.	(b)	page 214	oxygen diffusion
83.	(d)	page 213	room air nitrogen
84.	(e)	page 214	arterial PO_2
85.	(c)	page 214	arterial PCO_2
86.	(b)	page 219	airway obstruction
87.	(d)	page 219	airway obstruction
88.	(d)	page 220	aspiration of vomitus
89.	(a)	page 220	saliva contents
90.	(e)	page 220	vomitus contents
91.	(a)	page 219	laryngeal spasm
92.	(e)	pages 170-172, 220-221	respiratory assessment
93.	(c)	page 222	respiratory assessment
94.	(d)	page 172, g/d	accessory muscles
95.	(d)	pages 222-223	respiratory assessment
96.	(a)	page 226	airway maneuvers
97.	(d)	page 226	airway maneuvers
98.	(c)	page 228	oropharyngeal airway
99.	(e)	page 231	nasopharyngeal airway
100.	(d)	pages 232-234	EOA
101.	(b)	page 238	ET tube
102.	(e)	page 234	EOA
103.	(e)	page 236	EGTA
104.	(e)	page 236	EGTA
105.	(e)	page 241	airway management
106.	(c)	page 241	medication administration via ET tube
107.	(c)	page 237	straight/Miller laryngoscope blade
108.	(a)	page 237	curved/MacIntosh laryngoscope blade
109.	(e)	page 240	forceps for intubation
110.	(b)	page 238	straight/Miller laryngoscope blade
111.	(c)	page 237	curved/MacIntosh laryngoscope blade
112.	(d)	page 237	curved/MacIntosh laryngoscope blade
113.	(b)	page 238	straight/Miller laryngoscope blade
114.	(b)	page 239	intubation of children
115.	(d)	page 240	stylet use in intubation
116.	(a)	page 242	hyperventilation and intubation
117.	(d)	page 242	time allowed for intubation
118.	(e)	pages 242, 246	intubation
119.	(c)	page 247	endobronchial intubation
120.	(e)	page 247	endobronchial intubation
121.	(e)	page 245	endotracheal intubation
122.	(b)	page 261	nasotracheal intubation
123.	(a)	page 261	nasotracheal intubation

QUESTION	ANSWER	PAGE REFERENCE *Paramedic Emergency Care, 3/e*	
124.	(c)	page 263	nasotracheal intubation
125.	(b)	page 270	cricoid cartilage
126.	(a)	page 270	thyroid cartilage
127.	(d)	page 270	cricothyroid membrane
128.	(d)	pages 272-274	transtracheal jet insufflation
129.	(e)	pages 272-274	transtracheal jet insufflation
130.	(a)	pages 269-271	cricothyrotomy
131.	(d)	pages 275-276	suction
132.	(a)	page 277	nasal cannula
133.	(b)	page 277	nasal cannula
134.	(c)	page 278	simple face mask
135.	(d)	page 278	simple face mask
136.	(e)	page 278	nonrebreather mask
137.	(d)	page 278	nonrebreather mask
138.	(c)	page 279	venturi mask
139.	(c)	page 279	venturi mask
140.	(a)	page 279	mouth-to-mouth/stoma ventilation
141.	(e)	page 280	mouth-to-mask ventilation
142.	(c)	page 280	bag-valve-mask device
143.	(d)	page 280	demand valve device
144.	(e)	page 281	demand valve device
145.	(e)	pages 205-283	gastric distention
146.	(a)	page 289	electrolytes
147.	(b)	page 290	cations
148.	(c)	page 290	anions
149.	(a)	page 291	chief extracellular ion
150.	(b)	page 292	chief intracellular ion
151.	(d)	page 291	sodium
152.	(c)	page 292	chloride
153.	(e)	page 292	magnesium
154.	(e)	page 292	bicarbonate
155.	(b)	page 292	calcium
156.	(c)	page 292	potassium
157.	(a)	page 292	phosphate
158.	(b)	page 292	bicarbonate
159.	(d)	page 291	positively charged ions
160.	(b)	page 292	negatively charged ions
161.	(c)	page 292	isotonic fluids
162.	(d)	page 292	hypotonic fluids
163.	(b)	page 292	hypertonic fluids
164.	(e)	page 293	solute/solvent movement
165.	(d)	page 293	solute/solvent movement
166.	(d)	page 293	diffusion
167.	(a)	page 293	osmosis
168.	(a)	page 293	osmosis
169.	(d)	page 293	diffusion
170.	(a)	page 293	osmosis

QUESTION	ANSWER	PAGE REFERENCE *Paramedic Emergency Care, 3/e*	
171.	(b)	page 293	active transport
172.	(d)	page 294	facilitated diffusion
173.	(a)	page 294	hemoglobin
174.	(b)	page 294	erythrocytes
175.	(d)	page 294	leukocytes
176.	(e)	page 294	plasma
177.	(b)	page 294	erythrocytes
178.	(c)	page 294	thrombocytes
179.	(d)	page 294	leukocytes
180.	(c)	page 294	thrombocytes
181.	(a)	page 295	ABO blood classification system
182.	(c)	page 296	blood type B
183.	(a)	page 296	blood type O
184.	(d)	page 296	blood type A
185.	(b)	page 296	blood type AB
186.	(e)	page 296	universal blood donor
187.	(d)	page 295	universal blood recipient
188.	(e)	page 296	universal blood donor
189.	(e)	page 296	Rh negative blood
190.	(b)	page 297	colloid IV solutions
191.	(d)	page 298	crystalloid IV solutions
192.	(d)	page 298	crystalloid IV solutions
193.	(b)	page 297	colloid IV solutions
194.	(d)	page 298	crystalloid IV solutions
195.	(b)	page 297	colloid IV solutions
196.	(d)	page 299	crystalloid IV solutions
197.	(e)	page 298	hypertonic solutions
198.	(a)	page 298	isotonic solutions
199.	(c)	page 299	hypotonic solutions
200.	(e)	page 298	hypertonic solutions
201.	(a)	page 298	isotonic solutions
202.	(c)	page 299	hypotonic solutions
203.	(c)	page 299	acid-base balance
204.	(d)	pages 299-301	pH/acid-base balance
205.	(a)	pages 299-301	pH/acid-base balance
206.	(b)	pages 299-301	pH/acid-base balance
207.	(c)	pages 299-301	pH/acid-base balance
208.	(a)	page 300	acidosis
209.	(b)	page 300	alkalosis
210.	(a)	pages 299-301	pH/acid-base balance
211.	(e)	pages 299-301	pH/acid-base balance regulation
212.	(a)	pages 299-301	carbonate buffer system
213.	(c)	pages 299-301	renal buffer system
214.	(b)	pages 299-301	carbonate buffer system
215.	(e)	page 302	carbonic acid

QUESTION	ANSWER	PAGE REFERENCE *Paramedic Emergency Care, 3/e*	
216.	(a)	page 302	respirations and pH/acid-base balance
217.	(b)	page 302	respirations and pH/acid-base balance
218.	(b)	page 302	respiratory acidosis
219.	(a)	page 302	respiratory alkalosis
220.	(e)	page 301	renal regulation of pH/acid-base balance
221.	(b)	page 301	renal regulation of pH/acid-base balance
222.	(c)	page 301	renal regulation of pH/acid-base balance
223.	(e)	page 303	metabolic acidosis
224.	(a)	page 303	metabolic alkalosis
225.	(e)	page 303	shock
226.	(d)	page 303	Fick principle
227.	(e)	page 303	perfusion
228.	(c)	page 305	baroreceptors
229.	(c)	page 304	sympathetic nervous system and shock
230.	(e)	page 308	cellular oxygenation and shock
231.	(d)	page 309	inadequate cellular perfusion
232.	(a)	page 310	beta-blocker medications and shock
233.	(a)	page 310	decompensated shock
234.	(b)	page 309	peripheral effects of shock
235.	(e)	page 310	hypovolemic shock
236.	(e)	page 311	neurogenic shock
237.	(c)	page 312	blood pressure and pulse locations
238.	(e)	page 312	blood pressure and pulse locations
239.	(b)	page 312	blood pressure and pulse locations
240.	(b)	page 314	signs and symptoms of shock
241.	(a)	page 318	PASG
242.	(a)	page 319	PASG
243.	(b)	page 320	IV treatment
244.	(c)	page 320	IV treatment
245.	(e)	page 322	macrodrip administration sets
246.	(b)	page 322	microdrip administration sets
247.	(b)	page 322	IV catheter sizes
248.	(a)	page 325	extravasation
249.	(d)	page 327	pyrogenic reaction
250.	(a)	page 327	thrombophlebitis
251.	(e)	page 327	pyrogenic reaction signs and symptoms
252.	(a)	page 327	thrombophlebitis signs and symptoms
253.	(b)	page 329	treatment of shock trauma

TEST SECTION THREE ANSWER KEY

The page numbers following each answer indicate that subject's reference page within Brady's *Paramedic Emergency Care*, 3rd edition. If you do not have access to Brady's text, use the index of the text you do have to obtain information for review of that subject.

Following some answers is "g/d." This indicates that you should consult the glossary of your paramedic text or a medical dictionary.

Following some answers is "ACLS." This indicates that you should consult the American Heart Association's text book of Advanced Cardiac Life Support guidelines.

QUESTION	ANSWER	PAGE REFERENCE *Paramedic Emergency Care, 3/e*	
1.	(c)	page 332	drug sources
2.	(a)	page 332	drug sources
3.	(b)	page 332	drug sources
4.	(d)	page 332	drug sources
5.	(c)	page 333	controlled substance act
6.	(e)	page 333	narcotic classification
7.	(b)	page 333	narcotic classification
8.	(a)	page 333	narcotic classification
9.	(d)	page 334	official drug names
10.	(e)	page 334	chemical drug names
11.	(b)	page 334	generic drug names
12.	(c)	page 334	trade or proprietary drug names
13.	(a)	page 334	slang drug names
14.	(c)	page 334	drug solutions
15.	(b)	page 335	tinctures
16.	(d)	page 335	suspensions
17.	(a)	page 335	solid drug forms
18.	(a)	page 336	synergism
19.	(c)	page 336	cumulative action
20.	(d)	page 336	antagonism
21.	(b)	page 336	potentiation
22.	(e)	page 336	therapeutic action
23.	(b)	page 336	untoward effects
24.	(a)	page 336	side effects
25.	(e)	page 336	hypersensitivity
26.	(d)	page 336	idiosyncrasy
27.	(c)	page 336	contraindications
28.	(a)	page 336	refractory
29.	(e)	g/d	titration
30.	(b)	page 381	diuretic
31.	(c)	g/d	vasopressor

QUESTION	ANSWER	PAGE REFERENCE *Paramedic Emergency Care, 3/e*	
32.	(c)	page 360	enteral drugs
33.	(d)	page 352	parenteral drugs
34.	(d)	page 352	parenteral drugs
35.	(b)	page 337	drug absorption factors
36.	(e)	page 338	drug elimination
37.	(e)	page 339	pharmacodynamics
38.	(c)	page 340	parasympathetic nervous system
39.	(d)	page 340	sympathetic nervous system
40.	(c)	page 340	parasympathetic nervous system
41.	(d)	page 340	sympathetic nervous system
42.	(a)	page 347	acetylcholine stimulation
43.	(d)	page 345	beta 2 receptor stimulation
44.	(e)	page 345	dopaminergic stimulation
45.	(c)	page 345	beta 1 receptor stimulation
46.	(b)	page 345	alpha 1 receptor stimulation
47.	(c)	page 345	beta 1 receptor stimulation
48.	(d)	page 345	adrenergic drugs
49.	(b)	g/d	antiadrenergic drugs
50.	(b)	page 345	sympatholytic drugs
51.	(d)	page 345	sympathomimetic drugs
52.	(d)	page 347	parasympathomimetic/cholinergic drugs
53.	(e)	page 347	parasympathetic blockers
54.	(a)	page 363	inotropic
55.	(e)	page 363	chronotropic
56.	(a)	pages 348-350	pound and kilogram conversions
57.	(b)	pages 348-350	pound and kilogram conversions
58.	(e)	pages 348-350	pound and kilogram conversions
59.	(c)	pages 348-350	pound and kilogram conversions
60.	(b)	pages 350-352	drug dose calculations
61.	(c)	pages 350-352	drug dose calculations
62.	(b)	pages 350-352	drug dose calculations
63.	(a)	pages 350-352	drug dose calculations
64.	(d)	pages 350-352	drug dose calculations
65.	(c)	pages 350-352	drug dose calculations
66.	(a)	pages 350-352	drug dose calculations
67.	(c)	page 362	epinephrine
68.	(c)	page 362	epinephrine 1:10,000
69.	(e)	page 362	epinephrine 1:10,000
70.	(d)	page 362	epinephrine 1:10,000
71.	(e)	page 362	epinephrine 1:10,000
72.	(e)	page 362	epinephrine 1:10,000
73.	(a)	page 364	norepinephrine
74.	(c)	ACLS	isoproterenol
75.	(a)	ACLS	isoproterenol
76.	(d)	ACLS	isoproterenol

QUESTION	ANSWER	PAGE REFERENCE *Paramedic Emergency Care, 3/e*	
77.	(e)	ACLS	isoproterenol
78.	(c)	page 365	dopamine
79.	(d)	page 365	dopamine
80.	(b)	page 365	dopamine
81.	(e)	page 365	dopamine
82.	(b)	page 365	dopamine
83.	(a)	page 365	dobutamine
84.	(a)	page 369	lidocaine
85.	(a)	page 369	lidocaine
86.	(a)	page 369	lidocaine
87.	(e)	page 369	lidocaine
88.	(b)	page 369	lidocaine
89.	(e)	page 369	lidocaine
90.	(b)	page 369	lidocaine
91.	(e)	page 369	lidocaine
92.	(a)	page 373	bretylium tosylate
93.	(a)	page 373	bretylium tosylate
94.	(c)	page 373	bretylium tosylate
95.	(a)	page 373	bretylium tosylate
96.	(c)	page 372	procainamide
97.	(b)	page 372	procainamide
98.	(c)	page 375	verapamil
99.	(e)	page 375	verapamil
100.	(d)	page 375	verapamil
101.	(b)	page 375	verapamil
102.	(b)	page 374	adenosine
103.	(d)	page 374	adenosine
104.	(e)	page 374	adenosine
105.	(b)	page 374	adenosine
106.	(e)	page 374	adenosine
107.	(a)	page 374	adenosine
108.	(a)	page 374	adenosine
109.	(a)	page 374	adenosine
110.	(c)	page 376	atropine
111.	(c)	page 376	atropine
112.	(d)	page 376	atropine
113.	(a)	page 376	atropine
114.	(e)	page 376	atropine
115.	(d)	page 377	sodium bicarbonate
116.	(d)	page 377	sodium bicarbonate
117.	(b)	page 377	sodium bicarbonate
118.	(a)	page 377	sodium bicarbonate
119.	(a)	page 378	morphine sulfate
120.	(d)	page 378	morphine sulfate
121.	(d)	page 378	morphine sulfate
122.	(e)	page 378	morphine sulfate
123.	(c)	page 378	morphine sulfate

QUESTION	ANSWER	PAGE REFERENCE *Paramedic Emergency Care, 3/e*	
124.	(b)	page 378	morphine sulfate
125.	(d)	page 379	Nitronox
126.	(b)	page 379	Nitronox
127.	(e)	page 379	Nitronox
128.	(c)	page 380	furosemide
129.	(a)	page 380	furosemide
130.	(e)	page 380	furosemide
131.	(c)	page 380	furosemide
132.	(a)	page 380	furosemide
133.	(b)	page 381	nitroglycerin
134.	(d)	page 381	nitroglycerin
135.	(b)	page 381	nitroglycerin
136.	(a)	page 383	calcium chloride
137.	(c)	page 384	epinephrine 1:1000
138.	(e)	page 384	epinephrine 1:1000
139.	(b)	page 384	epinephrine 1:1000
140.	(c)	page 385	aminophylline
141.	(d)	page 385	aminophylline
142.	(d)	page 385	aminophylline
143.	(a)	page 386	racemic epinephrine
144.	(a)	page 387	albuterol
145.	(e)	page 389	$D_{50}W$
146.	(e)	page 389	$D_{50}W$
147.	(e)	page 389	$D_{50}W$
148.	(a)	page 389	$D_{50}W$
149.	(d)	page 389	thiamine
150.	(e)	page 389	thiamine
151.	(c)	page 392	steroid preparations
152.	(e)	page 392	methylprednisolone
153.	(c)	page 391	diazepam
154.	(b)	page 391	diazepam
155.	(d)	page 391	diazepam
156.	(e)	page 391	diazepam
157.	(c)	page 393	oxytocin
158.	(a)	page 393	magnesium sulfate
159.	(3)	ACLS	magnesium sulfate
160.	(b)	page 394	diphenhydramine
161.	(b)	page 394	diphenhydramine
162.	(c)	page 394	diphenhydramine
163.	(e)	page 395	ipecac
164.	(d)	page 396	activated charcoal
165.	(c)	page 396	naloxone
166.	(e)	page 396	naloxone
167.	(d)	page 396	naloxone
168.	(a)	ACLS	high-dose epinephrine
169.	(b)	ACLS	high-dose epinephrine
170.	(c)	ACLS	high-dose epinephrine

TEST SECTION FOUR ANSWER KEY

The page numbers following each answer indicate that subject's reference page within Brady's *Paramedic Emergency Care*, 3rd edition. If you do not have access to Brady's text, use the index of the text you do have to obtain information for review of that subject.

Following some answers is "g/d." This indicates that you should consult the glossary of your paramedic text or a medical dictionary.

QUESTION	ANSWER	PAGE REFERENCE *Paramedic Emergency Care, 3/e*	
1.	(a)	page 404	priorities in trauma care
2.	(e)	page 406	kinetics of trauma
3.	(b)	page 407	kinetics of trauma
4.	(a)	pages 406-411	kinetics of trauma
5.	(d)	page 406	kinetics of trauma
6.	(c)	page 406	kinetics of trauma
7.	(b)	page 407	deceleration trauma
8.	(a)	page 408	deceleration trauma
9.	(c)	page 412	kinetics of down and under travel
10.	(e)	page 412	kinetics of up and over travel
11.	(c)	page 414	kinetics of ejection
12.	(b)	page 414	kinetics of lateral impact
13.	(d)	page 416	kinetics of rear impact
14.	(e)	page 416	kinetics of rotational impact
15.	(e)	page 410	restraint systems
16.	(c)	page 418	alcohol, drugs, and trauma
17.	(e)	page 420	motorcycle trauma
18.	(a)	page 420	motorcycle trauma
19.	(b)	g/d	most commonly fractured bone
20.	(e)	page 422	deceleration trauma and falls
21.	(d)	page 426	projectile trauma
22.	(d)	page 435	facial anatomy
23.	(b)	page 435	cranial anatomy
24.	(c)	page 436	spinal column
25.	(d)	page 436	spinal column
26.	(b)	page 436	spinal column
27.	(b)	page 437	spinal column
28.	(a)	page 437	spinal column
29.	(a)	page 437	spinal column
30.	(c)	page 438	meninges
31.	(a)	page 438	meninges
32.	(d)	page 438	meninges
33.	(c)	page 435	neurons and regeneration
34.	(d)	page 440	cribriform plate

QUESTION	ANSWER	PAGE REFERENCE *Paramedic Emergency Care, 3/e*	
35.	(a)	page 440	dermatomes
36.	(b)	page 442	dermatomes
37.	(c)	page 442	dermatomes
38.	(e)	page 442	dermatomes
39.	(d)	page 442	circle of Willis
40.	(e)	page 442	orbits
41.	(b)	page 442	the eye
42.	(c)	page 442	the eye
43.	(e)	page 442	the eye
44.	(a)	page 442	the eye
45.	(c)	page 442	the eye
46.	(b)	page 442	the eye
47.	(a)	page 442	the eye
48.	(d)	page 442	the eye
49.	(d)	page 442	the eye
50.	(b)	page 442	the eye
51.	(e)	page 443	the ear
52.	(d)	page 443	the ear
53.	(b)	page 443	the cochlea
54.	(c)	page 443	the semicircular canals
55.	(a)	page 443	the tympanic membrane
56.	(e)	page 443	vertigo
57.	(c)	page 444	facial/cranial injuries
58.	(e)	page 445	skull fracture
59.	(e)	page 445	skull fracture
60.	(c)	page 445	skull fracture
61.	(a)	page 447	axial loading forces
62.	(c)	page 447	axial loading forces
63.	(b)	page 447	zygomatic fracture
64.	(e)	page 448	contrecoup injury
65.	(d)	page 448	contusion/concussion
66.	(b)	page 448	contusion/concussion
67.	(a)	page 448	contusion/concussion
68.	(a)	page 449	epidural hematoma
69.	(c)	page 449	intracerebral hemorrhage
70.	(a)	page 450	epidural hematoma
71.	(b)	page 449	subdural hematoma
72.	(b)	page 449	subdural hematoma
73.	(c)	page 450	increased ICP
74.	(e)	page 450	increased ICP
75.	(a)	page 450	increased ICP
76.	(d)	page 450	increased ICP
77.	(d)	page 450	spinal cord injury
78.	(a)	g/d	paraplegia
79.	(d)	g/d	paresis

QUESTION	ANSWER	PAGE REFERENCE *Paramedic Emergency Care, 3/e*	
80.	(e)	page 451	paresthesia
81.	(c)	g/d	hemiplegia
82.	(b)	g/d	quadriplegia
83.	(e)	page 450	spinal injury
84.	(c)	page 452	neurogenic shock
85.	(e)	page 452	cervical spine injury
86.	(c)	page 453	retinal detachment
87.	(e)	page 453	corneal abrasion
88.	(b)	page 453	acute retinal artery occlusion
89.	(a)	page 452	hyphema
90.	(d)	page 452	conjunctival hemorrhage
91.	(a)	page 452	hyphema
92.	(c)	page 452	vitreous humor loss
93.	(d)	page 462	cervical spine immobilization
94.	(d)	page 461	endotracheal stimulation
95.	(c)	page 454	Biot's breathing
96.	(e)	page 454	central neurogenic hyperventilation
97.	(d)	page 454	Cheyne-Stokes respirations
98.	(b)	page 454	ataxic respirations
99.	(a)	page 454	apneustic respirations
100.	(c)	page 460	treatment of increased ICP
101.	(a)	page 382	treatment of increased ICP/ spinal trauma
102.	(b)	page 460	treatment of increased ICP
103.	(c)	page 456	signs and symptoms of increased ICP
104.	(d)	page 458	Glasgow Coma Scale
105.	(c)	page 458	pain responses
106.	(a)	page 458	pain responses
107.	(d)	page 753	decorticate posturing
108.	(b)	page 753	decerebrate posturing
109.	(d)	page 753	decerebrate posturing
110.	(e)	page 467	treatment of neck trauma
111.	(c)	page 322	PASG and head trauma
112.	(d)	page 462	IV treatment of head trauma
113.	(e)	page 457	anisocoria
114.	(a)	page 180	dysconjugate gaze
115.	(d)	pages 450, 457	pupillary changes
116.	(e)	page 457	pupillary changes
117.	(b)	page 464	spinal immobilization
118.	(b)	page 759	mannitol
119.	(e)	page 466	furosemide
120.	(e)	page 466	methylprednisolone
121.	(d)	page 760	seizures and head trauma
122.	(a)	page 472	muscles of respiration
123.	(b)	page 473	pleural space
124.	(d)	page 476	atelectasis
125.	(e)	page 476	lung contusion

QUESTION	ANSWER	PAGE REFERENCE *Paramedic Emergency Care, 3/e*	
126.	(e)	page 476	rib fracture
127.	(b)	page 476	flail chest segment
128.	(c)	page 476	paradoxical movement
129.	(b)	page 476	traumatic asphyxia
130.	(a)	page 478	closed pneumothorax
131.	(c)	page 478	paper bag syndrome
132.	(a)	page 477	traumatic asphyxia
133.	(c)	page 478	tension pneumothorax
134.	(e)	page 478	signs and symptoms, simple/hemothorax
135.	(a)	page 478	signs and symptoms, tension pneumothorax
136.	(b)	page 478	signs and symptoms, hemothorax
137.	(d)	page 478	signs and symptoms, tension pneumothorax
138.	(b)	page 480	signs and symptoms, myocardial contusion
139.	(d)	page 603	visceral pericardium
140.	(b)	page 603	parietal pericardium
141.	(c)	page 473	pericardial sac
142.	(e)	page 480	pericardial tamponade
143.	(c)	page 480	signs and symptoms, pericardial tamponade
144.	(a)	page 480	signs and symptoms of thoracic aneurysm
145.	(b)	page 490	transport position for chest trauma
146.	(e)	page 490	treatment of chest trauma
147.	(a)	page 490	treatment of chest trauma
148.	(d)	page 493	needle decompression of the chest
149.	(b)	page 493	needle decompression of the chest
150.	(a)	page 494	treatment of myocardial contusion
151.	(e)	page 494	treatment of pericardial tamponade
152.	(e)	page 494	treatment of an impaled chest
153.	(d)	page 474	retroperitoneal organs
154.	(b)	page 474	LUQ abdomen
155.	(a)	page 474	RUQ abdomen
156.	(c)	page 474	RLQ abdomen
157.	(e)	page 474	LLQ abdomen
158.	(c)	page 152	common iliac arteries
159.	(d)	page 481	GSW wound to abdomen
160.	(e)	page 482	peritonitis
161.	(b)	pages 186, 772	rebound tenderness
162.	(d)	page 186	abdominal guarding
163.	(a)	page 495	treatment of closed abdomen
164.	(d)	page 496	treatment of evisceration
165.	(e)	page 496	treatment of impaled abdomen
166.	(c)	page 501	number of bones in human skeleton
167.	(b)	page 501	metaphysis
168.	(a)	page 501	epiphysis
169.	(c)	page 501	diaphysis
170.	(c)	page 501	axial skeleton
171.	(d)	page 501	appendicular skeleton
172.	(c)	page 503	skeletal A and P

QUESTION	ANSWER	PAGE REFERENCE *Paramedic Emergency Care, 3/e*	
173.	(a)	page 501	skeletal A and P
174.	(e)	page 503	skeletal A and P
175.	(c)	page 503	skeletal A and P
176.	(b)	page 504	ligaments
177.	(d)	page 505	tendons
178.	(c)	page 505	medial malleolus
179.	(d)	page 505	lateral malleolus
180.	(c)	page 504	acetabulum
181.	(c)	page 506	muscle tissue
182.	(e)	page 504	scapula
183.	(d)	page 504	radius
184.	(c)	page 504	humerus
185.	(b)	page 504	clavicle
186.	(a)	page 504	ulna
187.	(c)	page 504	carpals
188.	(d)	page 504	metacarpals
189.	(e)	page 504	phalanges
190.	(a)	page 505	tarsals
191.	(b)	page 505	metatarsals
192.	(e)	page 505	phalanges
193.	(b)	page 149	ilium
194.	(c)	page 149	iliac crest
195.	(a)	page 149	ischium
196.	(d)	page 149	sacrum
197.	(e)	page 149	pubic bones
198.	(a)	page 502	patella
199.	(b)	page 502	tibia
200.	(d)	page 502	fibula
201.	(b)	page 507	strain injury
202.	(d)	page 507	sprain injury
203.	(c)	page 509	green stick fracture
204.	(a)	page 508	transverse fracture
205.	(e)	page 508	oblique fracture
206.	(d)	page 509	impacted fracture
207.	(b)	page 508	comminuted fracture
208.	(e)	page 510	signs and symptoms of fractures
209.	(e)	page 508	fracture complications
210.	(e)	page 511	signs and symptoms of hip injury
211.	(a)	page 511	anterior hip dislocation
212.	(b)	page 511	posterior hip dislocation
213.	(c)	page 512	inferior shoulder dislocation
214.	(b)	page 512	posterior shoulder dislocation
215.	(a)	page 512	anterior shoulder dislocation
216.	(a)	page 152	popliteal pulse
217.	(e)	page 152	posterior tibial pulse
218.	(b)	page 152	dorsalis pedis pulse
219.	(b)	page 152	dorsalis pedis pulse

QUESTION	ANSWER	PAGE REFERENCE *Paramedic Emergency Care, 3/e*	
220.	(d)	page 516	splinting
221.	(c)	page 516	splinting
222.	(b)	page 517	treatment of pelvic fracture
223.	(a)	page 518	traction splint
224.	(a)	page 518	splinting
225.	(b)	page 525	A & P of the skin
226.	(c)	page 523	epidermis
227.	(d)	page 524	dermis
228.	(a)	page 524	subcutaneous tissue
229.	(a)	page 524	subcutaneous tissue
230.	(d)	page 523	sebum
231.	(b)	page 526	ecchymosis
232.	(a)	page 526	erythema
233.	(e)	page 526	contusion
234.	(a)	page 526	hematoma
235.	(c)	page 526	abrasion
236.	(d)	page 526	incision
237.	(a)	page 526	hematoma
238.	(b)	page 526	laceration
239.	(b)	page 536	eschar
240.	(e)	page 527	avulsion
241.	(d)	page 528	amputation
242.	(a)	page 528	crushing amputation
243.	(c)	page 528	clean amputation
244.	(d)	page 534	partial-thickness burn
245.	(b)	page 533	second-degree burn
246.	(c)	page 534	third-degree burn
247.	(d)	page 533	first- and second-degree burns
248.	(c)	page 534	full-thickness burn
249.	(c)	page 533	third-degree burn
250.	(c)	page 535	rule of nines
251.	(e)	page 535	rule of nines
252.	(c)	page 535	rule of nines
253.	(b)	page 535	rule of nines
254.	(a)	page 535	rule of nines
255.	(c)	page 535	rule of nines
256.	(c)	page 535	rule of nines
257.	(a)	page 535	palmar surface area percentage
258.	(e)	page 535	pediatric rule of nines
259.	(d)	page 535	pediatric rule of nines
260.	(e)	page 535	pediatric rule of nines
261.	(e)	page 535	pediatric rule of nines
262.	(d)	page 535	pediatric rule of nines
263.	(c)	page 535	pediatric rule of nines
264.	(a)	page 535	pediatric rule of nines
265.	(a)	page 536	systemic complications of burns
266.	(e)	page 536	circumferential burns

QUESTION	ANSWER	PAGE REFERENCE *Paramedic Emergency Care, 3/e*	
267.	(e)	page 532	inhalation injury
268.	(a)	page 532	inhalation injury
269.	(e)	page 529	electrical burns
270.	(d)	page 542	burn severity categorization
271.	(d)	page 542	burn severity categorization
272.	(b)	page 542	burn severity categorization
273.	(c)	page 548	burns and fluid resuscitation

EMT-PARAMEDIC NATIONAL STANDARDS REVIEW SELF TEST, 3/E

TEST SECTION FIVE ANSWER KEY

The page numbers following each answer indicate that subject's reference page within Brady's *Paramedic Emergency Care*, 3rd edition. If you do not have access to Brady's text, use the index of the text you do have to obtain information for review of that subject.

Following some answers is "g/d." This indicates that you should consult the glossary of your paramedic text or a medical dictionary.

QUESTION	ANSWER	PAGE REFERENCE *Paramedic Emergency Care, 3/e*	
1.	(e)	page 213	right lung anatomy
2.	(d)	page 213	left lung anatomy
3.	(c)	page 216	carbon dioxide
4.	(d)	page 217	hypoxia
5.	(b)	page 217	CO_2 production
6.	(b)	page 217	CO_2 elimination
7.	(d)	page 218	CO_2 elimination
8.	(d)	page 218	respiratory regulation
9.	(b)	page 218	respiratory regulation
10.	(e)	page 218	respiratory chemoreceptors
11.	(e)	page 218	respiratory chemoreceptors
12.	(c)	page 218	respiratory chemoreceptors
13.	(d)	page 218	hypoxic drive
14.	(d)	page 219	sighing
15.	(c)	page 219	hiccoughing
16.	(a)	page 219	coughing
17.	(a)	page 219	fever/respiratory effect
18.	(a)	page 219	anxiety/respiratory effect
19.	(b)	page 219	sleep/respiratory effect
20.	(a)	page 219	hypoxia/respiratory effect
21.	(b)	page 219	drugs/respiratory effect
22.	(a)	page 575	orthopnea
23.	(c)	page 577	chest palpation
24.	(a)	page 579	friction rub
25.	(d)	page 578	rhonchi
26.	(c)	page 578	wheezing
27.	(b)	page 578	stridor
28.	(e)	page 579	rales
29.	(c)	page 576	hypoxia
30.	(c)	page 582	treatment of obstructed airway
31.	(b)	page 582	treatment of obstructed airway
32.	(a)	page 583	treatment of obstructed airway
33.	(e)	page 583	treatment of obstructed airway

QUESTION	ANSWER	PAGE REFERENCE *Paramedic Emergency Care, 3/e*	
34.	(c)	page 582	treatment of obstructed airway
35.	(c)	page 582	treatment of obstructed airway
36.	(a)	page 582	treatment of obstructed airway
37.	(b)	page 585	chronic bronchitis
38.	(d)	page 585	COPD
39.	(a)	page 585	emphysema
40.	(e)	page 585	COPD
41.	(e)	page 585	COPD
42.	(a)	page 584	emphysema
43.	(b)	page 586	chronic bronchitis
44.	(c)	page 586	asthma
45.	(e)	page 585	COPD
46.	(e)	page 584	COPD
47.	(c)	page 586	asthma
48.	(e)	page 594	COPD
49.	(e)	page 595	pneumonia
50.	(d)	page 595	pneumonia
51.	(d)	page 595	treatment of pneumonia
52.	(e)	page 596	carbon monoxide poisoning
53.	(a)	page 596	carbon monoxide poisoning
54.	(b)	page 596	carbon monoxide poisoning
55.	(d)	page 596	carbon monoxide poisoning
56.	(b)	page 596	carbon monoxide poisoning
57.	(e)	page 597	pulmonary embolism
58.	(c)	page 597	pulmonary embolism
59.	(e)	page 597	pulmonary embolism
60.	(a)	page 597	hyperventilation syndrome
61.	(d)	page 597	hyperventilation syndrome
62.	(b)	page 726	exocrine glands
63.	(d)	page 726	endocrine glands
64.	(d)	page 729	pancreas
65.	(c)	page 726	pituitary gland
66.	(b)	page 729	adrenal glands
67.	(e)	page 728	parathyroid glands
68.	(d)	page 728	thyroid gland
69.	(a)	page 730	ovaries
70.	(c)	page 730	testes
71.	(b)	page 730	adrenal glands
72.	(c)	page 727	pituitary gland/ADH
73.	(a)	page 730	ovaries and testes/FSH
74.	(c)	page 727	pituitary gland/oxytocin
75.	(a)	page 730	ovaries and testes/LH
76.	(d)	page 729	adrenal glands
77.	(a)	page 726	pituitary gland
78.	(c)	page 728	thyroid gland
79.	(a)	page 729	alpha cells/glucagon
80.	(d)	page 729	beta cells/insulin

QUESTION	ANSWER	PAGE REFERENCE *Paramedic Emergency Care, 3/e*	
81.	(b)	page 773, g/d	glycogen
82.	(c)	page 729, g/d	glucose
83.	(a)	page 729	glucagon
84.	(d)	page 731	insulin
85.	(a)	page 732	hyperglycemia/diabetic ketoacidosis
86.	(d)	page 733	hypoglycemia
87.	(a)	page 732	hyperglycemia/diabetic ketoacidosis
88.	(d)	page 733	hypoglycemia
89.	(a)	page 732	hyperglycemia/diabetic ketoacidosis
90.	(a)	page 732	hyperglycemia/diabetic ketoacidosis
91.	(e)	page 735	hypoglycemia/hyperglycemia/ diabetic ketoacidosis
92.	(d)	page 735	hypoglycemia
93.	(e)	page 735	hypoglycemia/hyperglycemia/ diabetic ketoacidosis
94.	(a)	page 735	hyperglycemia/diabetic ketoacidosis
95.	(d)	page 735	hypoglycemia
96.	(e)	page 735	hypoglycemia/hyperglycemia/ diabetic ketoacidosis
97.	(d)	page 735	hypoglycemia
98.	(a)	page 735	hyperglycemia/diabetic ketoacidosis
99.	(a)	page 733	hyperglycemia/diabetic ketoacidosis
100.	(a)	page 732	hyperglycemia/diabetic ketoacidosis
101.	(d)	page 735	hypoglycemia
102.	(a)	page 733	hyperglycemia/diabetic ketoacidosis
103.	(e)	page 736	hypoglycemia/hyperglycemia/ diabetic ketoacidosis
104.	(c)	page 732	hyperglycemia
105.	(b)	page 732	polyphagia
106.	(e)	page 732	polyuria
107.	(b)	page 732	polydipsia
108.	(e)	page 741	neuron
109.	(b)	page 741	dendrite
110.	(c)	page 741	axon
111.	(d)	page 741	nerve impulse transportation
112.	(c)	page 744	brain stem
113.	(c)	page 744	cerebrum
114.	(a)	page 745	cerebellum
115.	(b)	page 745	medulla oblongata
116.	(a)	page 744	speech center
117.	(c)	page 744	personality center
118.	(a)	page 746	spinal cord
119.	(b)	page 746	afferent impulses
120.	(c)	page 747	efferent impulses
121.	(b)	page 747	cranial nerves
122.	(e)	page 747	peripheral nerves
123.	(b)	page 747	somatic sensory nerves

QUESTION	ANSWER	PAGE REFERENCE *Paramedic Emergency Care, 3/e*	
124.	(c)	page 748	brachial plexus
125.	(c)	page 755	altered level of consciousness/coma
126.	(b)	page 755	altered level of consciousness/coma
127.	(e)	page 755	altered level of consciousness/coma
128.	(a)	page 755	altered level of consciousness/coma
129.	(b)	page 755	altered level of consciousness/coma
130.	(d)	page 755	altered level of consciousness/coma
131.	(e)	page 757	$D_{50}W$
132.	(b)	page 758	naloxone
133.	(d)	page 758	thiamine
134.	(e)	page 760	seizures
135.	(c)	page 760	grand mal seizures
136.	(e)	page 761	hysterical seizures
137.	(d)	page 761	petit mal seizures
138.	(b)	page 761	psychomotor seizures
139.	(d)	page 761	petit mal seizures
140.	(c)	page 760	grand mal seizures
141.	(b)	page 760	focal motor seizures
142.	(a)	g/d	Jacksonian seizures
143.	(b)	page 761	psychomotor seizures
144.	(d)	g/d	gustatory aura
145.	(e)	g/d	tactile aura
146.	(c)	page 760	olfactory aura
147.	(b)	page 760	clonic phase of grand mal seizure
148.	(d)	page 760	hypertonic phase of grand mal seizure
149.	(e)	page 760	autonomic discharge phase of seizure
150.	(c)	page 760	tonic phase of grand mal seizure
151.	(e)	page 760	grand mal seizure
152.	(c)	page 762	vasovagal syncope
153.	(e)	page 762	seizure management
154.	(c)	page 763	status epilepticus
155.	(d)	page 763	status epilepticus
156.	(b)	page 763	status epilepticus, cause in adults
157.	(d)	page 763	status epilepticus management
158.	(a)	page 763	diazepam
159.	(e)	page 763	diazepam
160.	(e)	page 764	cerebrovascular accident (CVA)
161.	(b)	page 764	cerebrovascular accident (CVA)
162.	(c)	page 765	cerebrovascular accident (CVA)
163.	(c)	page 765	transient ischemic attack (TIA)
164.	(a)	page 765	signs and symptoms of CVA/TIA
165.	(e)	page 766	treatment of CVA/TIA
166.	(b)	page 767	treatment of CVA/TIA
167.	(e)	page 772	digestive enzymes
168.	(c)	page 773	bile/fat digestion
169.	(c)	page 773	liver/glycogen storage

QUESTION	ANSWER	PAGE REFERENCE *Paramedic Emergency Care, 3/e*	
170.	(d)	page 777	either hemorrhagic or nonhemorrhagic causes of abdominal pain
171.	(b)	page 776	nonhemorrhagic causes of abdominal pain
172.	(a)	page 780	hemorrhagic causes of abdominal pain
173.	(c)	page 777	appendicitis
174.	(e)	page 780	renal calculus
175.	(b)	page 780	pyelonephritis
176.	(d)	page 779	aortic aneurysm
177.	(e)	page 778	upper GI bleeding
178.	(d)	page 778	lower GI bleeding
179.	(c)	page 783	positioning for abdominal pain
180.	(a)	page 784	rebound tenderness
181.	(c)	page 784	rebound tenderness
182.	(b)	page 785	orthostatic vital signs, "tilt test"
183.	(d)	page 785	orthostatic vital signs, "tilt test"
184.	(e)	page 774	the kidneys
185.	(b)	page 780	renal failure
186.	(a)	page 781	uremia
187.	(a)	page 781	renal failure
188.	(d)	page 781	ascites
189.	(e)	page 781	renal failure
190.	(c)	page 781	renal failure
191.	(e)	page 790	dialysis patient management
192.	(e)	page 779	kidney stone
193.	(a)	page 779	kidney stone
194.	(a)	page 780	urinary tract infection
195.	(b)	page 787	renal dialysis
196.	(e)	page 788	complications of renal dialysis

TEST SECTION SIX ANSWER KEY

The page numbers following each answer indicate that subject's reference page within Brady's *Paramedic Emergency Care*, 3rd edition. If you do not have access to Brady's text, use the index of the text you do have to obtain information for review of that subject.

Following some answers is "g/d." This indicates that you should consult the glossary of your paramedic text or a medical dictionary.

Following some answers is "ACLS." This indicates that you should consult the American Heart Association's text book of Advanced Cardiac Life Support guidelines

QUESTION	ANSWER	PAGE REFERENCE *Paramedic Emergency Care, 3/e*	
1.	(b)	page 603	epicardium
2.	(d)	page 603	myocardium
3.	(a)	page 603	endocardium
4.	(e)	page 603	myocardial muscle
5.	(c)	page 604	atria/ventricles
6.	(d)	page 604	normal flow of blood through the heart
7.	(d)	page 607	coronary arteries
8.	(c)	page 608	lumen
9.	(b)	page 608	tunica imtima
10.	(e)	page 608	tunica media
11.	(a)	page 608	tunica adventitia
12.	(c)	page 608	arteries
13.	(a)	page 608	capillaries
14.	(b)	page 608	veins
15.	(b)	page 609	diastole
16.	(a)	page 609	systole
17.	(b)	page 609	diastole
18.	(b)	page 609	diastole
19.	(a)	page 609	systole
20.	(b)	page 609	diastole
21.	(e)	page 609	stroke volume
22.	(d)	page 609	preload
23.	(c)	page 609	afterload
24.	(e)	page 609	Starling's law
25.	(d)	page 610	stroke volume
26.	(c)	page 610	cardiac output
27.	(c)	page 610	cardiac nervous system effects
28.	(e)	page 349	alpha cardiac effects
29.	(d)	page 349	beta cardiac effects
30.	(a)	page 612	sodium

QUESTION	ANSWER	PAGE REFERENCE *Paramedic Emergency Care, 3/e*	
31.	(c)	page 612	potassium
32.	(d)	page 612	calcium
33.	(c)	page 615	automaticity
34.	(a)	page 615	excitability
35.	(e)	page 615	conductivity
36.	(b)	page 615	electrical conduction rates
37.	(d)	page 616	SA node rate
38.	(c)	page 616	AV node rate
39.	(b)	page 616	idioventricular rate
40.	(b)	page 616	ECG monitoring
41.	(b)	page 619	time measurement/ECG
42.	(e)	page 619	time measurement/ECG
43.	(a)	page 619	P wave
44.	(e)	page 619	atrial T wave
45.	(c)	page 619	QRS
46.	(b)	page 620	T wave
47.	(d)	page 620	P–R interval
48.	(e)	page 625	refractory period
49.	(e)	page 625	absolute refractory period
50.	(d)	page 625	absolute refractory period
51.	(e)	page 625	relative refractory period
52.	(c)	page 625	relative refractory period
53.	(d)	page 627	P–R interval
54.	(b)	page 627	QRS complex
55.	(b)	page 677	second-degree AV block type I
56.	(c)	page 679	second-degree AV block type II
57.	(d)	page 649	A-fib with bundle branch block
58.	(c)	page 672	AV sequential pacemaker
59.	(a)	page 655	junctional rhythm
60.	(e)	pages 641, 662	sinus rhythm with 1 PAC and 1 PVC
61.	(e)	pages 641, 662	sinus rhythm with 1 PAC and 1 PVC
62.	(a)	page 647	A-flutter with variable (4:1, 3:1) response
63.	(c)	page 653	sinus rhythm with 1 PJC
64.	(a)	page 633	sinus tachycardia with elevated S–T segment
65.	(a)	page 633	sinus tachycardia with BBB and inverted T wave
66.	(e)	page 689	There is no ectopy present; however, the elevated S–T segment is a strong indication of ischemia. There are clinical symptoms of AMI, and thus there is indication for prophylactic lidocaine.

QUESTION	ANSWER	PAGE REFERENCE *Paramedic Emergency Care, 3/e*	
67.	(e)	page 692	There is no mentation change or hypotension, no complaints of chest pain or shortness of breath. Pharmacological treatment of this patient is not indicated.
68.	(a)	page 643	Synchronized cardioversion is indicated. An atraumatic, well hydrated, 70-year-old female should not receive a fluid challenge. Dopamine will not correct the cause of her hypotension (tachycardia). Her blood pressure will not support NTG or MS. With transport only, she will suffer an AMI.
69.	(a)	page 662	This patient is having a painless AMI. The underlying rate of this rhythm is 80/min. The PVCs are malignant and require lidocaine.
70.	(b)	page 689	angina
71.	(a)	page 689	angina
72.	(b)	page 689	angina
73.	(c)	page 689	angina
74.	(c)	page 690	myocardial infarction
75.	(e)	page 690	most common cause of AMI
76.	(a)	page 691	most common cause of death from AMI
77.	(d)	page 690	most common site of AMI
78.	(a)	page 690	subendocardial infarction
79.	(e)	page 690	transmural infarction
80.	(e)	page 691	most common complication of AMI
81.	(b)	page 692	management of AMI
82.	(e)	page 694	causes of left ventricular failure
83.	(b)	page 694	left ventricular failure
84.	(d)	page 695	lung sounds and left ventricular failure
85.	(b)	page 695	rhonchi and left ventricular failure
86.	(c)	page 696	wheezes and left ventricular failure
87.	(a)	page 695	rales and left ventricular failure
88.	(d)	page 696	jugular vein distention
89.	(c)	page 696	vital signs of left ventricular failure
90.	(d)	page 696	left ventricular failure management
91.	(a)	page 697	cause of right ventricular failure
92.	(d)	page 698	right ventricular failure, venous congestion
93.	(c)	page 699	management of right ventricular failure

QUESTION	ANSWER	PAGE REFERENCE *Paramedic Emergency Care, 3/e*	
94.	(e)	page 699	cardiogenic shock
95.	(e)	page 699	cardiogenic shock
96.	(a)	page 699	management of cardiogenic shock
97.	(c)	page 701	sudden death
98.	(d)	page 701	management of cardiac arrest
99.	(a)	page 703	EMD
100.	(d)	page 705	abdominal aneurysm
101.	(a)	page 706	thoracic aneurysm
102.	(b)	page 705	abdominal aneurysm
103.	(d)	page 705	aortic aneurysm
104.	(a)	page 706	aortic aneurysm
105.	(e)	page 709	malignant hypertension
106.	(b)	page 709	malignant hypertension
107.	(b)	page 709	hypertension-related emergencies
108.	(d)	page 712	Lanoxin
109.	(a)	page 712	Inderal
110.	(c)	page 715	precordial thump
111.	(d)	page 715	precordial thump
112.	(b)	page 715	defibrillation
113.	(e)	page 716	defibrillation
114.	(e)	page 717	adult defibrillation energy levels
115.	(e)	page 717	adult defibrillation energy levels
116.	(d)	page 717	adult defibrillation energy levels
117.	(c)	page 717	pediatric defibrillation energy levels
118.	(b)	page 717	pediatric defibrillation energy levels
119.	(b)	page 717	synchronized cardioversion
120.	(c)	page 717	synchronized cardioversion
121.	(c)	page 645, ACLS	energy levels for synchronized cardioversion
122.	(d)	ACLS	energy levels for synchronized cardioversion
123.	(a)	page 717	carotid sinus massage
124.	(b)	page 717	carotid sinus massage
125.	(e)	page 717	carotid sinus massage
126.	(e)	page 721	transcutaneous cardiac pacing (TCP)

Figure 6-14 (sample)

Figure 6-15 Runaway pacemaker with effective capture at a rate of 100/min and accelerating.

Figure 6-16 Accelerated junctional rhythm (rate of 80/min) with elevated S–T segments.

Figure 6-17 Complete (third-degree) AV block, junctional escape (rate of 60/min) with elevated S–T segments.

Figure 6-18 Sinus tachycardia with inverted T waves and multifocal R-on-T PVCs that degenerates into ventricular tachycardia.

Figure 6-19 Mobitz I second-degree AV block (Wenkebach) with inverted T waves, in a 4:3 ratio with a ventricular rate of 60/min.

Figure 6-20 Sinus rhythm (rate of 80/min) with a BBB, elevated S–T segments, and muscle tremor ECG interference.

Figure 6-21 Pacemaker rhythm with effective capture at a rate of 72–80/min (underlying A-fib).

Figure 6-22 Sinus dysrhythmia with inverted T waves, at a rate of 70/min.

Figure 6-23 Sinus bradycardia at a rate of 50/min, with elevated S–T segments.

Figure 6-24 Sinus rhythm with a BBB, elevated S–T segments, 2 PACs, and an underlying rate of 80/min.

Figure 6-25 Sinus rhythm with a BBB, elevated S–T segments, 2 unifocal PVCs, and an underlying rate of 80/min.

Figure 6-26 Sinus rhythm with inverted T waves, trigeminal PACs, and an underlying rate of 80/min.

Figure 6-27 Controlled A-fib at a rate of 90/min, with elevated S–T segments.

Figure 6-28 Atrial tachycardia (or SVT) at a rate of 180/min, with inverted T waves.

Figure 6-29 Sinus tachycardia with inverted T waves, 4 PACs, and an underlying rate of 110/min.

Figure 6-30 Sinus rhythm with a BBB, elevated S–T segments, 2 multifocal PVCs, and an underlying rate of 80/min.

Figure 6-31 Sinus rhythm with a prolonged PRI (first-degree AV block), a BBB, elevated S–T segments, 1 PAC, and an underlying rate of 80/min.

Figure 6-32 Sinus bradycardia with a prolonged PRI (first-degree AV block) at a rate of 60/min.

Figure 6-33 Sinus bradycardia with inverted T waves at a rate of 40/min, with 1 interpolated ventricular complex.

Figure 6-34 Atrial flutter with elevated S–T segments, in a 4:1 ratio, and a ventricular rate of 80/min.

Figure 6-35 Sinus tachycardia with elevated S–T segments, bigeminal PACs, and an underlying rate of 110/min.

Figure 6-36 Sinus rhythm, with 1 PVC and 1 PAC, and an underlying rate of 80/min.

Figure 6-37 Sinus rhythm with a BBB, elevated S–T segments, 1 PAC, 2 unifocal PVCs, and an underlying rate of 80/min.

Figure 6-38 Sinus rhythm with bigeminal PJCs and an underlying rate of 100/min.

Figure 6-39 Sinus rhythm with elevated S–T segments, 1 couplet of PVCs, and an underlying rate of 80/min.

Figure 6-40 Sinus rhythm with 1 PJC, and an underlying rate of 80/min.

Figure 6-41 Sinus rhythm without ectopy that develops a BBB with depressed S–T segments, at a rate of 90/min.

Figure 6-42 Atrial flutter with a variable ventricular response (4:1, 2:1, 3:1), at a rate of 120/min.

Figure 6-43 Sinus bradycardia with inverted T waves, 2 unifocal interpolated ventricular complexes, and an underlying rate of 40/min.

Figure 6-44 Sinus rhythm with a prolonged PRI (first-degree AV block), a BBB, elevated S–T segments, at a rate of 60/min, without ectopy.

Figure 6-45 Sinus rhythm with a prolonged PRI (first-degree AV block), inverted T waves, 4 PACs, and an underlying rate of 80/min.

Figure 6-46 Atrial flutter with a 3:1 response, a ventricular rate of 100/min, without ectopy.

Figure 6-47 Sinus rhythm with quadrigeminal unifocal PVCs, and an underlying rate of 90/min.

Figure 6-48 Uncontrolled atrial fibrillation with elevated S–T segments at a rate of 110–120/min, without ectopy.

Figure 6-49 Atrial fibrillation with a BBB, elevated S–T segments, at a bradycardic ventricular response rate of 40–50/min, without ectopy.

Figure 6-50 Sinus rhythm with a BBB, elevated S–T segments, bigeminal unifocal PVCs, and an underlying rate 80/min.

Figure 6-51 Sinus rhythm with a 3-beat run of V-tach (salvo of 3 PVCs), and an underlying rate of 70/min.

Figure 6-52 Sinus rhythm with inverted T waves, trigeminal unifocal PVCs, and an underlying rate of 80/min.

Figure 6-53 Sinus rhythm with inverted T waves, 3 multifocal PVCs, and an underlying rate of 80/min.

Figure 6-54 Sinus tachycardia with 1 couplet and 2 multifocal PVCs, and an underlying rate of 100/min.

Figure 6-55 Sinus rhythm with a BBB (PRI of 0.12), inverted T waves, at a rate of 80/min, without ectopy.

Figure 6-56 Demand pacemaker rhythm with effective capture that is overtaken by the patient's sinus dysrhythmia with a prolonged PRI (first-degree AV block), inverted T waves; paced rate of 60/min, sinus rate of 80/min, and muscle tremor ECG interference.

Figure 6-57 Sinus rhythm with inverted T waves, 2 PACs, and an underlying rate of 80/min.

Figure 6-58 Mobitz I second-degree AV block (Wenkebach) with elevated S–T segments, in a 4:3 ratio with a ventricular rate of 60/min, without ectopy.

Figure 6-59 Complete (third-degree) AV block with a junctional escape at a rate of 60/min, without ectopy.

Figure 6-60 Idioventricular rhythm at a rate of 40/min.

Figure 6-61 Pacemaker rhythm with ineffective capture; pacer spikes 80/min, ventricular capture of 30-40/min.

Figure 6-62 Junctional bradycardia with bigeminal interpolated ventricular escape beats, and a junctional rate of 30/min ("PVCs" would come earlier in the cycle).

Figure 6-63 Ventricular tachycardia at a rate of 190–200/min.

Figure 6-64 Junctional rhythm with inverted T waves, at a rate of 50/min, without ectopy.

Figure 6-65 Mobitz II second-degree (classical) AV block in a 2:1 ratio with a ventricular rate of 50/min, without ectopy.

TEST SECTION SEVEN ANSWER KEY

The page numbers following each answer indicate that subject's reference page within Brady's *Paramedic Emergency Care*, 3rd edition. If you do not have access to Brady's text, use the index of the text you do have to obtain information for review of that subject.

Following some answers is "g/d." This indicates that you should consult the glossary of your paramedic text or a medical dictionary.

QUESTION	ANSWER	PAGE REFERENCE *Paramedic Emergency Care, 3/e*	
1.	(d)	pages 794-797	antigen-antibody reaction
2.	(e)	pages 794-797	antigen entry routes
3.	(c)	page 797	histamine effects
4.	(a)	page 797	histamine effects
5.	(b)	page 797	histamine effects
6.	(c)	page 798	signs and symptoms of anaphylaxis
7.	(e)	page 798	signs and symptoms of anaphylaxis
8.	(b)	page 798	urticaria
9.	(d)	page 798	airway management of anaphylaxis
10.	(d)	page 799	IV fluids and anaphylaxis
11.	(a)	page 799	medications and anaphylaxis
12.	(a)	page 800	epinephrine 1:1000
13.	(e)	page 800	epinephrine 1:10,000
14.	(c)	page 807	most common route of poisoning
15.	(e)	page 807	onset of poison effects
16.	(a)	page 812	contraindications for ipecac
17.	(b)	page 812	ingestion of commercial toilet cleaner
18.	(a)	page 813	adult dose of activated charcoal
19.	(c)	page 813	pediatric dose of activated charcoal
20.	(b)	page 812	ingestion of bleach
21.	(b)	page 813	pediatric ipecac dose
22.	(d)	page 813	adult ipecac dose
23.	(e)	page 818	toxic inhalation
24.	(c)	page 822	treatment of bites and stings
25.	(b)	page 822	brown recluse spider
26.	(a)	page 824	black widow spider
27.	(b)	page 824	brown recluse spider
28.	(c)	page 825	scorpion sting
29.	(a)	page 824	black widow spider bite
30.	(d)	page 827	coral snake
31.	(c)	page 827	treatment of snake bite
32.	(a)	pages 548-550	treatment of dry lime exposure
33.	(e)	pages 548-550	treatment of chemical exposure
34.	(b)	pages 548-550	treatment of phenol exposure

QUESTION	ANSWER	PAGE REFERENCE *Paramedic Emergency Care, 3/e*	
35.	(c)	pages 548-550	treatment of sodium metal exposure
36.	(d)	page 829	signs and symptoms of organophosphate poisoning
37.	(d)	page 830	treatment of organophosphate poisoning
38.	(e)	page 831	opiates/narcotics
39.	(d)	page 831	treatment of Darvon overdose
40.	(b)	page 835	alcohol withdrawal syndrome
41.	(c)	page 842	components of the immune system
42.	(c)	page 843	interstitial fluid
43.	(a)	page 843	lymph
44.	(c)	page 843	the primary immune system organ/ spleen
45.	(e)	page 594	pneumonia
46.	(e)	page 851	meningitis
47.	(a)	page 851	tuberculosis
48.	(c)	page 855	hepatitis A
49.	(b)	page 855	hepatitis B
50.	(b)	page 855	hepatitis B
51.	(c)	page 851	signs and symptoms of meningitis
52.	(d)	pages 851, 858	signs and symptoms of tuberculosis/AIDS
53.	(d)	page 858	AIDS transmission
54.	(b)	page 859	Kaposi's sarcoma
55.	(d)	page 859	Pneumocystis carinii
56.	(e)	page 857	sexually transmitted diseases
57.	(d)	page 860	universal precautions
58.	(e)	page 865	thermoregulatory control center location
59.	(e)	page 866	heat elimination mechanisms
60.	(c)	page 867	heat preservation/generation
61.	(d)	page 865	heat loss from evaporation
62.	(b)	page 865	heat loss from conduction
63.	(c)	page 865	heat loss from convection
64.	(e)	page 869	heat-related illnesses
65.	(a)	page 870	heat-related illnesses
66.	(a)	page 870	hyperkalemia and heat stroke
67.	(b)	page 869	treatment of heat cramps
68.	(d)	page 870	treatment of heat stroke
69.	(d)	page 864	normal oral temperature
70.	(a)	page 871	normal rectal temperature
71.	(c)	page 871	pyrexia
72.	(b)	page 873	mild hypothermia/Fahrenheit degrees
73.	(d)	page 873	mild hypothermia/centigrade degrees

QUESTION	ANSWER	PAGE REFERENCE *Paramedic Emergency Care, 3/e*	
74.	(d)	page 873	severe hypothermia/Fahrenheit degrees
75.	(b)	page 873	severe hypothermia/centigrade degrees
76.	(c)	page 873	J wave/Osborn wave
77.	(e)	page 874	treatment of mild hypothermia
78.	(d)	page 874	treatment of severe hypothermia (with vital signs)
79.	(e)	page 876	treatment of severe hypothermia (without vital signs)
80.	(e)	page 874	metabolic causes of hypothermia
81.	(a)	page 876	pathophysiology of frostbite
82.	(c)	page 877	treatment of frostbite
83.	(c)	page 878	fresh-water is hypotonic
84.	(a)	page 879	salt-water is hypertonic
85.	(d)	page 878	fresh-water aspiration
86.	(b)	page 879	salt-water aspiration
87.	(e)	page 879	successful resuscitation factors
88.	(d)	page 879	treatment of near-drowning
89.	(e)	page 885	diving injuries
90.	(e)	page 886	diving injuries
91.	(d)	page 888	signs and symptoms of decompression sickness, Type I
92.	(e)	page 888	signs and symptoms of decompression sickness, Type II
93.	(c)	page 889	treatment of decompression sickness
94.	(e)	page 890	pulmonary overpressure accidents
95.	(c)	page 890	treatment of pulmonary overpressure accidents
96.	(a)	page 890	signs and symptoms of pneumomediastinum
97.	(b)	page 890	treatment of pneumomediastinum
98.	(e)	page 880	alpha, beta, gamma, and X-rays
99.	(a)	page 880	radiation injury sources
100.	(e)	page 881	radiation exposure factors
101.	(b)	page 895	elderly physiologic changes
102.	(c)	page 896	cardiac hypertrophy
103.	(a)	page 896	osteoporosis
104.	(e)	page 900	spondylolysis
105.	(b)	page 896	kyphosis
106.	(d)	page 896	fibrosis
107.	(a)	page 898	cataracts
108.	(c)	page 898	glaucoma
109.	(a)	page 898	elderly alcoholism, depression, suicide

QUESTION	ANSWER	PAGE REFERENCE *Paramedic Emergency Care, 3/e*	
110.	(e)	page 899	trauma and the geriatric patient
111.	(b)	page 900	treatment of geriatric trauma
112.	(c)	page 902	geriatric medical emergencies
113.	(a)	page 905	senility, dementia, and OBS
114.	(e)	page 905	etiologies mistaken for OBS
115.	(a)	page 905	other geriatric emergencies
116.	(b)	page 909	geriatric abuse
117.	(b)	page 919	normal condition of fontanelle
118.	(a)	page 919	increased ICP condition of fontanelle
119.	(c)	page 919	dehydration condition of fontanelle
120.	(c)	page 914	neonate age range
121.	(b)	page 940	SIDS
122.	(d)	page 940	SIDS
123.	(d)	page 940	SIDS
124.	(d)	page 940	SIDS
125.	(a)	page 923	child abuse
126.	(e)	page 924	the child abuser
127.	(c)	page 924	signs and symptoms of child abuse
128.	(b)	page 924	historical signs of child abuse
129.	(b)	page 924	sexual abuse and neglect
130.	(e)	page 927	pediatric seizures
131.	(a)	page 927	pediatric seizures
132.	(a)	page 928	infant dose of diazepam
133.	(b)	page 928	child dose of diazepam
134.	(d)	page 935	pediatric dehydration
135.	(b)	page 936	sepsis
136.	(c)	page 929	Reyes syndrome
137.	(d)	pages 929, 937	meningitis and septicemia
138.	(a)	page 929	meningitis
139.	(a)	page 929	meningitis
140.	(c)	page 929	Reyes syndrome
141.	(e)	pages 929, 936	meningitis/septicemia/Reyes syndrome
142.	(c)	page 931	croup (laryngotracheobronchitis)
143.	(a)	page 933	bronchiolitis
144.	(d)	pages 931–933	bronchiolitis/epiglottitis/croup
145.	(c)	pages 931–932	croup
146.	(b)	page 932	epiglottitis
147.	(b)	page 931-932	epiglottitis
148.	(e)	pages 931-932	epiglottitis and croup
149.	(e)	pages 932-933	bronchiolitis/epiglottitis/croup
150.	(c)	page 951	initial pediatric defibrillation
151.	(d)	page 951	subsequent pediatric defibrillation
152.	(e)	page 945	pediatric endotracheal intubation
153.	(e)	page 959	pelvic inflammatory disease (PID)

QUESTION	ANSWER	PAGE REFERENCE *Paramedic Emergency Care, 3/e*	
154.	(d)	page 959	causes of PID
155.	(b)	page 959	signs and symptoms of PID
156.	(a)	page 961	sexual assault
157.	(c)	page 961	sexual assault evidence preservation
158.	(a)	page 966	functions of the placenta
159.	(a)	page 970	primagravida
160.	(b)	page 970	primapara
161.	(d)	page 970	multipara
162.	(c)	page 970	multigravida
163.	(d)	page 970	gravida/para
164.	(e)	page 972	pregnancy and medical disorders
165.	(d)	page 974	miscarriage
166.	(a)	page 975	ectopic pregnancy
167.	(b)	page 975	most common site of ectopic pregnancy
168.	(b)	page 975	ectopic pregnancy
169.	(d)	page 975	abruptio placenta
170.	(c)	page 977	placenta previa
171.	(c)	page 977	placenta previa
172.	(c)	page 977	treatment of obstetric emergencies
173.	(d)	page 978	preeclampsia
174.	(e)	page 978	eclampsia
175.	(d)	page 978	treatment of preeclampsia
176.	(e)	page 979	treatment of eclampsia
177.	(b)	page 979	supine-hypotensive syndrome
178.	(a)	page 979	treatment of supine-hypotensive syndrome
179.	(c)	page 982	stages of labor
180.	(b)	page 982	stages of labor
181.	(d)	page 981	stages of labor
182.	(e)	page 982	stages of labor
183.	(a)	page 982	signs and symptoms of imminent delivery
184.	(a)	g/d	cephalic presentation
185.	(b)	page 988	breech presentation
186.	(e)	page 983	preparation for delivery
187.	(a)	page 983	management of unruptured amniotic sac
188.	(c)	page 987	management of wrapped umbilical cord
189.	(a)	page 987	umbilical clamp placement
190.	(b)	page 987	umbilical clamp placement
191.	(c)	page 987	postpartum hemorrhage
192.	(e)	page 987	cephalopelvic disproportion
193.	(c)	page 988	abnormal deliveries

QUESTION	ANSWER	PAGE REFERENCE *Paramedic Emergency Care, 3/e*	
194.	(b)	page 991	multiple births
195.	(b)	page 992	uterine rupture
196.	(e)	page 992	uterine inversion
197.	(d)	page 992	treatment of uterine inversion
198.	(e)	page 992	pulmonary embolus
199.	(d)	page 998	infant suctioning
200.	(e)	page 998	infant positioning after birth
201.	(b)	page 999	APGAR assessment times
202.	(c)	page 1000	APGAR mnemonic
203.	(a)	page 1000	APGAR mnemonic
204.	(d)	page 1000	APGAR mnemonic
205.	(d)	page 1000	APGAR mnemonic
206.	(e)	page 1000	treatment of the premature infant
207.	(a)	page 1001	meconium
208.	(e)	page 1001	infant CPR requirements
209.	(e)	page 1015	pathologies/behavior disorders
210.	(c)	page 1018	open-ended questions
211.	(b)	page 1019	suicidal risk factors
212.	(d)	page 1020	suicidal/homicidal patients
213.	(a)	page 1018	general management of behavioral or psychiatric disorders

GUIDE TO COMMON MEDICAL ABBREVIATIONS AND SYMBOLS

Many of the following abbreviations were used in this text. For the purpose of self-improvement, this is a more comprehensive list. It contains abbreviations that should be familiar to the professional paramedic and which she/he should use on a regular basis. When an abbreviation is based on the Latin or Greek form of a word, the Latin or Greek word is presented in italic print in parentheses.

The appropriate use of upper- and lowercase letters is an important distinction in medical abbreviations. For example: "cc" is the abbreviation for cubiccentimeter, whereas "CC" is the abbreviation for chief complaint; "Ca" is the abbreviation for calcium, whereas "CA" is the abbreviation for cancer.

ā .... before (*ante*)
AAOX3 .... alert and oriented to person/place/time
abdo .... abdomen
ABG .... arterial blood gases
ac .... before meals (*ante cibum*)
AC .... anticubital fossa
ACLS .... advanced cardiac life support
ad lib .... as desired (*ad libitum*)
A-fib .... atrial fibrillation
A-flutter .... atrial flutter
AIDS .... acquired immune deficiency syndrome
ALS .... advanced life support
AMA .... against medical advice
AMI .... acute myocardial infarction (heart attack)
amp .... ampule
ARC .... AIDS-related complex
ASA .... acetylsalicylic acid (aspirin)
ASHD .... arteriosclerotic heart disease
AV .... atrioventricular

BA .... blood alcohol level
BBB .... bundle branch block
BBS .... bilateral breath sounds
BCP .... birth control pills
b.i.d. .... twice a day (*bis in die*)
BLS .... basic life support
BM .... bowel movement
bp or b/p .... blood pressure
BS .... breath sounds or blood sugar
BVM .... bag–valve–mask device

$\bar{c}$ .... with (*cum*)
Ca .... calcium
CA^{++} .... cancer (carcinoma)
$CaCl_2$ .... calcium chloride
CAD .... coronary artery disease
cap(s) .... capsule(s)
CBC .... complete blood count
cc .... cubic centimeter
CC .... chief complaint
CHF .... congestive heart failure
Cl^- .... chloride
cm .... centimeter
CNS .... central nervous system
c/o .... complains of
CO .... carbon monoxide
CO_2 .... carbon dioxide
conx .... conscious
COPD or COLD .... chronic obstructive pulmonary (lung) disease
cp .... chest pain
CSF .... cerebrospinal fluid
CSM .... carotid sinus massage
CVA .... cerebrovascular accident or costovertebral angle

D_5W .... dextrose 5 percent in water
$D_{50}W$ .... dextrose 50 percent in water
D & C .... dilation and curettage
dc or d/c .... discontinue
dig .... digitalis
DIP .... distal interphalangeal (joint)
Dx .... diagnosis or dislocation

ea .... each
ECG or EKG .... electrocardiogram
ECT .... electroconvulsive therapy
EEG .... eletroencephalogram
ED .... Emergency Department
EGTA .... esophageal gastric tube airway
EJ .... external jugular vein
EMD .... electromechanical dissociation
EMS .... Emergency Medical Services
EMT .... Emergency Medical Technician
EMT–B .... Emergency Medical Technician—Basic level
EMT–I .... Emergency Medical Technician—Intermediate
EMT–P .... Emergency Medical Technician—Paramedic
EOA .... esophageal obturator airway
ER .... Emergency Room
et .... and (*et*)
ET .... endotracheal tube
ETOH .... alcohol (ethyl alcohol)

FBS .... fasting blood sugar
fib .... fibrillation
FROM .... full range of motion
Fx .... fracture

GI .... gastrointestinal
g, Gm, or gm .... gram
gr .... grain
GSW .... gunshot wound
gtt .... drops (*guttae*)
GTT .... glucose tolerance test
GU .... genitourinary
GYN .... gynecological

H_2O .... water
HA .... headache
HCO_3^- .... bicarbonate
Hct .... hematocrit
Hctz .... hydrochlorothiazide
HEENT .... head, ears, eyes, nose, throat
Hgb .... hemoglobin
HIV .... human immunodeficiency virus
HPTN .... hypertension
hs .... at bedtime (*hora somni*)
Hx .... history

ICP .... intercranial pressure
ICS .... intercostal space
IM .... intramuscular
IP .... interphalangeal (joint)
IPPB .... intermittent positive pressure breathing
IV .... intravenous
IVP .... IV push or intravenous pyelogram

JVD .... jugular vein distention

K^+ .... potassium (kalium)
Kg or kg .... kilogram
KVO .... keep vein open

Ⓛ .... left
l .... liter
lac .... laceration
LAD .... left axis deviation or left anterior descending (coronary artery)
LBB .... long backboard
LBBB .... left bundle branch block
LGL .... Lown–Ganong–Levine syndrome
LLQ .... left lower quadrant

LMP .... last menstrual period
LOC .... level of consciousness (not loss of conx)
LPM .... liters per minute
LR .... lactated ringers
LUQ .... left upper quadrant

MAO .... monoamine oxidase (MAO inhibitor medications)
MAST .... military antishock trousers
MCI .... mass casualty incident
MCL .... mid-costal line or mid-clavicular line
MCP .... metacarpal–phalangeal (joint)
mEq .... milliequivalents
mg .... milligram
Mg^{++} .... magnesium
MI .... myocardial infarction (heart attack)
MICU .... mobile intensive care unit
min .... minute
ml .... milliliter
mm .... millimeter
MOI .... mechanism of injury
MS .... morphine sulfate
MTP .... metatarsal–phalangeal (joint)
mv .... milivolt

N_2O .... nitrous oxide
n/a .... not applicable
Na^+ .... sodium (*L. natrium*)
$NaHCO_3^-$ .... sodium bicarbonate
nc .... nasal cannula
nka or NKA .... no known allergies
noc or noct .... night (*nocte*)
NPO .... nothing by mouth (nothing *per os*)
NRB .... nonrebreather mask
NS .... normal saline
NT .... nontender
NTG .... nitroglycerine
n/v .... nausea and vomiting
n/v/d .... nausea, vomiting, and diarrhea

OB .... obstetrics
OBS .... organic brain syndrome
OD .... overdose or right eye (*occulus dexter*)
os .... mouth (*os* or *ora*)
OS .... left eye (*occulus sinister*)
OU .... both eyes (*occulus uterque*)
oz .... ounce

$\overline{p}$ .... after (*post*)
PAC .... premature atrial contraction

PASG .... pneumatic antishock garment
PAT .... paroxysmal atrial tachycardia
pc .... after meals (*post cibum*)
pCO_2 .... concentration of carbon dioxide in arterial blood
PCP .... pneumocystis carinii pneumonia or phencyclidine hydrochloride
PE .... pulmonary embolism
PEARL .... pupils equal and reactive to light
ped .... pedestrian or foot
pedi .... pediatric
pH .... hydrogen concentration
PID .... pelvic inflammatory disease
PIP .... proximal interphalangeal (joint)
PJC .... premature junctional contraction
PND .... paroxysmal nocturnal dyspnea
PNS .... peripheral nervous system
po .... by mouth (*per os*)
pO_2 .... concentration of oxygen in arterial blood
ppm .... parts per million
prn .... as needed (*pro re nata*)
PSVT .... paroxysmal supraventricular tachycardia
Ptl .... pharyngeo-tracheal lumen (airway)
PTOA .... prior to our arrival
PMH .... past medical history
PVC .... premature ventricular contraction

$\bar{q}$ .... every (*quaque*)
qd .... every day (*quaque die*)
q.i.d. .... four times a day (*quarter in die*)

Ⓡ .... right
rad .... radiation absorbed dose
RAD .... right axis deviation
RBBB .... right bundle branch block
RBC .... red blood cell
RL .... ringers lactate
RLQ .... right lower quadrant
R/O .... rule out
ROM .... range of motion
RUQ .... right upper quadrant
Rx .... recipe or prescription

$\bar{s}$ .... without (*sans*)
sc or sq .... subcutaneous
SIG .... label (*signa* or *signetur*)
sl .... sublingual
SOB .... shortness of breath
ss .... a half (*semi* or *semisse*)
s/sx .... signs and symptoms

stat .... immediately (*statim*)
SVT .... supraventricular tachycardia
Sx .... symptoms

TIA .... transient ischemic attack
t.i.d. .... three times a day (*ter in die*)
TKO .... to keep open
T.O. .... telephone order
trans .... transport
Tx .... treatment (not trans)

UA .... urinalysis
U/A .... upon arrival
uggt or uggts .... microdrop or microdrops
unconx .... unconscious
URI .... upper respiratory infection
UTI .... urinary tract infection

VF or V-fib .... ventricular fibrillation
V.O. .... verbal order
vs .... vital signs
VT .... ventricular tachycardia

WBC .... white blood cell
w/d .... warm and dry
w/d/pink .... warm, dry, and pink
WNL .... within normal limits
WPW .... Wolfe–Parkinson–White syndrome
ws .... watt seconds

y/o .... years old

COMMON MEDICAL SYMBOLS

α	alpha
≈	approximately
△	change
>	greater than
<	less than
♀	female
♂	male
×	multiply by
2°	secondary to
β	beta
⊕	positive for
⊖	negative for
Ⓗ	husband
Ⓕ	father
Ⓜ	mother
Ⓦ	wife
o‿‿⌋	lying
	sitting
	standing